Lecture Notes in Applied and Computational Mechanics

Volume 58

Series Editors

Prof. Dr.-Ing. Friedrich Pfeiffer
Prof. Dr.-Ing. Peter Wriggers

Lecture Notes in Applied and Computational Mechanics

Edited by F. Pfeiffer and P. Wriggers

Further volumes of this series found on our homepage: springer.com

Vol. 58 Zavarise, G., Wriggers, P. (Eds.)
Trends in Computational Contact Mechanics
354 p. 2011 [978-3-642-22166-8]

Vol. 57 Stephan, E.,
Wriggers, P.
Modelling, Simulation and Software Concepts for Scientific-Technological Problems
251 p. 2011 [978-3-642-20489-0]

Vol. 54: Sanchez-Palencia, E.,
Millet, O., Béchet, F.
Singular Problems in Shell Theory
265 p. 2010 [978-3-642-13814-0]

Vol. 53: Litewka, P.
Finite Element Analysis of Beam-to-Beam Contact
159 p. 2010 [978-3-642-12939-1]

Vol. 52: Pilipchuk, V.N.
Nonlinear Dynamics: Between Linear and Impact Limits
364 p. 2010 [978-3-642-12798-4]

Vol. 51: Besdo, D., Heimann, B., Klüppel, M.,
Kröger, M., Wriggers, P., Nackenhorst, U.
Elastomere Friction
249 p. 2010 [978-3-642-10656-9]

Vol. 50: Ganghoffer, J.-F., Pastrone, F. (Eds.)
Mechanics of Microstructured Solids 2
102 p. 2010 [978-3-642-05170-8]

Vol. 49: Hazra, S.B.
Large-Scale PDE-Constrained Optimization in Applications
224 p. 2010 [978-3-642-01501-4]

Vol. 48: Su, Z.; Ye, L.
Identification of Damage Using Lamb Waves
346 p. 2009 [978-1-84882-783-7]

Vol. 47: Studer, C.
Numerics of Unilateral Contacts and Friction
191 p. 2009 [978-3-642-01099-6]

Vol. 46: Ganghoffer, J.-F., Pastrone, F. (Eds.)
Mechanics of Microstructured Solids
136 p. 2009 [978-3-642-00910-5]

Vol. 45: Shevchuk, I.V.
Convective Heat and Mass Transfer in Rotating Disk Systems
300 p. 2009 [978-3-642-00717-0]

Vol. 44: Ibrahim R.A., Babitsky, V.I., Okuma, M. (Eds.)
Vibro-Impact Dynamics of Ocean Systems and Related Problems
280 p. 2009 [978-3-642-00628-9]

Vol.43: Ibrahim, R.A.
Vibro-Impact Dynamics
312 p. 2009 [978-3-642-00274-8]

Vol. 42: Hashiguchi, K.
Elastoplasticity Theory
432 p. 2009 [978-3-642-00272-4]

Vol. 41: Browand, F., Ross, J., McCallen, R. (Eds.)
Aerodynamics of Heavy Vehicles II: Trucks, Buses, and Trains
486 p. 2009 [978-3-540-85069-4]

Vol. 40: Pfeiffer, F.
Mechanical System Dynamics
578 p. 2008 [978-3-540-79435-6]

Vol. 39: Lucchesi, M., Padovani, C., Pasquinelli, G., Zani, N.
Masonry Constructions: Mechanical Models and Numerical Applications
176 p. 2008 [978-3-540-79110-2]

Vol. 38: Marynowski, K.
Dynamics of the Axially Moving Orthotropic Web
140 p. 2008 [978-3-540-78988-8]

Vol. 37: Chaudhary, H., Saha, S.K.
Dynamics and Balancing of Multibody Systems
200 p. 2008 [978-3-540-78178-3]

Vol. 36: Leine, R.I.; van de Wouw, N.
Stability and Convergence of Mechanical Systems with Unilateral Constraints
250 p. 2008 [978-3-540-76974-3]

Vol. 35: Acary, V.; Brogliato, B.
Numerical Methods for Nonsmooth Dynamical Systems: Applications in Mechanics and Electronics
545 p. 2008 [978-3-540-75391-9]

Vol. 34: Flores, P.; Ambrósio, J.; Pimenta Claro, J.C.;
Lankarani Hamid M.
Kinematics and Dynamics of Multibody Systems with Imperfect Joints: Models and Case Studies
186 p. 2008 [978-3-540-74359-0

Vol. 33: Nies ony, A.; Macha, E.
Spectral Method in Multiaxial Random Fatigue
146 p. 2007 [978-3-540-73822-0]

Trends in Computational Contact Mechanics

Giorgio Zavarise, Peter Wriggers (Eds.)

Springer

Prof. Giorgio Zavarise
University of Salento
Department of Innovation
Engineering
Via per Monteroni - Edificio
"La Stecca"
73100 Lecce
Italy

Prof. Dr.-Ing. habil. Peter Wriggers
Leibniz Universitaet Hannover
Institut fuer Kontinuumsmechanik
Appelstr. 11
30167 Hannover
E-mail: wriggers@ikm.uni-hannover.de
http://www.ikm.uni-hannover.de

ISBN: 978-3-642-22166-8 e-ISBN: 978-3-642-22167-5

DOI 10.1007/ 978-3-642-22167-5

Lecture Notes in Applied and Computational Mechanics ISSN 1613-7736

e-ISSN 1860-0816

Library of Congress Control Number: 2011930522

Typeset & Cover Design: Scientific Publishing Services Pvt. Ltd., Chennai, India.

Printed on acid-free paper

9 8 7 6 5 4 3 2 1 0

springer.com

Preface

Contact mechanics is a science that has a great impact on everyday life and is present in many different fields. These include civil, mechanical and environmental engineering, but also medicine since locomotion as well as functional joints do not work without friction. In one application friction is needed – like traction of car tyres – and in another application friction produces wear and costs – like in bearings. Thus it is of the utmost interest to have reliable and efficient methods and associated analysis tools that can be applied to a vast range of contact problems.

Using the power of today's computers many complex contact problems can be solved with numerical simulation tools. Despite the progress that has been reached with respect to the implementation of contact algorithms in commercial codes, vivid research is still going on in the area of contact mechanics. Thus, within the last years, computational contact mechanics has been a topic of intense research. The aim of the development is to devise robust solution schemes and new discretization techniques, which can be applied to different problem classes in engineering and science.

These are wide-ranging and include computational aspects of discretization techniques using finite and boundary element methods. Special solution algorithms for single- and multi-processor computing environments are of great interest for efficient solutions. Furthermore, multi-scale approaches have been applied successfully to contact problems and multi-field formulations were used for thermo-mechanical or electro-thermo-mechanical applications involving contact. Discrete element models include always contact and pose a challenge for the numerical treatment due to the high number of particles. Finally, problems like rolling wheels and tyres need special contact formulations and special algorithmic approaches.

Technical applications incorporate different interface problems. Examples are failure processes in heterogeneous materials, textile and laminated composites, interaction between road and tyres, hip implants or artificial knee joints as well as spraying of particles on surfaces and impact analysis of cars.

The present book summarizes work in the area of computational contact mechanics that was presented at the 1st International Conference on Computational

Contact Mechanics in Lecce, Italy. The authors discuss different theoretical methodologies, algorithms for the solution of contact problems and apply these to different engineering problems.

Hannover and Lecce, April 2011
P. Wriggers and G. Zavarise

Table of Contents

Contact Modelling in Entangled Fibrous Materials

Damien Durville

Abstract An approach to model contact-friction interactions between beams within assemblies of fibers is presented in this paper in order to simulate the mechanical behaviour of entangled structures at the scale of individual fibers using the finite element method. The determination of contact elements associating pairs of material particles is based on the detection of proximity zones between beams and on the construction of intermediate geometries approximating the actual contact zone, and allowing to consider contact along zones of non-zero lengths. The penalty method for contact is improved by adjusting the penalty parameter for each contact zone, thus stabilizing contact algorithms and allowing to handle high numbers of contact elements. Applications to samples of textile materials involving few hundreds of fibers are presented to demonstrate the abilities of the method. The presented examples are related to the simulation of woven fabrics – computation of the initial configuration and application of test loadings – and the identification of the transverse mechanical behaviour of a twisted textile yarn.

1 Introduction

Entangled fibers are involved in different types of structures and materials, ranging from biological tissues to technical textiles used as reinforcements in composites. The characteristic features of the nonlinear mechanical behaviour of media constituted by fibers rely mostly on contact-friction interactions developed between individual fibers. The finite element simulation can be usefully employed to better understand elementary mechanisms ruling the global behaviour of such structures and identifying their mechanical properties. The approach presented in this paper is aimed to this purpose. Its goal is to consider samples of reduced size of fibrous

Damien Durville
LMSSMat, Ecole Centrale Paris/CNRS UMR8579, Grande Voie des Vignes,
92290 Chatenay-Malabry, France; e-mail: damien.durville@ecp.fr

G. Zavarise & P. Wriggers (Eds.): Trends in Computational Contact Mechanics, LNACM 58, pp. 1–22.
springerlink.com

materials, taking into account all individual fibers contained in these samples, and modeling contact-friction interactions between them. The mechanical problem is set in the form of determining the equilibrium of an assembly of fibers, submitted to large displacements and finite strains. A kinematically enriched beam model is used to represent fibers, and assuming quasistatic loadings, an implicit scheme is employed to solve the problem.

The issue of modelling contact between beams, at the core of our approach, has been addressed in various ways in the literature, in a finite element framework. Usual techniques to determine contact between deformable surfaces seem hardly applicable to the case of beams. These techniques, based on master/slave strategies, check contact at nodes on one surface, associating to each node a target on the opposite surface, frequently using the normal direction to the first surface. Because crossings between beams can occur anywhere with respect to the finite element discretization, checking contact at nodes is often not sufficient in the case of beams. To overcome this difficulty, some authors [4–6] use a minimum distance criterion to characterize the location of contact. This approach has the advantage to determine accurately the location of contacts between beams and to provide a symmetrical treatment of interacting beams, and is particularly suitable for situations where the contact limits to only one point within each contact zone. However, in cases where contact is to be considered as a continuous phenomenon along a contact zone of non-zero length, the notion of minimum distance loses its relevance.

In order to handle the various contact configurations that can be encountered depending on the angle formed between interacting beams, we propose another way of determining contact, based on the determination of proximity zones between beams and on the construction of intermediate geometries in each proximity zone. The intermediate geometry, whose role is to approximate the geometry of the actual unknown contact zone, is used as a support for the contact discretization. Contact elements are generated between pairs of material particles on the surfaces of interacting beams that are predicted to enter into contact at some discrete locations defined on the intermediate geometry. This way contact is determined considering symmetrically both interacting beams with respect to the intermediate geometry, and a discretization size for contact can be defined. Another advantage of the introduction of proximity zones is to reduce the cost of the contact search by splitting it into two different tasks: the first one at a global level consisting only of associating coarsely pairs of close parts of beams, and the second one performed more accurately at the local level to associate particles for contact elements.

The high density of contacts involved in entangled structures is a challenge for the convergence of nonlinear algorithms dedicated to contact friction. In order to stabilize contact algorithms, two main improvements to the penalty method are implemented: a quadratic regularization for small penetrations to smooth transitions between contacting and non-contacting status, and a local adjustment of the penalty parameter for each contact zone to control the maximum penetration. The combination of these two ingredients is essential to make algorithms converge for cases involving up to nearly 100 000 contact elements.

The paper is organized as follows. Section 2 summarizes the way the global problem is set, and describes the 3D beam model used to represent fibers, based on three vector fields (nine degrees of freedom) according to Antman's theory with two unconstrained directors. Section 3 presents the approach to determine contact between beams based on the construction of proximity zones and intermediate geometries. Section 4 is dedicated to the mechanical models for contact-friction interactions, and to some algorithmic aspects. Three applications are then presented in Section 5 to demonstrate the capabilities of the approach. The first presented application is a test of sliding between two orthogonal beams to illustrate the taking into account of a moving frictional contact. The second one deals with the simulation of samples of woven fabrics, made of 480 fibers, from the determination of the unknown initial configuration of such structures, until the characterization by biaxial and shear loading tests. The third example refers to the identification of the transversal behaviour of a twisted textile yarn constituted of 250 fibers, and crushed between moving rigid tools.

2 Mechanical Equilibrium of an Assembly of Entangled Fibers

2.1 *Principle of Virtual Work*

In order to simulate the mechanical behaviour of an entangled structure, we consider it as an assembly of N fibers submitted to various loadings and undergoing large displacements, and we characterize the global displacement solution $\mathbf{u}$ by the following principle of virtual work, using a full-Lagrangian formulation:

Find $\mathbf{u}$ kinematically admissible, such that $\forall \mathbf{v}$ kinematically admissible

$$\sum_{I=1}^{N}\left(\int_{\Omega_0^I} \mathrm{Tr}\left(\mathbf{s}(\mathbf{u})\frac{\mathbf{DE}}{\mathbf{Du}}\cdot\mathbf{v}\right)\mathbf{d}\boldsymbol{\omega} + \sum_{J=I}^{N} W_{\mathrm{cf}}^{IJ}(\mathbf{u},\mathbf{v})\right) = \sum_{I=1}^{N} W_{\mathrm{ext}}^{I}(\mathbf{v}). \qquad (1)$$

In the above expression, the internal work for each fiber is expressed as function of the second Piola–Kirchhoff stress tensor $\mathbf{s}$ and the Green–Lagrange strain tensor $\mathbf{E}$; W_{cf}^{IJ} is the virtual work of contact-friction interactions between fibers I and J, and W_{ext}^{I} is the virtual of external loads applied to the fiber I. While the expression of contact-friction interactions is the main subject of the paper, the beam model employed to represent the behaviour of each fiber is now briefly described.

2.2 *3D Beam Model*

The model used to account for the mechanical response of fibers describes the kinematics of each cross-section by the means of three kinematical vectors, following

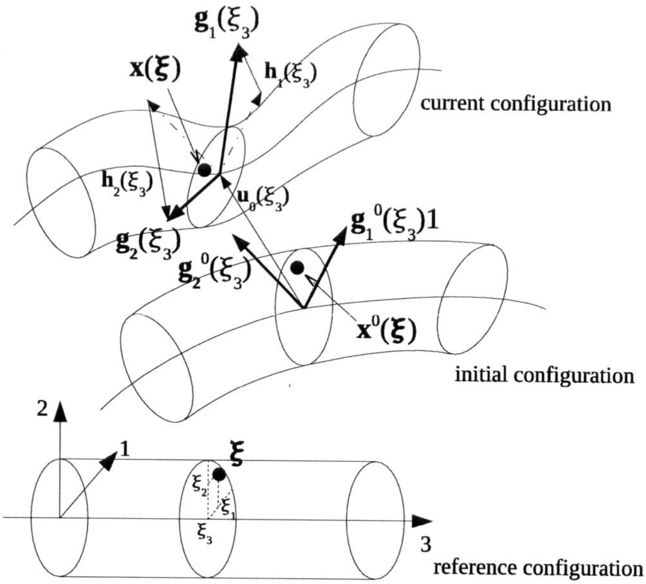

Fig. 1 Kinematical beam model.

Antman's theory [1]. According to this model, the position $\mathbf{x}$ of any particle of the beam, identified by its coordinates (ξ_1, ξ_2, ξ_3) in the reference material configuration is assumed to be expressed as follows:

$$\mathbf{x}(\boldsymbol{\xi}) = \mathbf{x}_0(\xi_3) + \xi_1 \mathbf{g}_1(\xi_3) + \xi_2 \mathbf{g}_2(\xi_3). \tag{2}$$

In this equation, ξ_1 and ξ_2 stand for the transverse coordinates of the particle in the cross-section, and ξ_3 for its curvilinear abscissa along the beam axis (see Figure 1). The three kinematical vectors involved in the model, defined on the centroidal line of the beam, are the position of the center of the cross-section, $\mathbf{x}_0(\xi_3)$, and two directors of the cross-section, $\mathbf{g}_1(\xi_3)$ and $\mathbf{g}_2(\xi_3)$. Accordingly, the displacement of any particle ξ of the beam is expressed as

$$\mathbf{u}(\boldsymbol{\xi}) = \mathbf{u}_0(\xi_3) + \xi_1 \mathbf{h}_1(\xi_3) + \xi_2 \mathbf{h}_2(\xi_3). \tag{3}$$

where $\mathbf{u}_0(\xi_3)$ is the displacement of the center of the cross-section, and $\mathbf{h}_1(\xi_3)$ and $\mathbf{h}_2(\xi_3)$ are the variations of the cross-section directors. The variations of cross-section directors are unconstrained, and both the norms of these directors and the relative angle between them may vary. According to this first order kinematical model, cross-sections are assumed to remain plane but may deform depending on the variations of cross-sections directors. For instance, intially circular cross-sections can take can any elliptical shape depending on the applied mechanical loading. In particular, the Poisson effect – contraction of the cross-section generated by an axial stretch – is naturally considered by this model.

The use of three kinematical vector fields (nine degrees of freedom) to describe the kinematics of each cross-section allows the derivation of a full 3D strain tensor, including in particular planar deformations of cross-sections. Standard constitutive laws can thus be employed to express 3D stresses in the beam as function of strains. Cases of specific behaviours, such as for example reduced bending or torsional stiffnesses for some particular fibers, can be considered either by using an orthotropic constitutive law, or by changing artificially the values of the moments of inertia used in the calculation of stresses.

The 3D beam model is implemented under a nonlinear finite strain framework.

3 Geometrical Handling of Contacts within an Assembly of Fibers

The case of entangled structures presents a specific context for the contact detection characterized by a high number of interacting fibers, and by large relative displacements between fibers making the arrangement of fibers continuously changing.

The goal of the contact detection is to predict locations where contact is likely to occur, and to provide with pairs of entities between which non-penetrating conditions can be formulated. Because of the high number of fibers possibly considered, the global algorithm for detecting contact must be fast.

3.1 Continuous Geometrical Approach of Contact between Fibers

3.1.1 Considerations of Various Contact Configurations

Diversified situations of contacts depending on the angle between fibers and on the extension of the contact zone can be encountered in a general assembly of fibers. When fibers cross each other with an angle close to 90°, the contact is almost pointwise. Yet for fibers forming locally a small angle, contact can be viewed as a continuous phenomenon along a contact surface that can be assimilated to a line of given length. In cases where two fibers are almost parallel or where a fiber is wound around another, the contact may be continuous all along the fibers.

For the pairing task in the contact detection, that consists in associating parts of the structure that are likely to come into contact, we seek a unique strategy able to handle the various contact configurations that can be encountered in a general assembly of fibers. The minimum distance criterion, employed in most of methods dealing with contact between beams to determine contact points, is effective in case of crossing between fibers, where the contact zone can be reduced to a point. However, if contact is continuous along a zone of non-zero length, the distance between fibers is almost constant in this region; the search of a minimum distance loses its relevance and another way of associating points in contact is required.

Various strategies are available for the consideration of contact between deformable surfaces. To couple contact entities, most of them take points on one of the surfaces, and determine corresponding target points on the opposite surface, using for example the normal direction to the first surface for this search. Such strategies, that demonstrate efficiency for deformable surfaces, seem nevertheless hard to adapt to the case of contact between deformable beams. Besides the fact it provides a non-symmetrical treatment of both interacting beams, the use of normal directions (planes orthogonal to the centerline of beams) raises difficulties in regions with high curvatures.

3.1.2 Pairing of Portions of Fibers

In order to preserve the continuous aspect of contact along a zone, instead of coupling directly points through a minimum distance criterion, we suggest to associate first pairs of portions of fibers that are close to each other. By this way, we try to consider simultaneously the geometry of both fibers in contact in order to better approximate the geometry of the actual contact zone. To this end, we proceed as follows. First we determine proximity zones between fibers in the whole assembly, defining a proximity zone as a pair of portions of fibers that are stated to be close to each other. Next, for each proximity zone, we determine an intermediate geometry, defined as the average between the two line segments constituting the proximity zone. This intermediate geometry can be viewed as an approximation of the actual contact zone. Instead of defining contact on one beam with respect to the other, contact is now defined on this intermediate geometry with respect to both beams. Because it depends on both geometries, normal directions determined from this intermediate geometry are better suited to the search of contact than those determined only from one of both beams.

3.2 Proximity Zones

As the number of contacts in entangled structures may be high, the determination of proximity zones meets a first need of reducing the cost of the contact search by operating first a coarse localization of contact. The purpose of this task is to delimit pairs of intervals on beam centerlines that are estimated to be close to each other. It is performed considering any possible pair of beams in the assembly. For a pair of beams (I,J), test points generated only to evaluate the distance between beams, are distributed on the first beam I according to a discretization size l_{zp}, and the closest distance to the other beam is calculated for each of these test points. The k-th test point on the first beam being identified by its curvilinear abscissa $s_k^I = kl_{zp}$, we search the curvilinear abscissa of the closest point on the centerline opposite beam,

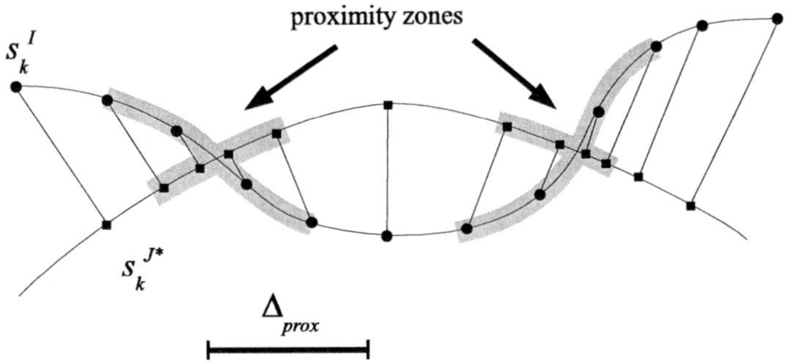

Fig. 2 Determination of proximity zones.

denoted s_k^{J*} (Figure 2). Considering the centerline of beams in their discretized form by finite elements, the closest point is either an orthogonal projection on a finite element or a node.

Giving a proximity criterion Δ_{prox}, we define the k-th proximity zone between beams I and J, denoted Z_k^{IJ}, in the following way:

$$\mathsf{Z}_k^{IJ} = \left([s_{k_1}^I, s_{k_2}^I], [s_{k_1}^{J\,*}, s_{k_2}^{J\,*}]\right), \text{ such that}$$

$$\forall k \in [k_1, k_2], \mathrm{dist}(\mathbf{x}(s_k^I), \mathbf{x}(s_k^{J*})) \leq \Delta_{\mathrm{prox}}, \tag{4}$$

where $\mathrm{dist}(\cdot,\cdot)$ stands for the distance between two points.

The process of determination of proximity zones provides a set of pairs of intervals.

3.3 Intermediate Geometries

For each proximity zone, we define an intermediate geometry as the average of the two line segments delimited on the centerlines of beams. The position of a point on the intermediate geometry related to the zone Z_k^{IJ}, and identified by its relative curvilinear abscissa ζ is calculated as

$$\mathbf{x}_{\mathrm{int}}^{IJ}(\zeta) = \frac{1}{2}\left(\mathbf{x}(s_{k_1}^I + \zeta(s_{k_2}^I - s_{k_1}^I)) + \mathbf{x}(s_{k_1}^{J\,*} + \zeta(s_{k_2}^{J\,*} - s_{k_1}^{J\,*}))\right), \quad \zeta \in [0,1], \tag{5}$$

and the tangent vector to the intermediate geometry $\mathbf{t}_{\mathrm{int}}^{IJ}(\zeta)$ is obtained straightforward by derivating this expression.

The intermediate geometry can be viewed as a first approximation of the geometry of the unknown actual contact zone. It will be used as a referential with respect to which contact between the two beams will be analyzed.

3.4 Discretization by Contact Elements

Owing to its symmetrical position with respect to the two interacting beams, the intermediate geometry affords a proper location from which to consider contact. Despite the contact zone is geometrically considered as continuous, we chose to approach contact along this zone by the means of discrete elements. These contact elements are defined at discrete locations on the intermediate geometry, and are constituted by the two material particles located on the surfaces of both beams that can be predicted to enter into contact at these discrete locations. Taking the intermediate geometry as referential to consider contact allows us to formulate the contact detection in the form of the question: Which particles on interacting beams are likely to come into contact at a given point of the intermediate geometry?

Contact elements are generated according to the following procedure. First, we determine the discrete locations on the intermediate geometry where the contact is to be checked. The number of contact elements to be generated is calculated depending on the length of the intermediate geometry and on a given discretization size. This discretization size is chosen according to the smaller finite element length on both beams, denoted $h_{\min}$, and to the polynomial degree of shape functions. If quadratic shape functions are used, the discretization size is fixed so as to get two contact elements per finite element, in order to be consistent with the number of constraints these finite elements can support. Denoting L_{int} the length of the intermediate geometry, the number of contact elements N_c is calculated as

$$N_c = \left[2 \frac{L_{\text{int}}}{h_{\min}} \right], \tag{6}$$

where $[\cdot]$ stands for the floor function, and the relative abscissa ζ_k of contact elements on the intermediate geometry are calculated as

$$\zeta_k = \frac{k-1}{N_c - 1}, \quad k = 1, \ldots, N_c. \tag{7}$$

Then, the determination of the pair of particles likely to come into contact at each point $\mathbf{x}^{IJ}_{\text{int}}(\zeta_k)$ on the intermediate geometry is performed in two steps. First, the two beam cross-sections candidate to contact at the relative abscissa ζ_k on the intermediate geometry, are selected at the intersections between the orthogonal plane to the intermediate geometry and the centerlines of both beams (see Figure 3). The curvilinear abscissae s^I_k and s^J_k of these cross-sections are characterized by

$$\begin{aligned} \left(\mathbf{x}^I(s^I_k) - \mathbf{x}^{IJ}_{\text{int}}(\zeta_k), \mathbf{t}^{IJ}_{\text{int}}(\zeta_k)\right) &= 0, \\ \left(\mathbf{x}^J(s^J_k) - \mathbf{x}^{IJ}_{\text{int}}(\zeta_k), \mathbf{t}^{IJ}_{\text{int}}(\zeta_k)\right) &= 0. \end{aligned} \tag{8}$$

In a second step, the two particles candidate to contact constituting the contact element are positionned on the outline of these cross-sections. Each of these particles, identified by their coordinates $\boldsymbol{\xi}^I_k$ and $\boldsymbol{\xi}^J_k$ in the reference configuration, is

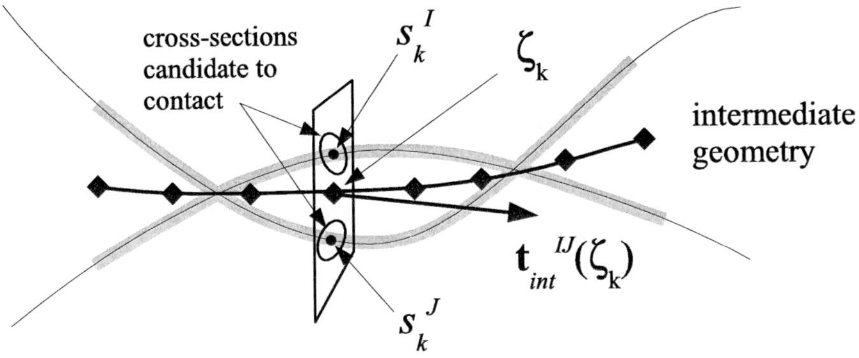

Fig. 3 Intermediate geometry.

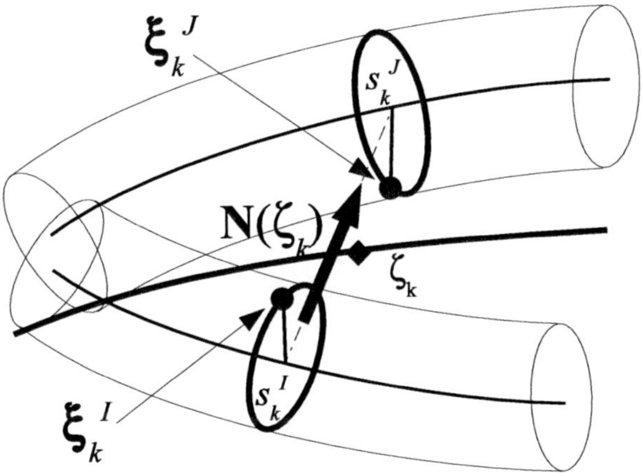

Fig. 4 Positions of particles candidate to contact on the outline of selected cross-sections.

placed at the intersection between the projection onto the cross-section of the direction between the centers of cross-sections and the outline of this cross-section (see Figure 4).

The global procedure associates to each contact test location identified by its relative abscissa ζ_c on the intermediate geometry, a contact element, denoted $E_c(\zeta_c)$, and defined as the pair of material particles predicted to enter into contact at this location :

$$E_c(\zeta_c) = (\xi_k^I, \xi_k^J). \tag{9}$$

3.5 Kinematical Contact Conditions

Non-penetration conditions are commonly expressed for each contact element by defining a gap function between contact particles and prescribing this function to remain positive. A normal direction according to which the distance between particles is measured, is required to evaluate this gap function. The role of this normal direction is predominant since it determines the direction according to which the contact is considered between the two beams. This normal direction is evaluated in differents ways depending on the angle θ between the two interacting beams. If the angle is greater than a given criterion θ_{cross}, reflecting a situation of crossing between the two beams, the normal direction is taken as the normalized vector product between the tangent vectors to both beams. When the angle is smaller than a criterion θ_{parall}, indicating that beams are nearly parallel, the normal direction is taken as the direction between the centers of the cross-sections candidate to contact. For intermediate angles, the normal direction is calculated as a linear combination between the last two expressions. This can be summarized as follows:

$$\text{if } |\theta| > \theta_{\text{cross}}, \quad \mathbf{N}(\zeta_k) = \frac{\mathbf{t}^I(\xi^I_{k3}) \times \mathbf{t}^J(\xi^J_{k3})}{\|\mathbf{t}^I(\xi^I_{k3}) \times \mathbf{t}^J(\xi^J_{k3})\|},$$

$$\text{if } |\theta| < \theta_{\text{parall}}, \quad \mathbf{N}(\zeta_k) = \frac{\mathbf{x}^I(\xi^I_{k3}) - \mathbf{x}^J(\xi^J_{k3})}{\|\mathbf{x}^I(\xi^I_{k3}) - \mathbf{x}^J(\xi^J_{k3})\|},$$

$$\text{if } \theta_{\text{parall}} \leq |\theta| \leq \theta_{\text{cross}}, \tag{10}$$

$$\mathbf{N}(\zeta_k) = \frac{\theta_{\text{cross}} - |\theta|}{\theta_{\text{cross}} - \theta_{\text{parall}}} \cdot \frac{\mathbf{x}^I(\xi^I_{k3}) - \mathbf{x}^J(\xi^J_{k3})}{\|\mathbf{x}^I(\xi^I_{k3}) - \mathbf{x}^J(\xi^J_{k3})\|} + \frac{|\theta| - \theta_{\text{parall}}}{\theta_{\text{cross}} - \theta_{\text{parall}}} \cdot \frac{\mathbf{t}^I(\xi^I_{k3}) \times \mathbf{t}^J(\xi^J_{k3})}{\|\mathbf{t}^I(\xi^I_{k3}) \times \mathbf{t}^J(\xi^J_{k3})\|}.$$

The kinematical condition to be fulfilled at each contact element is expressed by means of the gap function denoted g_N as follows:

$$g_N(\zeta_k) = \left(\mathbf{x}^I(\boldsymbol{\xi}^I_k) - \mathbf{x}^J(\boldsymbol{\xi}^J_k), \mathbf{N}(\zeta_k)\right) \geq 0. \tag{11}$$

4 Mechanical Models for Contact and Friction

4.1 Quadratic Regularization of the Penalty Method for Contact

Dealing with an entangled structure with a high number of contacts, the convergence for the contact problem is practically unreachable using the standard penalty method. The high number of contacts, the softness of fibers and the connections between them favor unstable situations where oscillations of contact in one region have repercussions in neighboughring regions. For these reasons, contact needs to be stabilized through proper adaptations of the penalty method.

Using a quadratic regularization of the penalty is a first way to stabilize the contact algorithm. Such a regularization consists in employing a quadratic function for very small gaps, below a given regularization threshold p_{reg}, and to express the normal reaction R_N as function of the gap:

$$\begin{cases} \text{if } g_N > 0, \quad R_N = 0 \\ \text{if } -p_{\mathrm{reg}} \le g_n \le 0, \quad R_N = \dfrac{k_N}{2p_{\mathrm{reg}}} g_N^2, \\ \text{if } g_n < -p_{\mathrm{reg}}, \quad R_N = -k_N g_N - \dfrac{k_N}{2} p_{\mathrm{reg}}. \end{cases} \tag{12}$$

The quadratic regularization appears to be particularly effective for contact elements holding very low contact forces, and whose contact status may easily change from one iteration to another. In this case, the quadratic regularization of the penalty smooths the change of contact stiffness between contacting and non-contacting status. However, in order to be really effective, a small, but significant, proportion of contact elements has to be concerned by this regularization, and so needs to have penetrations lower than the threshold p_{reg}.

4.2 Local Adjustment of the Penalty Parameter

The need to control the penetration to ensure the effectiveness of the penalty regularization leads to the second improvement of the penalty, by adjusting locally the penalty coefficient in order to limit the maximum penetration. Because resultant contact forces can be very different from one contact zone to another, and can also vary widely, penetrations of very different orders would be expected if a unique and constant penalty coefficient was used for all contact zones. To avoid such circumstances, the penalty coefficient is adjusted for each proximity zone so as the maximum penetration within this zone is equal to a given maximum allowed penetration, denoted $p_{\max}$. Penalty parameters are iteratively adjusted during the solution for each loading step. At the i-th iteration of this process, the penalty parameter k_c^i is adjusted for each proximity zone in function of the previous parameter k_c^{i-1} in the following way:

$$k_c^i = \frac{g_{N,\max}}{p_{\max}} k_c^{i-1}, \tag{13}$$

where $g_{N,\max}$ is the maximum penetration measured on the proximity zone. Similar techniques to adjust the penalty parameter have already been suggested [2]. One difference here is that the parameter is not adjusted at the level of each contact element, but more globally for a set of contact elements contained in a proximity zone.

4.3 Regularized Coulomb's Law for Friction

As far as tangential reactions at contact elements are concerned, a regularized Coulomb's law accounting for a small reversible relative displacement before the gross sliding occurs is formulated in an incremental way. Knowing the reversible tangential displacement at the previous step $\mathbf{g}_{T,\text{rev}}^{n-1}$, and the increment of relative tangential displacement at the current step $\Delta \mathbf{g}_T^n$, a trial reversible relative tangential displacement $\mathbf{g}_{T,\text{rev}}^{n,\text{tr}}$ for the current step is computed as follows:

$$\mathbf{g}_{T,\text{rev}}^{n,\text{tr}} = \mathbf{g}_{T,\text{rev}}^{n-1} + \Delta \mathbf{g}_T^n. \tag{14}$$

The current reversible tangential displacement is evaluated according to

$$\text{if } \|\mathbf{g}_{T,\text{rev}}^{n,\text{tr}}\| \leq g_{T,\max}, \quad \mathbf{g}_{T,\text{rev}}^{n} = \mathbf{g}_{T,\text{rev}}^{n,\text{tr}}, \tag{15}$$

$$\text{else} \quad \mathbf{g}_{T,\text{rev}}^{n} = g_{T,\max} \frac{\mathbf{g}_{T,\text{rev}}^{n,\text{tr}}}{\|\mathbf{g}_{T,\text{rev}}^{n,\text{tr}}\|}, \tag{16}$$

where $g_{T,\max}$ is the maximum allowed reversible tangential displacement.

4.3.1 Transfer of History Variables Related to the Friction Model

The reversible part of the friction law requires a transmission of the history variable $\mathbf{g}_{T,\text{rev}}^{n}$ from one step to the next. However, the fact that contact elements have no continuity in time raises difficulties regarding this transfer. Since this information cannot be attached to contact elements, whose constituting particles are constantly changing, the vector of reversible tangential displacement is stored for each particle of both beams at the end of each loop on the determination of contact. For the next iteration on contact determination (either within the same step or at the begining of the next step), for any generated contact element, the value of the vector of reversible tangential displacement is first interpolated for each contact particle from the values stored at the previous iteration on the beam holding the particle, and then interpolated for the contact element between the vectors determined at both particles.

4.4 Algorithmic Aspects

The global problem is solved using an implicit method, within a quasi-static framework. It involves nonlinearities of various kinds that need different algorithms to be solved, depending essentially on whether the concerned nonlinear quantities can be easily linearized or not.

Consistent with our approach of the contact problem, the global problem can be divided into two nested nonlinear subproblems, the first one dealing with the statement of linearized kinematical contact conditions (Eq. 11), and the second one dedicated to solving the mechanical problem satisfying these conditions. For the latter problem, under fixed unilateral contact conditions, most nonlinear quantities can be differentiated, and a Newton–Raphson type algorithm can be employed to solve simultaneously the nonlinearities related to the contact status, the sliding status, and the nonlinear terms involved in the internal virtual work of beams.

A linearization of the first level problem seems very hard to obtain since the two nonlinear handled entities, namely the contact elements and the normal directions for contact, are defined through geometrical constructions, and not directly by differentiable equations. Newton-like algorithms can therefore not be employed. To account for the nonlinear character of this problem, iterations on the determination of the two entities are simply made using two nested fixed point algorithms. This leads to the following algorithm, made of three nested loops, to solve the problem for each loading step:

First level loop – fixed point algorithm on the determination of contact elements

Second level loop – fixed point algorithm on the determination of normal directions for contact

Third level loop – Newton–Raphson algorithm to solve

- contact status,
- sliding status,
- finite strains nonlinearities.

Although the convergence of the two first level fixed point iterations is hard to prove, from our experience, three iterations for each of these two loops is generally sufficient to obtain a good convergence on the global solution for not too large loading increments.

5 Applications

5.1 Test of Alternate Sliding between Two Beams

A test of sliding between beams is performed to illustrate the effectiveness of the transfer of history variables related to the friction model to follow sliding phenomena. Two 4 millimeter long crossing beams, with a radius of 0.1 mm, and a Young modulus equal to 4000 MPa, are considered. A force of 0.1 N is applied at both ends of one of the beam, while the other beam is maintained vertically at ends (Figure 5). An alternate translation of 0.75 mm in their longitudinal direction, divided in 25 increments, is applied at both ends of each beams. A friction coefficient of 0.2 is considered, with a maximum allowed reversible tangential displacement of 0.5 micrometers. The maximum penetration allowed for each proximity zone, used to adjust the penalty ocefficient for contact, is equal to 5×10^{-4} mm, that is to say 0.5% of the radius of the beam.

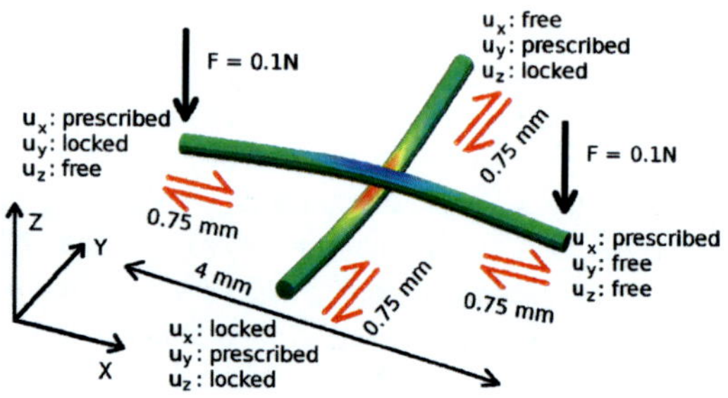

Fig. 5 Description of the cyclic alternate loading applied to the beams.

Fig. 6 Deformed configurations at end of each alternate motion.

After a first translation, the full cycle of alternate displacements is repeated twice. For each alternate motion, there is first a sticking phase at contact between beams, before the gross sliding occurs. Each beam is discretized with 10 quadratic finite elements, and during the sliding phase, the location of contact goes accross two successive elements on each beam (Figure 6). The total number of iterations per loading step is between 13 and 23, depending on the step.

The resultant horizontal force applied to each of the beams is plotted in Figure 7. The curves for the two successive cycles exactly superimpose. They exhibit clearly a sticking and a sliding phase. The slight slope in the sliding part is presumably due to the curvature of the bent beams, which makes the orientation of the tangential plane at contact vary as the location of contact along the beams changes, while the force is measured in the horizontal plane. Small irregularities on the curve are related to the sliding from one finite element to the next, and may be due to the discontinuity in curvature at the junction between these two elements. However, the ability of the model to reproduce the transition between the sticking and the sliding status demonstrates the effectiveness of the transfer of history variables between contact elements which are discontinuous in time.

5.2 *Modeling of Woven Fabrics*

When woven textiles are employed as reinforcements to manufacture textile composites, their mechanical properties have to be known in particular to assess their formability. Indentifying local mechanisms responsible for the global response of

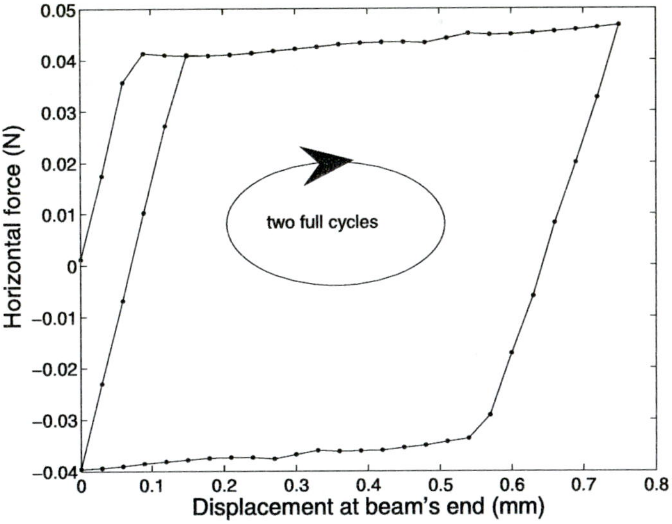

Fig. 7 Horizontal interaction force between the two sliding beams.

Fig. 8 Starting configuration for the calculation of the initial geometry.

such structures is also important to better understand these materials. The simulation at the scale of individual fibers can be very helpful in this context. To have a good representation of the media constituting these materials, a reasonable number of fibers has to be considered for each yarn.

5.2.1 Calculation of the Initial Geometry

As the layout of fibers within a woven fabric cannot be determined a priori, the initial geometry of the structure to study has to be computed by trying to reproduce the way yarns are intertwined together through the weaving process. As the process itself is not easy to simulate, the way we use to obtain this initial geometry is to make

Table 1 Characteristic features of the woven sample model.

Material characteristics	
Fiber diameter	0.052 mm
Young Modulus	73 000 MPa
Poisson ratio	0.3
Model characteristics	
Coefficient of reduction of bending stiffness	0.12
Friction coefficient	0.1
Maximum allowed penetration	2×10^{-4} mm
Number of fibers	480
Number of finite elements per fiber	40
Total number of DOFs	349 920
Number of contact elements	$\approx$ 90 000

yarns move gradually above or below each other at their crossings, depending on the chosen weaving pattern. To do this, starting from an initial configuration where all yarns lay on the same plane and interpenetrate each other (Figure 8), the direction of contact between fibers belonging to different yarns is temporarily chosen vertical and oriented according to the stacking order defined at each crossing by the weaving pattern. This way fibers of different yarns are gradually moved until there is no more penetration between yarns. Once this step is achieved, classical contact conditions are considered, and the equibrium of the fabrics is found, just applying small tensile loads at ends of yarns.

This procedure is illustrated on an example involving ten yarns, each of them constituted by 48 fibers. Starting from the same initial configuration, two weaving patterns – a plain weave and a twill weave – are applied. Main features of the model are summarized in Table 1. Fifteen steps are necessary to compute the initial configuration, with in average 30 total Newton iterations per step.

The final shapes of both weaves are shown in Figures 9 and 10. Cuts of the samples at some steps of this initial procedure (Figures 11 and 12) show the rearrangement of fibers within yarns.

5.2.2 Application of Test Loadings

Once the initial configuration has been computed, biaxial tensile tests and shear loading tests that are commonly performed to characterize woven fabrics, can be simulated through the application of appropriate boundary conditions.

Biaxial tensile tests, prescribing identical elongations in warp and weft directions, are simulated using 20 loading steps, with in average 45 Newton iterations per step. These tests exhibit a nonlinearity at the start of the loading curve (Figure 13). This nonlinearity is likely related to the possible compression of yarns cross-sections at crossings, allowed by the free spaces existing between fibers. This

Fig. 9 Computed initial configuration for the plain weave.

Fig. 10 Computed initial configuration for the twill weave.

nonlinearity is stronger for the plain weave than for the twill weave, probably due to the fact that fibers are initially more undulated in the plain weave than in the twill weave.

A cyclic shear loading test is performed on the plain weave sample, by applying alternate opposite displacements to edges of the sample in the weft direction, in 43 loading steps. Approximately 25 Newton iterations are needed in average for each step. This test demonstrates an hysteretic behaviour due to friction interactions between fibers (Figure 15).

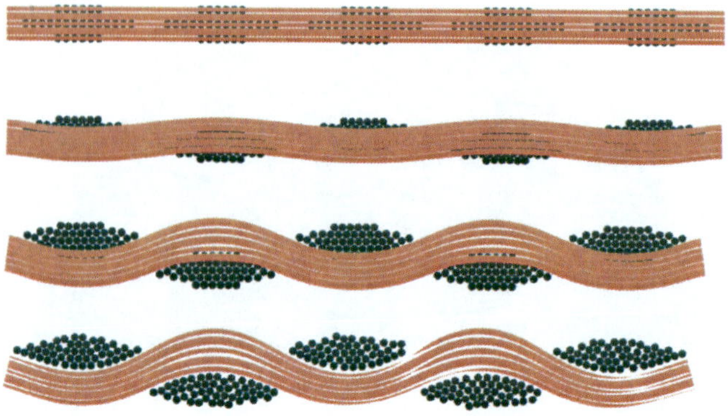

Fig. 11 Cuts of the plain weave sample during the computation of the initial configuration.

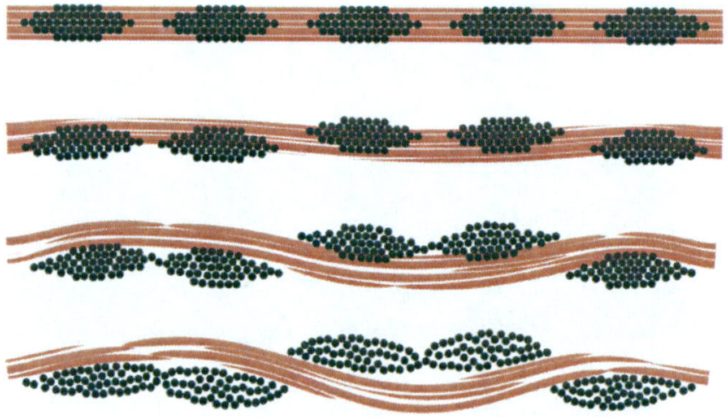

Fig. 12 Cuts of the twill weave sample during the computation of the initial configuration.

5.3 Identification of the Transverse Mechanical Behaviour of a Twisted Textile Yarn

A model is developed to study locally the effects of the transverse compression undergone by yarns at their crossings in woven fabrics, in order to better understand the influence of various parameters such as the tensile force, the torsion, and even the disorder in the initial layout of fibers, on their transverse behaviour.

So as to simulate a test experiment consisting in crushing a tightened twisted yarn (Figure 16) by means of two moving rigid tools, a finite element model (characteristics in Table 2) representing a yarn made of 250 fibers is considered. We start with a configuration where all fibers are parallel and in the form of a compact arrangement. A disorder is first introduced in this initial layout applying small random perturbations to the straight trajectories of fibers. As these random perturb-

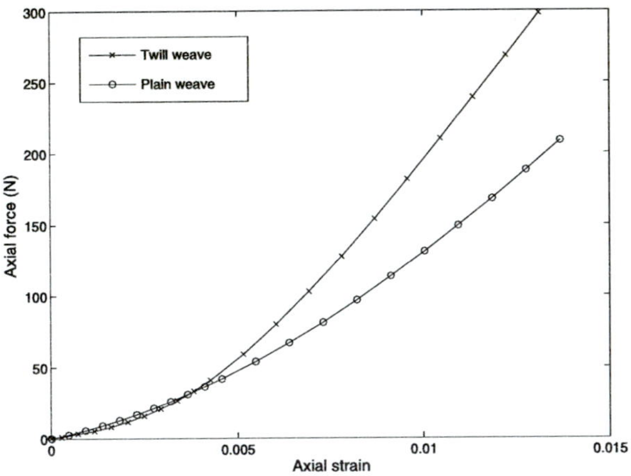

Fig. 13 Biaxial loading curves for the plain and twill weave samples.

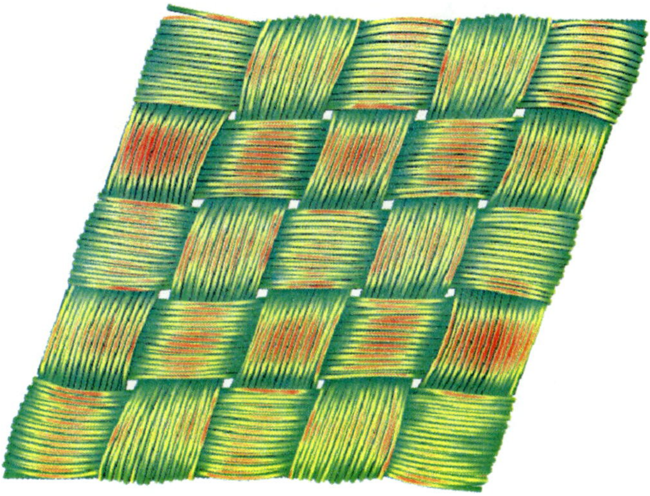

Fig. 14 Plain weave sample under a shear loading.

ations cause interpenetration between fibers, a first stage of simulation is needed to find a new equilibrium configuration fulfiling contact conditions between fibers, and an equilibrated disordered configuration is obtained. A tensile force is then applied and the yarn is twisted until a given torsion.

To simulate the transverse compression, contact conditions between fibers and two moving rigid tools are considered. The transverse compression is applied in 150 steps, approximately 30 Newton iterations being necessary to solve each step. Evolutions of the central cross-section of yarn are represented in Figure 17, showing some interesting features. Fibers are more tightened at the periphery of the yarn than

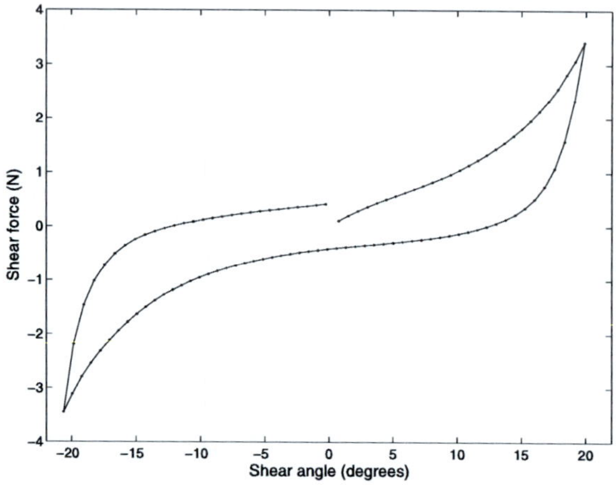

Fig. 15 Shear loading curve for the plain weave sample.

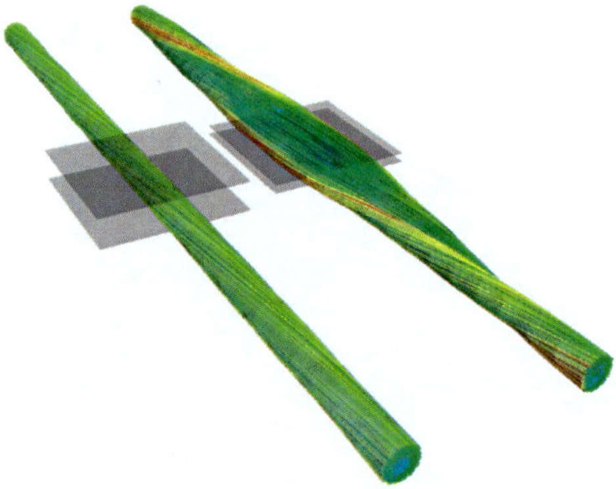

Fig. 16 Transverse compression of a twisted yarn between two moving rigid tools.

at the center, due to the elongation induced by the deflection of outside fibers. These outside fibers produce a confinement effect in relation with the global twisting, increasing the density of fibers at the periphery of the yarn. The global deformation induces large relative displacements between fibers that may be compared to flows of granular materials.

The global loading curve (Figure 18) shows some irregularities that could be related to slidings between groups of fibers corresponding to possible alignements between fibers appearing at some steps of the loading. This kind of curve has similarities with behaviours observed on generalized entangled media such as wools [3].

Table 2 Characteristic features of the twisted yarn model.

Material characteristics	
Fiber diameter	0.02 mm
Young Modulus	73 000 MPa
Poisson ratio	0.3
Model characteristics	
Coefficient of reduction of bending stiffness	0
Friction coefficient	0.05
Maximum allowed penetration	10^{-4} mm
Number of finite elements per fiber	15
Number of fibers	250
Number of DOFs	69 750
Number of contact elements	$\approx$ 17 000

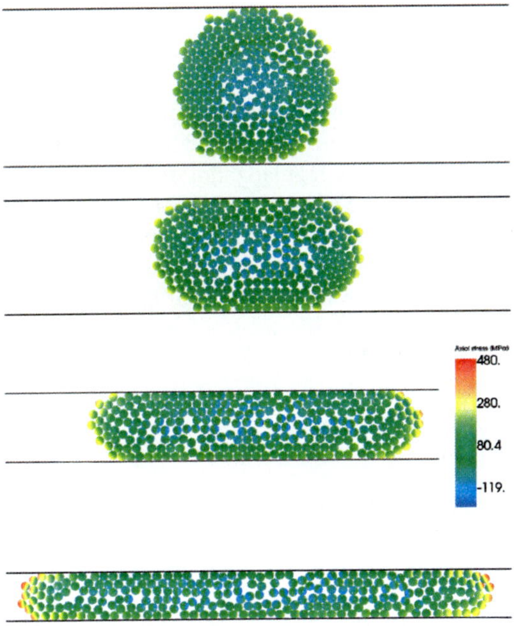

Fig. 17 Evolution of the yarn median cross-section during the transverse compression.

6 Conclusion

The geometrical approach to the detection of contact between beams based on the determination of intermediate geometries at the level of proximity zones allows to consider contact for beams arranged according to various layouts. The high numbers

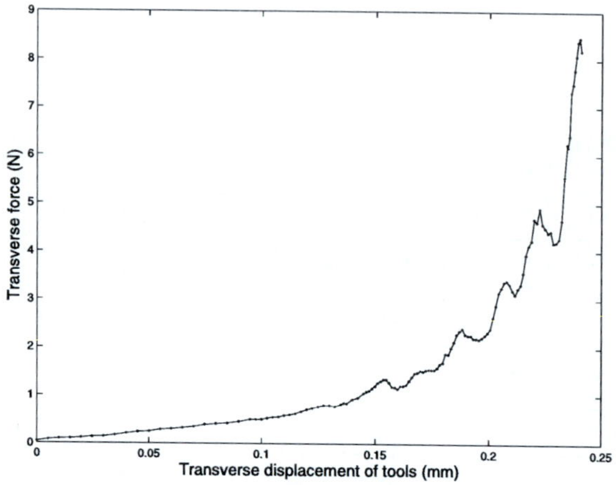

Fig. 18 Loading curve for the transverse compression of a twisted yarn.

of fibers and contacts that are to be handled to simulate the behaviour of entangled structures incite to improve the convergence of algorithms for contact and friction, in particular by using a quadratic regularization for the penalty method employed for contact and by adjusting the penalty parameter for each contact zone so as to control the maximum penetration. These improvements enable to solve problems with high numbers of contact elements using an implicit solution scheme. Applications to textile materials made of few hundreds of fibers demonstrate the ability of the method to explore the complex mechanical behaviour of entangled structures.

References

1. Antman, S.S.: Nonlinear Problems of Elasticity. Springer, New Yark (1995)
2. Chamoret, D., Saillard, P., Rassineux, A., Bergheau, J.-M.: New smoothing procedures in contact mechanics. Journal of Computational and Applied Mathematics 168(1-2), 107–116 (2004)
3. Durville, D.: Numerical simulation of entangled material mechanical properties. Journal of Material Sciences 40(22), 5941–5948 (2005)
4. Konyukhov, A., Schweizerhof, K.: Geometrical covariant approach for contact between curves representing beam and cable type structures. PAMM, Proc. Appl. Math. Mech. 8, 10299–10300 (2008)
5. Litewka, P., Wriggers, P.: Frictional contact between 3D beams. Computational Mechanics 28, 26–39 (2002)
6. Zavarise, G., Wriggers, P.: Contact with friction between beams in 3-D space. International Journal for Numerical Methods in Engineering 49(8), 977–1006 (2000)

3D Contact Smoothing Method Based on Quasi-C1 Interpolation

Maha Hachani and Lionel Fourment

Abstract This paper describes the effect of tool discretization on the simulation of metal forming processes, especially for processes where the contact area is quite small with respect to the component size. The smoothing of contact surfaces, which are defined by linear triangles, is based on a higher order quadratic interpolation of the curved surface. This interpolation is derived from the node positions and their normal vectors, as proposed by Nagata. The normal vectors are calculated at each node from the existing discretized surface by considering a patch of surrounding elements and using a consistent strategy. The efficiency and reliability of the resulting contact model are assessed on several examples, such as the indentation of a parallelepiped and the drawing of a wire.

1 Introduction

Finite element simulation of forming processes is now widely recognized as an efficient tool for designing actual forming processes in industry. However, for metal forming processes where the contact area is quite small with respect to the component size, like wire drawing, extrusion and rolling, there is still an extensive need for increasing the accuracy and robustness of the results while decreasing the computational time. In fact, despite using parallel computers during several weeks, results might not be accurate enough to properly reproduce the experiments.

In general, these processes involve a complex material flow resulting from contact over a very small area with either simple or very complex and uneven shapes. The numerical modelling of this contact problem requires more realistic descriptions of the discretized contact geometry than this which is usually proposed. Indeed, the description of such surfaces by straight elements (segments in 2D and facet in 3D)

Maha Hachani · Lionel Fourment
CEMEF – Centre for Material Forming, Mines ParisTech, CNRS UMR 7635, BP 207,
06904 Sophia Antipolis Cedex, France; e-mail: maha.hachani@mines-paristech.fr

G. Zavarise & P. Wriggers (Eds.): Trends in Computational Contact Mechanics, LNACM 58, pp. 23–40.
springerlink.com

generally require a very high density of the mesh in the contact zone, leading to complex and computationally inefficient calculations. The analysis shows that, very often, simulation results significantly depend on the mesh size in the contact area. So, a significant source of errors lies in the contact treatment, which is usually not accurate enough, even with a proper mesh refinement. The main aim of this study is to improve the contact treatment without increasing the computational cost of the finite element resolution, in other words, without decreasing the finite element mesh size or the time step.

Generally, the numerical simulation of contact problems closely depends on the robustness and precision with which contact interactions can be described, which is related to the manner in which the contact constraints are imposed. This issue is discussed in a large body of literature which shows the relative advantages and disadvantages of different formulations, such as the penalty method [1, 2], the Lagrange multipliers method [3] and the augmented Lagrangian method [4, 5]. In addition, the difficulty of contact algorithms arises from the non-linearity and non-smoothness of the interactions. In fact, large deformation problems often result into potentially large slidings. As points slide over long distances on the discretized surface, they are brought to cross boundary elements. This results into geometric numerical discontinuities and sudden changes in the material flow.

The specific goal of this work is to develop new techniques for reducing the non-smoothness of the contact interactions that arises from the finite element discretization of the contact surface, in order to make the numerical schemes more precise and robust. Various techniques have been proposed in the literature, such as NURBS, Bezier patches, Gregory patches and local Hermite diffuse interpolations, for instance in [6–9], respectively. In spite of their efficiency to provide good approximations of the original surface and to ensure a high continuity degree of the contact surface, their applications to complex 3D problems raise several difficulties. Most approaches are based on specific assumptions, while other require free parameters that have to be given a priori, which is hardly possible since the analytic properties of the original surface are generally unknown [10]. On the other hand, the high level of continuity of the smoothed surface is generally obtained from geometric informations that are derived from large element patches (sets of elements that surround a surface node or element, see Figure 1), but those patches are quite difficult to handle in a parallel computational environment, as its elements can belong to a large number of different processors. In order to avoid this difficult parallelization problem, it is decided to restrict the patches to first order neighbours (see Figure 1), in other word to the elements that contain the considered node. Such first order patches can be easily handled in a parallel framework, by just changing the way of scanning the elements.

Moreover, as the smoothed surfaces often are described by complex expressions (cubic, higher polynomials or rational functions), the contact search algorithm is made significantly more elaborated and computationally more expensive. In this study, the rather simple smoothing procedure proposed by Nagata [10] is selected, and a corresponding contact search algorithm is developed. This approach is applied

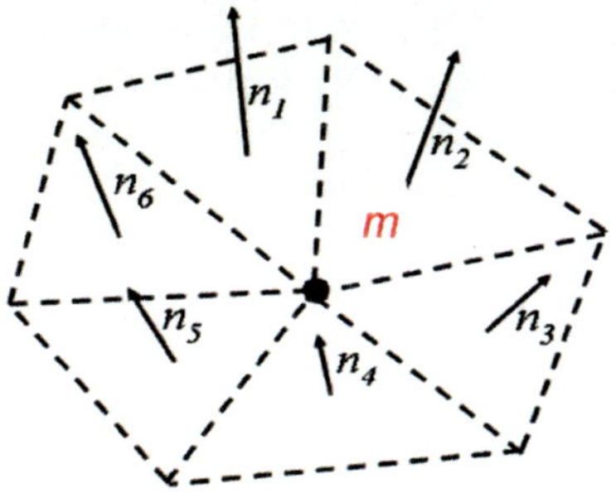

Fig. 1 First order patch ∂P_m where the centre is the considered node m.

to a specific and discriminating problem, the Hertz's indentation problem, and to a more complex metal forming problem, the wire drawing.

2 Finite Element Formulation

The proposed smoothing contact technique based on Nagata's idea is introduced in the 3D FEM code FORGE3® [11], so the corresponding Finite Element formulation is first briefly described.

2.1 Constitutive Equation and Integral Formulation

An updated Lagrangian formulation is used to describe the material deformation (6). Only dense materials are considered, and the elastic material deformation is neglected, so the incompressibility condition is written as

$$\mathrm{div}(\mathbf{v}) = 0 \tag{1}$$

where $\mathbf{v}$ is the velocity field. The rate form of the equilibrium equations and boundary conditions at time t are expressed by the virtual power principle, in term of the velocity $\mathbf{v}$ and p pressure fields satisfying the contact equations, for any virtual velocity $\mathbf{v}^*$ and pressure p^* fields

$$\begin{cases} \int_\Omega \mathbf{s} : \dot{\varepsilon}^* d\omega - \int_\Omega p \cdot \mathrm{div}(\mathbf{v}^*) d\omega - \int_{\partial\Omega_c} \tau \cdot \mathbf{v}^* dS = 0 \\ \int_\Omega p^* \cdot \mathrm{div}(\mathbf{v}) d\omega = 0 \end{cases} \tag{2}$$

where $\mathbf{s}$ is the deviatoric part of the stress tensor. The expression of $\mathbf{s}$ is given by the constitutive equation which can be elastic (Hook law), elasto-plastic (Prandtl–

Reuss criteria), viscoplastic (Norton–Hoff law) or elasto-viscoplastic. is the strain rate tensor. Ω, $\partial\Omega$ and $\partial\Omega_c$ respectively denote the domain occupied by the body, its boundary and the contact surface at each time t. τ is the tangential shear stress resulting from friction.

2.2 Finite Element Formulation

The $P1+/P1$ tetrahedral finite element is used to interpolate the pressure and velocity fields in a compatible way [12]. The pressure p is linearly interpolated on each tetrahedron, while the velocity is linearly expressed in term of the N_n^l shape functions with enrichment from the N_m^b bubble function, as

$$p = \sum_n P_n N_n^l \tag{3}$$

The parametric relation between the physical coordinate vector $\mathbf{x}$ and the local coordinate vector ξ, using the nodal coordinate $\mathbf{X}_n$ is expressed as follows:

$$\mathbf{x} = \sum_n \mathbf{X}_n N_n^l \tag{4}$$

The contact condition can be imposed at the nodes, as described below, and enforced using a penalty formulation. After finite element discretization, the complete set of equations (2) can be written in a symbolic way as

$$R(\mathbf{X}, \mathbf{V}, \mathbf{P}) = 0 \tag{5}$$

This system of equations has to be satisfied at any simulation time. An Eulerian implicit scheme is used time integration, and the configuration update is written as

$$\forall t \mathbf{x}^{t+\Delta t} = \mathbf{x}^t - \mathbf{v}^{t+\Delta t} \Delta t \tag{6}$$

2.3 Discrete Contact Treatment

To take into account the contact constraints that govern the interaction between Ω and the obstacle, the relative position of each material point of the boundary domain (coordinate $\mathbf{x}$) with respect to the tool has to be checked. The tools are considered as rigid and their surfaces are discretized by triangular facets. The signed distance is defined by

- $\delta(\mathbf{x}) > 0$ if the point is outside the tool;
- $\delta(\mathbf{x}) = 0$ if the point is on the tool surface;
- $\delta(\mathbf{x}) < 0$ means that the point has penetrated into the tool.

A nodal approach (node to facet) is used. For any node n, the contact condition is prescribed at the end of the time increment:

$$\delta_n^{t+\Delta t}(\mathbf{x}) \geq 0 \tag{7}$$

This condition is linearized with the assumption that the tool can locally be approximated by its tangential plane, which will be referred to as an *explicit* scheme:

$$\begin{aligned}\delta_n^{t+\Delta t} &= \delta(\mathbf{M}_n^{t+\Delta t}) \\ &= \delta(\mathbf{M}_n^t) + \frac{d\delta}{dt}(\mathbf{M}_n^t)\Delta t + O(\Delta t^2) \\ &\approx \delta_n^t + (\mathbf{v}_{\text{tool}}^t - \mathbf{v}^t)\cdot \mathbf{n}^t \Delta t\end{aligned} \tag{8}$$

where $\mathbf{v}_{\text{tool}}$ is the tool velocity and n_n^t is the surface normal vector at the point of the surface tool which is the closest to node n.

Finally, the discrete form of the contact potential is expressed as

$$\varphi_{\text{contact}}(\mathbf{v}_h) = \frac{1}{2}\rho \sum_{n\in\partial\Omega_c} \left[(\mathbf{V}_n - \mathbf{v}_{\text{tool}})\cdot \mathbf{n}_n - \left(\frac{\delta_n + \delta_{\text{pen}}^-}{\Delta t} \right) \right]^{+2} S_n \tag{9}$$

where ρ is a penalty coefficient and S_n is the surface associated to the node n and δ_{pen}^- is a small numerical coefficient.

2.4 Friction

The most common friction model used for metal forming applications is Coulomb's. The shear stress τ_f is proportional to the contact pressure $p_c = \sigma_n = (\sigma \cdot \mathbf{n}) \cdot \mathbf{n}$, as follows:

$$\tau_f = -\mu \sigma_n \frac{\Delta \mathbf{v}_s}{\|\Delta \mathbf{v}_s\|} \tag{10}$$

where μ is the Coulomb friction coefficient, $\Delta \mathbf{v}_s$ is the sliding velocity, $\Delta \mathbf{v}_s = (\mathbf{v} - \mathbf{v}_{\text{tool}}) - [(\mathbf{v} - \mathbf{v}_{\text{tool}}) \cdot \mathbf{n}] \cdot \mathbf{n}$.

According to the smallness of the time steps in metal forming applications, a simple explicit time integration scheme is used in the FORGE® software. The friction surface is determined by the nodes that are in contact at the beginning of time increment t, while the contact pressure σ_n is determined from the stress tensor values that are calculated at $t - \Delta t$, the previous time step .With this explicit formulation, the friction conditions are determined at the previous time step, so any change brought to the contact treatment does bring any change to the friction treatment.

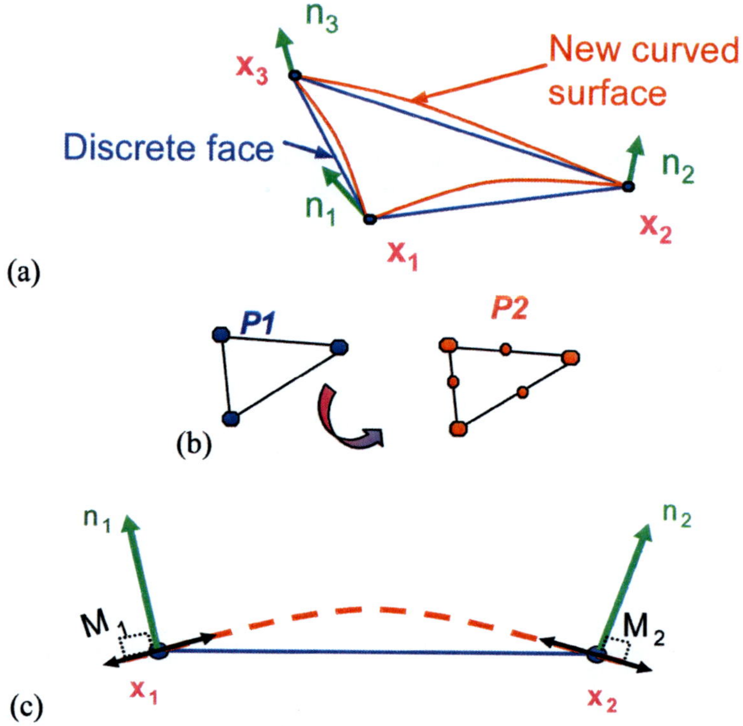

Fig. 2 Nagata patch: (a), (b): interpolation of a face procedure; (c): interpolation of a curved segment.

3 Smoothing Technique

3.1 Nagata's Method

The central idea of Nagata's algorithm [10] is to increase the interpolation order of the triangularized surface, from linear to quadratic (see Figure 2a), by only using local operators. It has been successfully applied to sheet metal forming problems [13]. The curvature of triangular (or quadrilateral) facets (see Figure 2) is recovered by only using the positions and normal vectors directions at the vertices of polyhedral meshes. This approach is simple and local, so it is both computationally inexpensive and easy to parallelize.

Each triangle is composed of three edges. In a first stage, each edge is independently interpolated by second order polynomial (see Figure 2c). Then, in a second stage, the surface of the triangle can be interpolated using the second degree polynomial that is defined by its trace on the quadratic segments (see Figure 2b).

The curvature of each segment is locally and independently recovered by computing the quadratic curve passing through the end nodes $\mathbf{M}_1$ and $\mathbf{M}_2$ (see Figure 2a) which coordinates are respectively $\mathbf{X}_1$ and $\mathbf{X}_2$, and that is orthogonal to the normal vectors $\mathbf{n}_1$ and $\mathbf{n}_2$, respectively at these points (see Figure 2a). This is achieved by a hierarchical h interpolation. A mid-segment node is added to the mesh edge, so allowing the second order interpolation (see Figure 2b). The equation of this curve is as follows:

$$\mathbf{x}(\xi) = \mathbf{x}_1 N_1(\xi) + \mathbf{x}_2 N_2(\xi) + \bar{\mathbf{x}} N_3(\xi) \tag{11}$$

where ξ is the local coordinate, which satisfies the condition $0 \leq \xi \leq 1$, N_1, N_2 are the linear interpolation functions ($N_1(\xi) = 1 - \xi; N_2(\xi) = \xi$) of the segment, and N_3 is the hierarchical quadratic function: $N_3(\xi) = 4\xi(1-\xi)$. $\bar{\mathbf{X}}$ is the position of the mid-edge node, which is unknown. It is determined by imposing that the curve is orthogonal to the normal vectors $\mathbf{n}_1$ and $\mathbf{n}_2$ given at nodes $\mathbf{X}_1$ and $\mathbf{X}_2$, respectively, in the least square sense, using the generalized inverse:

$$\bar{\mathbf{x}} = (\mathbf{A}^t\mathbf{A} + \varepsilon\mathbf{E})^{-1}\mathbf{A}^t\mathbf{B} \tag{12}$$

where

$$A = \begin{bmatrix} u_1\, v_1\, w_1 \\ u_2\, v_2\, w_2 \end{bmatrix}$$

where (u_1, v_1, w_1) and $(u_2 v_2 w_2)$ respectively are the coordinates of $\mathbf{n}_1$ and $\mathbf{n}_2$, $B_1 = -\frac{1}{4}(\overrightarrow{M_1M_2} \cdot \mathbf{n}_1)$, and $B_2 = \frac{1}{4}(\overrightarrow{M_1M_2} \cdot \mathbf{n}_2)$; $E$is the unit matrix and ε is a small numerical coefficient.

The different mid-edge points of the triangle, $\bar{\mathbf{X}}_4, \bar{\mathbf{X}}_5, \bar{\mathbf{X}}_6$, are calculated in the same way. After replacing each edge of the triangular facet by the new quadratic curve, the quadratic surface is defined by the following equation:

$$\begin{aligned} \mathbf{x}(\xi,\eta) = {} & \mathbf{x}_1 N_1(\xi,\eta) + \mathbf{x}_2 N_2(\xi,\eta) + \mathbf{x}_3 N_3(\xi,\eta) + \bar{\mathbf{X}}_4 N_4(\xi,\eta) \\ & + \bar{\mathbf{X}}_5 N_5(\xi,\eta) + \bar{\mathbf{X}}_6 N_6(\xi,\eta) \end{aligned} \tag{13}$$

with the original linear functions of the triangle:

$$\begin{cases} N_1(\xi,\eta) = 1 - \xi - \eta \\ N_2(\xi,\eta) = \xi \\ N_3(\xi,\eta) = \eta \end{cases}$$

and the hierarchical quadratic functions at the mid-edge nodes

$$\begin{cases} N_4(\xi,\eta) = 4(1 - \xi - \eta)\xi \\ N_5(\xi,\eta) = 4\xi\eta \\ N_6(\xi,\eta) = 4(1 - \xi - \eta)\eta \end{cases}$$

Singular points of the domain such as edges and corners are handled by taking into account multiple normal vectors at these points (see Figure 3): two for edges and

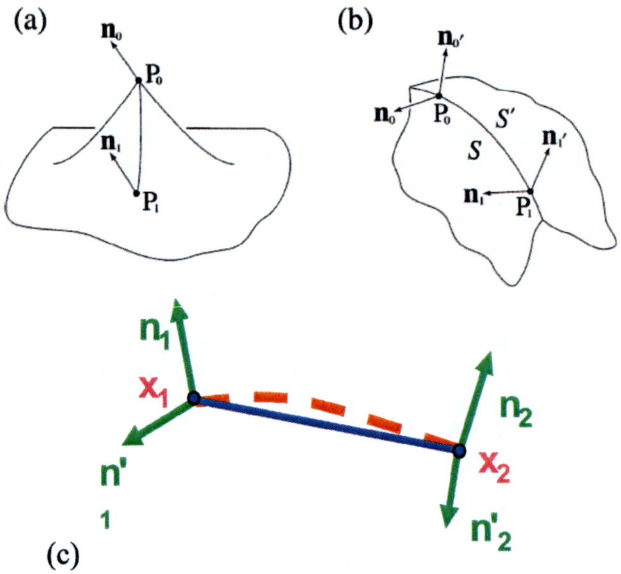

Fig. 3 (a), (b): Surface features: sharp edge; (c): interpolation of a curved sharp edge.

three for corners. The number of equations that the mid-edge point $\bar{\mathbf{X}}$ has to satisfy is consequently increased.

3.2 Validation

The validity and efficiency of the smoothing procedure are first evaluated by applying the algorithm to analytical shapes, a sphere and a cylinder. In these cases, the normal vectors directions are provided by the analytical values. In order to visualize the resulting quadratic facets, they are over-discretized by 16 linear facets, which nodes belong to the smoothed surface, as shown in Figure 4.

Table 1 qualitatively shows how this method is able to recover smoothed surfaces from initially coarsely discretized shapes.

3.3 Normal Vector Calculation

For actual surfaces that are only known from their tessellation, normal vectors are generally unknown at vertices, so they have to be calculated. It is observed that Nagata's method is quite sensitive to these values. If normal directions are not accurately calculated, then the smoothing procedure is not satisfactory and does not

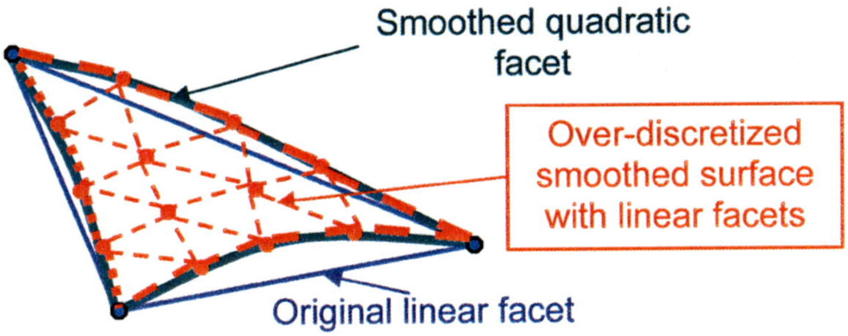

Fig. 4 New facet description for visualization.

Table 1 Quadratic interpolation of coarsely tessellated surfaces of a sphere and a cylinder using analytical normals. Left: original surface discretization using linear facets. Middle: smoothed geometry obtained with Nagata's method. Right: comparison between the original (lines) and smoothed (facets) surfaces.

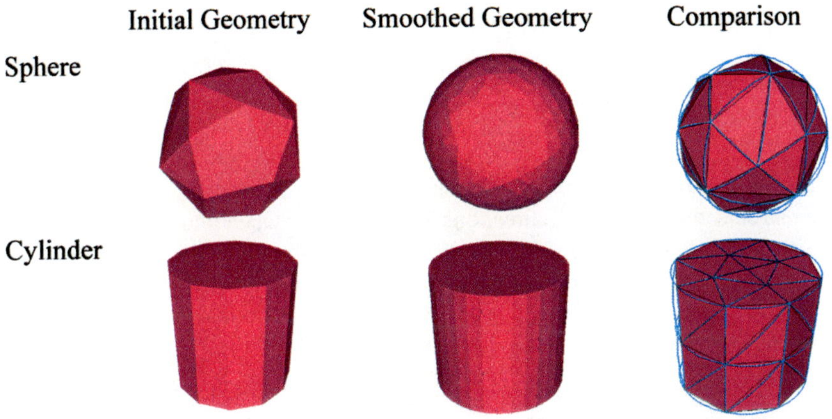

converge (by increasing the number of facets) toward the exact surface. This calculation is also aimed at detecting the different surface features, such as corners and sharp edges.

Several methods can be found in the literature. Some are based on angle criteria [9, 14]. They can be quite efficient but they significantly depend upon the selected angle threshold and tolerance, so they can be difficult to use for complex shapes. Other methods have been investigated, such as the calculation of a consistent normal [15] in the frame of flux evaluations, or as the SPR (Superconvergent Patch Recovery) method [16]. Normal directions are calculated from the available values at the centre of the triangles, and by considering a neighbourhood (or patch) of elements surrounding a node, as shown in Figure 1.

In the present study, the consistent normal is used. It was originally proposed in the frame of an Arbitrary Lagrangian or Eulerian (ALE) formulation, and was successfully used for surface mesh rezoning [15, 17]. For each boundary node, the consistent normal is regarded as the direction in which it is the most crucial to impose the material flux conservation condition. This condition is written for each facets f of a surface patch ∂P_m (see Figure 1). The flux is written as follows:

$$\int_{\partial P_m} (\mathbf{v}_{\text{may}} - \mathbf{v}_{\text{mat}}) \cdot \mathbf{n}^f dS = 0 \tag{14}$$

where $\mathbf{n}^f$ is the normal of the facet f, vmay and vmat respectively are the mesh and material velocities. This condition is imposed by minimizing the following expression:

$$\min(\Phi_m) = \min\left(\frac{1}{2}\left(\int_{\partial P_m} (\mathbf{v}_{\text{may}} - \mathbf{v}_{\text{mat}}) \cdot \mathbf{n}^f dS\right)^2\right) \tag{15}$$

which leads to solve the following system:

$$\frac{\partial \Phi_m}{\partial \mathbf{v}_{\text{may}}} = 0 \Rightarrow \mathbf{A} \cdot \mathbf{V}_{\text{may}} = \mathbf{B} \tag{16}$$

with

$$\mathbf{A}_{ij} = \sum_{f \in \partial P_m} \int \mathbf{n}_i N_m dS \int \mathbf{n}_j N_m dS \tag{17}$$

The invertible matrix A is expressed in its eigenvectors basis, which allows classifying and selecting the normal directions that have to be taken into account to satisfy the minimization problem (15). The eigenvectors provide the normal and tangent directions at the central vertex m. A criterion based on the eigenvalues λ_1, λ_2 and λ_3 ($\lambda_1 \geq \lambda_2 \geq \lambda_3$) intensities allows selecting the main normal, and to make a distinction between surface nodes, edges and corners, as follows:

- Surface (see Figure 5a): if $\lambda_2/\lambda_1 < \lambda_{\text{crit}}$: the normal of vertex $\mathbf{m}$ is $\overrightarrow{\mathbf{u}_1}$
- Edge (see Figure 5b): if $\begin{cases} \frac{\lambda_2}{\lambda_1} \geq \lambda_{\text{crit}} \\ \frac{\lambda_3}{\lambda_1} < \lambda_{\text{crit}} \end{cases}$: the normals of vertex $\mathbf{m}$ is $\overrightarrow{\mathbf{u}_1}$ and $\overrightarrow{\mathbf{u}_2}$, assuming that $\overrightarrow{\mathbf{u}_1}$ and $\overrightarrow{\mathbf{u}_2}$ are orthogonal.
- Corner (see Figure 5c): if $\begin{cases} \frac{\lambda_2}{\lambda_1} \geq \lambda_{\text{crit}} \\ \frac{\lambda_3}{\lambda_1} \geq \lambda_{\text{crit}} \end{cases}$: the normals of vertex $\mathbf{m}$ is $\overrightarrow{\mathbf{u}_1}$, $\overrightarrow{\mathbf{u}_2}$ and $\overrightarrow{\mathbf{u}_3}$, assuming that $\overrightarrow{\mathbf{u}_1}$, $\overrightarrow{\mathbf{u}_2}$ and $\overrightarrow{\mathbf{u}_3}$ are orthogonals.

The calculations presented in Table 1 are carried out again using the consistent normals instead of analytical values. Results are shown in Table 2. The obtained shapes are quite similar, but they are slightly not as good, which shows how the smoothing quality significantly depends on the accuracy of the normal directions.

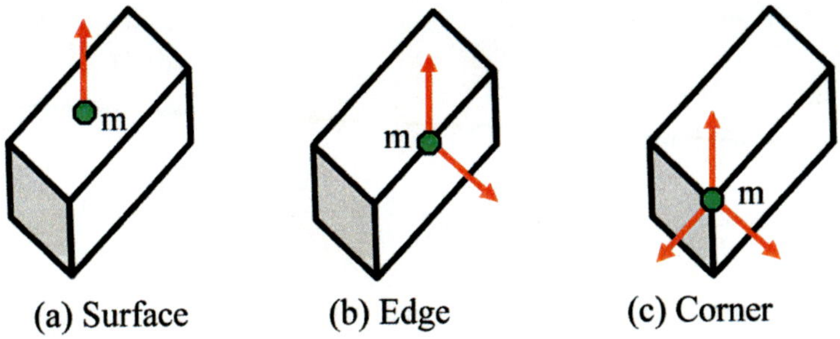

Fig. 5 Recognition of the surface features using consistent normals.

Table 2 Impact of using consistent normal in sphere smoothing.

	Initial Shape	Smoothed Sphere (Analytical normals)	Smoothed Sphere (Consistent normals)
Sphere	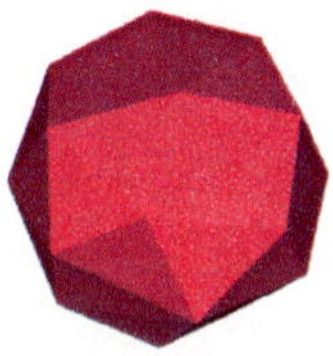		

4 Contact Algorithm for the Smoothed Surface

Any contact algorithm requires computing the orthogonal projection **P** of any spatial point **M** onto the contact surface (see Figure 6). After smoothing, this surface is made quadratic facets (13). Therefore, **M** is first projected on the linear facets of the contact surface, and then a Newton–Raphson algorithm is used to find the exact (ξ, η) local coordinates of **P** that satisfy the surface equation (13).

5 Applications

In order to validate the robustness of the proposed contact algorithm with smoothed surfaces and to examine its contribution to the forming processes simulation, a simple but discriminating example is first studied, and then a more realistic problems is considered.

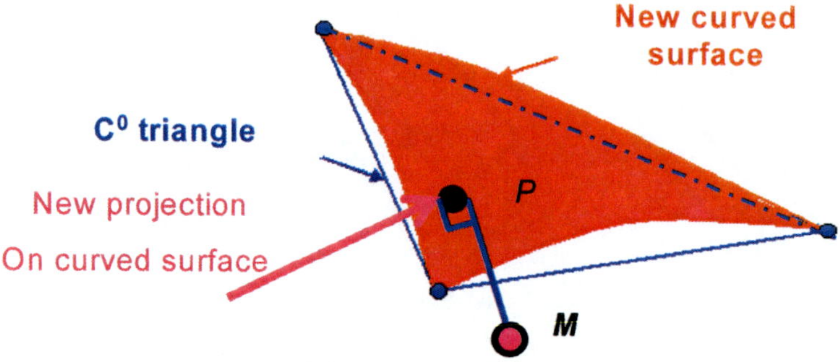

Fig. 6 Projection on a curved surface.

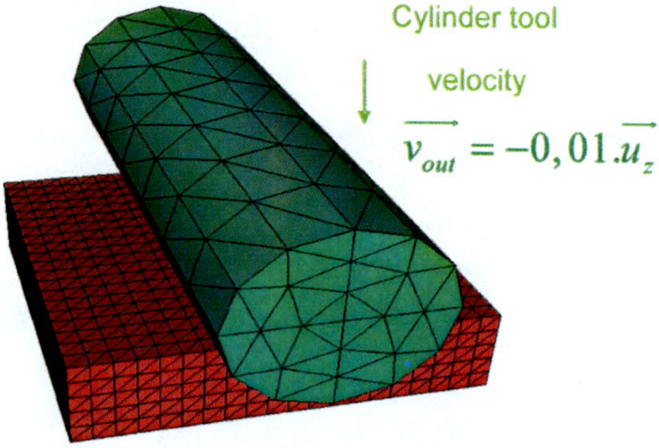

Fig. 7 Meshes for the indentation of a bulk parallelepiped.

5.1 Hertz's Contact: Indenting a Bulk Parallelepiped

5.1.1 Simulations Conditions

A rigid cylinder is indented onto an elastic bulk parallelepiped, which is defined by its Young's modulus $E = 2.10^5$ Pa, and Poison's ratio $\nu = 0.3$, with a vertical velocity of 0.01 mm·s^{-1} (see Figure 7). The thickness of the parallelepiped is 5 mm. The mesh size is approximately $h = 1$ mm. The cylinder radius is 5 mm and tool mesh size is approximately $h = 1.7$ mm. The time step is $\Delta t = 1$ s and the indentation depth is 2 mm.

5.1.2 Simulation Results

The contact areas resulting from the different calculations that have been carried out using (1) the original discretized tool, (2) the tool smoothed by the proposed

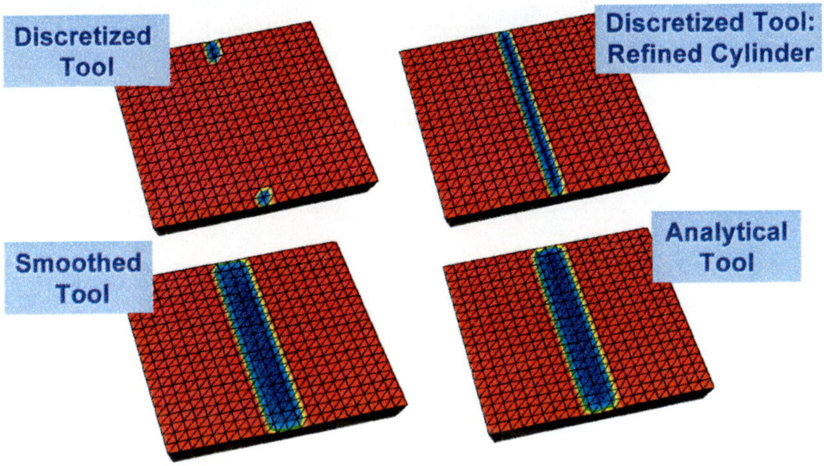

Fig. 8 Meshes for the indentation of a bulk parallelepiped.

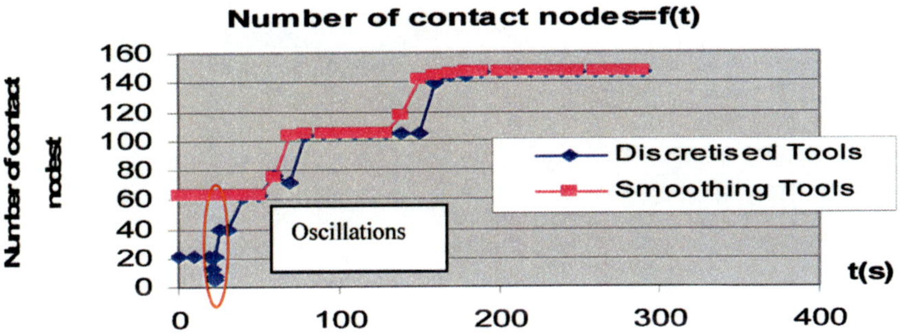

Fig. 9 Number of calculated contact nodes with the facetized and smoothed tools.

method, (3) the analytical tool, and (4) a more finely refined tool using a three times smaller mesh size for the cylinder: $h = 0.5mm$, show how smoothing allows a better calculation of the contact surface (see Figure 8). Figure 9 presents the evolution of the number of contact nodes. It points out how smoothing allows a better detection of contacts, provides larger and more regular contact surfaces, and removes time oscillations. The new contact model is also more accurate. The contact normal is almost perfectly calculated, as shown in Figure 10, which presents the maximum error between analytical and calculated values (18). Smoothing reduces this error from 25 to 5%.

$$\text{Error} = \max_{l \in \partial\Omega} \left(\sum_{i=1,3} |n_i^l - m_i^l| \right) \tag{18}$$

where n_i^l and m_i^l respectively are the calculated and analytical components in the ith space direction of the normal of node l.

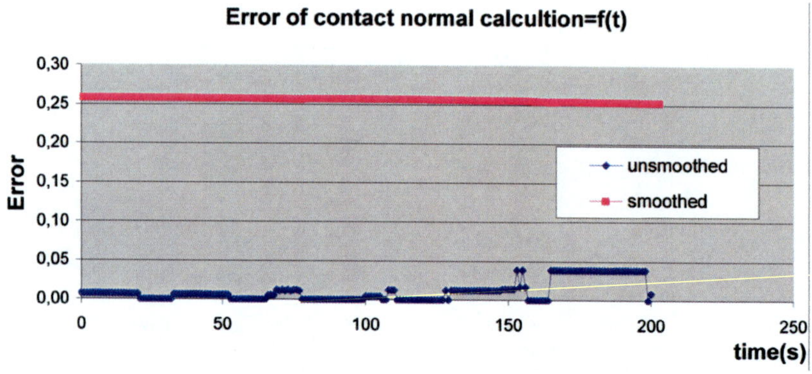

Fig. 10 Time evolution of the contact normal error using facetized and smoothed tool.

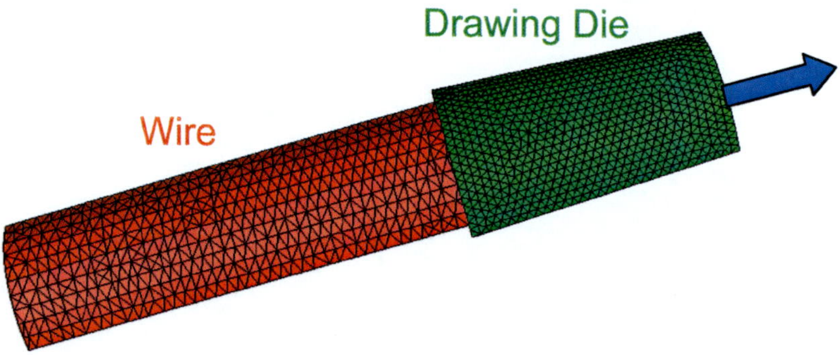

Fig. 11 Wire drawing test.

5.2 Wire Drawing

5.2.1 Simulation Conditions

The geometry and the finite element discretization of the wire drawing die are shown in Figure 11. The wire is pulled to pass through the die that has a reducing section. Its decreases the wire diameter while increasing its mechanical properties. The material is regarded as elastoviscoplastic. The contact between the wire and the die is assumed to be frictionless, which provides a good first approximation of the low friction of the actual process. The mesh size is approximately $h = 0.75$ mm and the tool mesh size is about $h = 0.5$ mm.

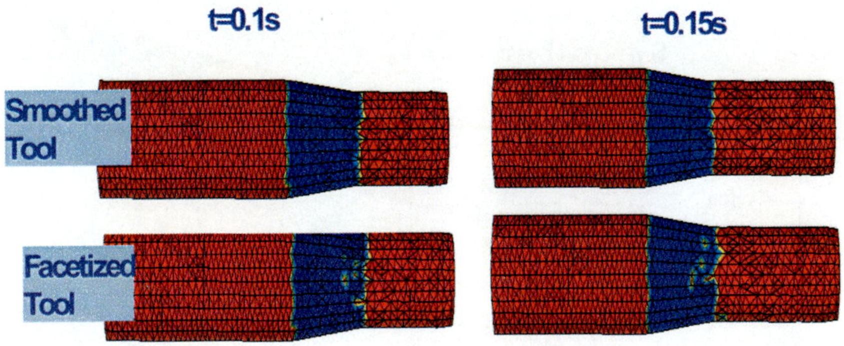

Fig. 12 Comparison of contact surface between facetized tool description and smoothed one.

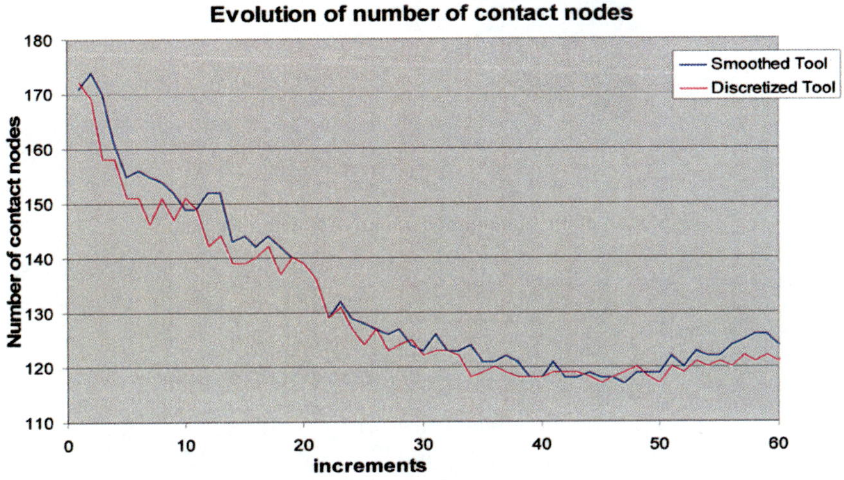

Fig. 13 Time evolution of the number of contact nodes.

5.2.2 Simulation Results

A detailed comparison between smoothed and unsmoothed simulations is presented. First, it is observed that, even though the die is accurately discretized with very small triangles, the smoothing procedure allows a better detection of the contact zone. Figures 12 and 13 respectively show the contact area and the evolution of contact nodes numbers. It can be seen that contact area is smoother, more regular and larger with smoothing. Figure 14 shows that the equivalent strains are lower and more homogenous. The maximum equivalent strain is about 0.75 using the standard algorithm and about 0.5 using the smoothing technique, which corresponds better with the expectations.

Moreover, the number of iterations for the resolution of the non linear system of equations is reduced when the surface smoothing procedure is used (Figure 15),

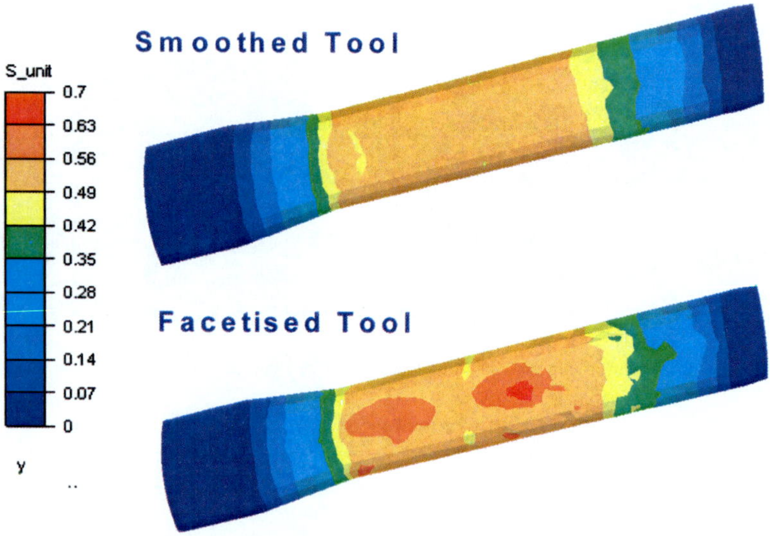

Fig. 14 Equivalent strain using facetized or smoothed tool.

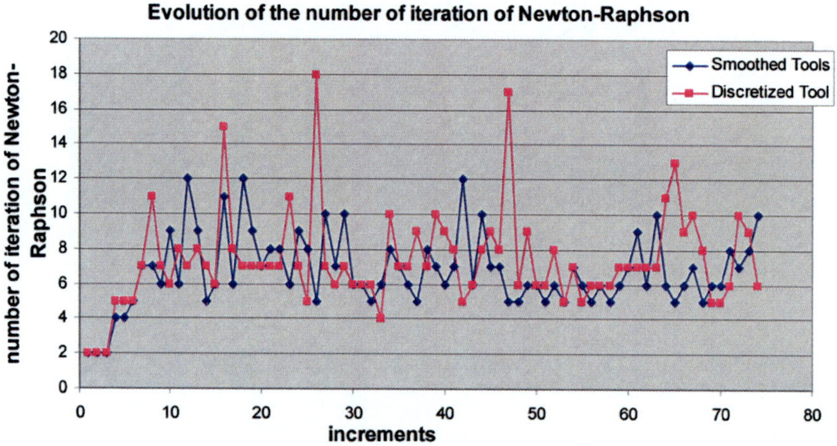

Fig. 15 Time evolution of the number of iterations of Newton–Raphson.

which can be explained by the fact that contact is better handled, which results in less time oscillations and consequently in less complex linear systems to solve. This fact explains that the CPU time of the total problem resolution is not significantly increased, even though a more expensive quadratic contact algorithm is used. The CPU time is 7 mins 55 s with smoothed contact and 7 mins 37 s without smoothing.

6 Conclusions

Based on a locally quadratic interpolation of the contact surface that combines the smoothing technique proposed by Nagata and the consistent normal computation initially proposed for ALE formulations, the developed contact algorithm provides significantly more precise simulation of metal forming processes with reduced contact area, with an additional computational cost that only represents few percents of the total cost. The obtained contact surfaces are more accurate and regular, without time oscillations, which results into more precise results in terms of equivalent strain rates and total strains that better fit with expectations. Last and not least for large 3D problems, this local approach can be easily extended to parallel calculations.

Acknowledgements This work has been carried out in the frame of the "Faible Contact" French project gathering several companies of the metal forming industry. Its financial and technical support is gratefully acknowledged.

References

1. Wriggers, P., Vu Van, T., Stein, E.: Finite element formulation of large deformation impact – Contact problems with friction. Computers and Structures 37, 319–331 (1990)
2. Oden, J.T., Kikuchi, N.: Contact Problems in Elasticity: A Study of Variational Inequalities and Finite Element Methods. SIAM, Philadelphia (1988)
3. Bathe, K.J., Chaudhary, A.: A solution method for planar and axisymmetric contact problems. International Journal for Numerical Methods in Engineering 21, 61–88 (1985)
4. Wriggers, P., Simo, J.C., Taylor, R.L.: Penalty and augmented Lagrangian formulations for contact problems. In: Proceedings of the NUMETA 1985 Conference. Elsevier, Amsterdam (1985)
5. Simo, J.C., Laursen, T.A.: An augmented Lagrangian treatment of contact problems involving friction. Computers and Structures 42, 97–116 (1992)
6. Stadler, M., Holzapfel, G.A., Korelc, J.: Cn continuous modeling of smooth contact surfaces using NURBS and application to 2D problems. International Journal for Numerical Methods in Engineering 57, 2177–2203 (2003)
7. Wriggers, P., Krstulovic-Opara, L., Korelc, J.: Smooth C1 interpolations for two-dimensional frictional contact problems. International Journal for Numerical Methods in Engineering 51, 1469–1495 (2001)
8. Puso, M.A., Laursen, T.A.: A 3D contact smoothing method using Gregory patches. International Journal for Numerical Methods in Engineering 54, 1161–1194 (2002)
9. Rassineux, A., Villon, P., Savignat, J.-M., Stab, O.: Surface remeshing by local Hermite diffuse interpolation. International Journal for Numerical Methods in Engineering 49, 31–49 (2000)
10. Nagata, T.: Simple local interpolation of surfaces using normal vectors. In: Proceedings of Computer Aided Geometric Design (2005)
11. Chenot, J.L., Fourment, L., Coupez, T., Ducloux, R., Wey, E.: Forge3R: A general tool for practical optimisation of forging sequence of complex 3-D parts in industry. In: Proceedings ICFT 1998, pp. 113–122. Institution of Mechanical Engineers, Birmingham (1998)

12. Coupez, T., Soyris, N., Chenot, J.L.: 3-D finite element modeling of the forging process with automatic remeshing. J. Mat. Processing Technology 27, 119–133 (2004)
13. Hama, T., Nagata, T., Teodosiu, C., Makinouchi, A., Takuda, H.: Finite-element simulation of springback in sheet metal forming using local interpolation for tool surfaces. International Journal of Mechanical Science 50, 175–192 (2008)
14. Frey, P.J., George, P.L.: Mesh Generation. Application to Finite Elements, Hermes (2002)
15. Guerdoux, S.: Numerical simulation of the Friction Stir Welding process. Ph.D. Thesis, Mines ParisTech (2007)
16. Zienkiewicz, O.C., Zhu, J.Z.: The Superconvergent Patch Recovery (SPR) and adaptive finite element. Computer Methods in Applied Mechanics and Engineering 101, 207–224 (1992)
17. Philippe, S., Fourment, L., Montmitonnet, P.: Application of the Arbitrary Lagrangian Eulerian formulation to the numerical simulation of stationary forming processes with dominant tangential material motion. In: Proceedings of Metal Forming 2008, pp. 22–24 (September 2008)

On a Geometrically Exact Theory for Contact Interactions

Alexander Konyukhov and Karl Schweizerhof

Abstract The focus of the contribution is on the developments concerning an unified geometrical formulation of contact algorithms in a covariant form for various geometrical situations of contacting bodies leading to contact pairs: surface-to-surface, line-to-surface, point-to-surface, line-to-line, point-to-line, point-to-point. The computational contact algorithm will be considered in accordance with the geometry of contact bodies in a covariant form. This combination forms *a geometrically exact theory of contact interaction*. The contribution focuses on an overview of the literature and then presents in a review type the contributions of the authors on the topic.

1 On Geometrical Approaches in Contact Mechanics

Contact interaction from a geometrical point of view can be seen as an interaction between deformable surfaces and, therefore, geometrical approaches can be exploited. However, there are only a few publications covering geometrical issues to some extent. Gurtin et al. [2] (1998) considered surface tractions on curvilinear interfaces describing them from a geometrical point of view. Jones and Papadopoulos [6] (2006) considered contact describing various mappings from the reference configuration employing the Lie derivative. Laursen and Simo [19] (1993) and Laursen [18] (1994) described some contact parameters via geometrical surface parameters. Heegaard and Curnier [4] (1996) considered geometrical properties of slip operators.

Alexander Konyukhov · Karl Schweizerhof
Karlsruhe Institute of Technology (KIT), Institute of Mechanics, Otto-Ammann-Platz 9, D-76131 Karlsruhe, Germany; e-mail: {alexander.konyukhov; karl.schweizerhof}@kit.edu

G. Zavarise & P. Wriggers (Eds.): Trends in Computational Contact Mechanics, LNACM 58, pp. 41–56.
springerlink.com

1.1 Bottleneck: Consistent Linearization

The iterative solution of, e.g., Newton type is a standard way to obtain the solution in computational contact mechanics. However, one of the difficult points is to obtain the full derivative of the functional necessary for a fast Newton solver – this procedure is known as linearization. Two approaches for linearization of the final functional representing the work of contact tractions can be distinguished in order to obtain consistent tangent matrices. *The direct approach* follows the following sequence: *functional – discretization – linearization* and *the covariant approach* follows the rule: *functional – linearization – discretization.*

The direct approach, historically motivated by the development of the finite element method for the analysis of solid mechanics structures assumes that the discretization is first involved in the process and linearization is provided with regard to the displacement vector **u** and, therefore, of the discretized system. This leads to the final results containing a set of approximation matrices: for surface to surface contact it is described in Wriggers and Simo [31] (1985), Wriggers et al. [32] (1990), Parisch [25] (1989), Parisch and Luebbing [26] (1997), Peric and Owen [27] (1992), Simo and Laursen [28] (1992), Laursen and Simo [19] (1993) and in books by Wriggers [34] and Laursen [20]; for anisotropic friction in Alart and Heege [1] (1995), for beam type contact in Wriggers and Zavarise [33] (1997), Zavarise and Wriggers [35] (2000), Litewka and Wriggers [22] (2002), Litewka and Wriggers [23] (2002). The complexity in the derivation for curved contact interfaces led to the use of codes containing an automatic derivation with mathematical software, see Heege and Alart [5] (1996), Stadler, Holzapfel and Korelc [30] (2003), Krstulovic-Opara, Wriggers and Korelc [17] (2002), for anisotropic friction in Montmitonnet and Hasquin [24] (1995) and for beams in Litewka [21] (2007).

Open questions and drawbacks of *the direct approach* can be summarized as follows:

1. A closed form for tangent matrices is available only for linear approximations of surfaces. For curved interfaces, either a form depending on approximations (mathematical software), or a form of simplified matrices (taken for linear approximations of the curved surfaces) is reported.
2. The structure of the derived matrices is very complicated and often intransparent. There is no clear mechanical interpretation of each part possible. Thus, simplifications are hardly manageable.
3. A specification of complex contact interface laws with properties explicitly depending on the surface geometry (e.g. *arbitrary* anisotropy) is not possible.
4. A contact description of many geometrical features (curved line-to-curved line, curved line-to-surface) is almost not possible because of the necessity of convective surface coordinates.
5. Geometrically motivated measures of contact interaction are coupled with convective variables in a specially defined coordinate system. However, in the direct approach, they can hardly be defined separately for various geometrical features (surfaces, edges, etc.).

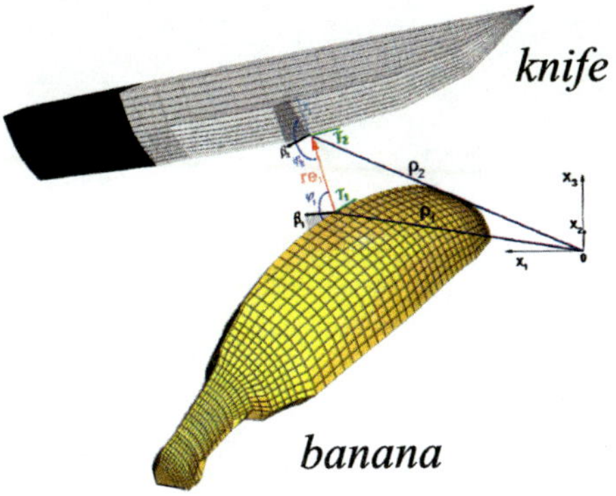

Fig. 1 Various geometrical situations in contact lead to different contact algorithms: surface-to-surface, line-to-surface, point-to-line, line-to-line and point-to-point.

The fully covariant approach, however, assumes only a local coordinate system associated with the deformed continuum (convective coordinates) and requires extensive application of covariant operations (derivatives, etc.). The approach started with the consideration of convective variables arising from the surface approximations directly for contact traction and displacements: see Simo et al. [29] (1986), Wriggers et al. [32] (1990), Laursen and Simo [19] (1993), Laursen [18] (1994). Two convective variables ξ^1, ξ^2 in a surface covariant basis ρ_1, ρ_2 are used as tangential measure.

This approach has many advantages:

1. Objectivity is straightforwardly observed because the surface coordinates ξ^i are used.
2. Geometrical interpretation of a measure – line on a surface; geometrical interpretation of a linearized measure – relative tangent velocity of a contact point can be directly included.
3. The number of history variables is minimal (two for surface interaction).
4. A complex constitutive law for tangent interaction can be easily formulated in a robust form for computation.
5. Expressions for contact tangent matrices are by far less complex within the fully covariant approach than for the direct approach.
6. Any type of surfaces geometry interpolation can be directly used – Lagrange and/or NURBS interpolation.

A fully covariant approach, though, is intended for the finite element method, but does not assume approximations from the beginning and it serves to describe all parameters necessary for the solution based on the geometry of the contacting bodies in the local coordinate system. The method, however, requires a number of preliminary transformations based on differential geometry of contacting objects (surfaces or curves) and extensive application of tensor analysis especially for differential operations and linearization.

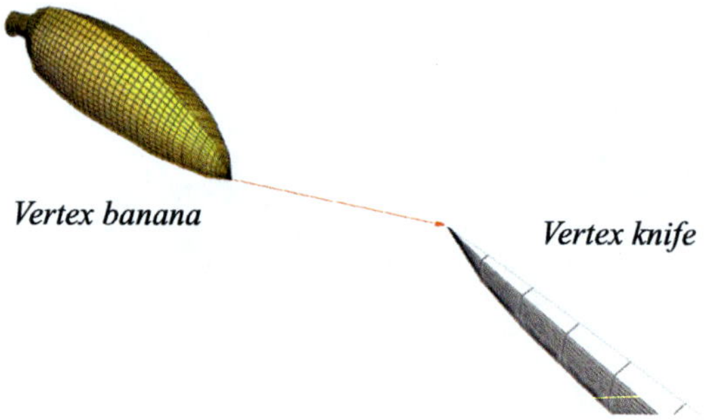

Fig. 2 Point-to-point (PTP) contact algorithm.

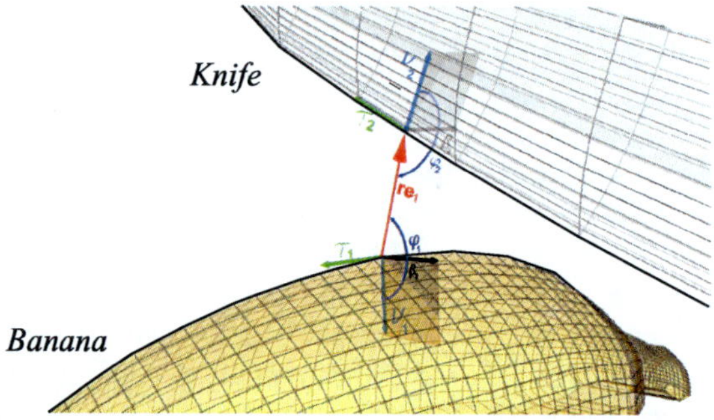

Fig. 3 Line-to-line (LTL) contact algorithm.

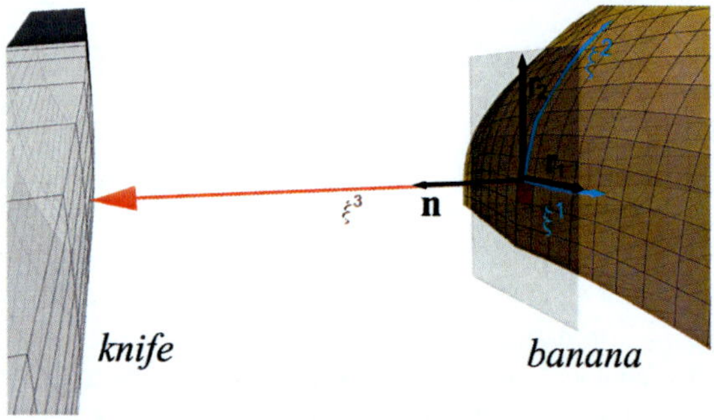

Fig. 4 Surface-to-surface (STS) contact algorithm.

2 Development of a Geometrically Exact Theory

In order to formulate goals and summarize the developments concerning a geometrically exact theory we consider a model contact problem with two bodies possessing smooth surfaces as well as various geometrical features such as edges and vertices – an example of this is a banana and a knife shown in Figure 1.

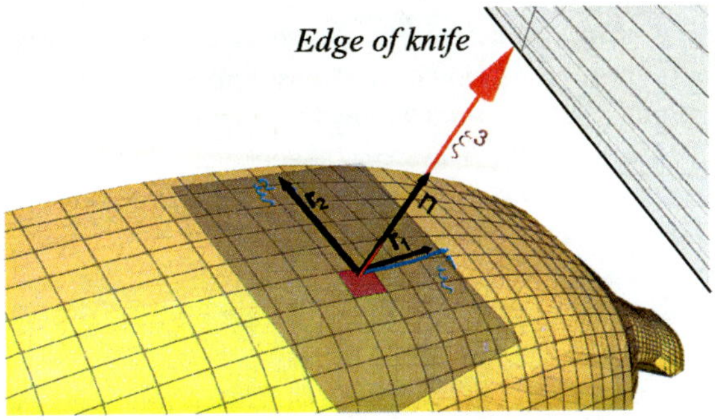

Fig. 5 Line-to-surface contact as surface-to-surface algorithm.

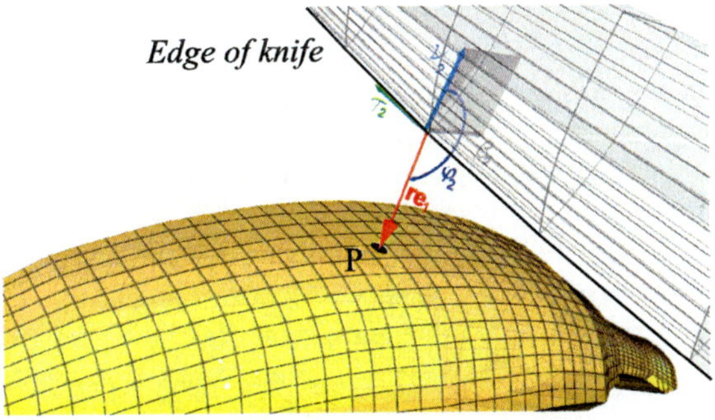

Fig. 6 Line-to-surface contact as point-to-line algorithm.

Considering all possible geometrical situations in which knife and banana can contact each other, the following hierarchical sequence of contact pairs appears.

Possible contact pairs:

1. Point-to-point contact pair, see Figure 2.
2. Point-to-line contact pair, see Figure 6.
3. Point-to-surface contact pair, see Figures 4 and 5.
4. Line-to-line contact pair, see Figure 3.
5. Line-to-surface contact pair, see Figures 5 and 6.
6. Surface-to-surface contact pair, see Figure 4.

2.1 Goals of a Geometrically Exact Theory

The following open problems can be stated as goals for a geometrically exact theory:

1. Development of a unified geometrical formulation of contact conditions in a covariant form for various geometrical situations of contacting bodies leading

to contact pairs: surface-to-surface, line-to-surface, point-to-surface, line-to-line, point-to-line, point-to-point (joint). The description will be fully based on the differential geometry of specific features forming a continuum, because it is carried out in the local coordinate systems attached to this feature: in the case of a surface in the Gaussian surface coordinate system; in the case of a curved line in the Serret–Frenet basis; in the case of a point in the coordinate system in a standard form as is known for rigid body rotation problem (e.g. via the Euler angles). *This general description forms a geometrically exact theory for contact interaction.*

2. A full set of contact pairs requires various closest point projection (CPP) procedures. Thus, fundamental problems of existence and uniqueness of closest point projection routines corresponding to the following situations have to be investigated: point-to-surface, point-to-line, line-to-line.
3. A solution of existence and uniqueness problems of closest points routines leads to "projection domains" as the "maximal searching domains".
4. Since contact interaction between arbitrary bodies is modeled via a corresponding set of contact pairs (surface-surface, surface-line, etc.), then the necessary transfer algorithm for history variables can be constructed correctly.
5. For the enforcement of contact conditions via various applicable methods: Lagrange multipliers methods, penalty methods or augmented Lagrange multipliers method derivation of a unified covariant description is needed. Consistent tangent matrices are given in closed covariant form possessing a clear geometrical structure.
6. Straightforward recipes for the implementation with any order of approximation for finite elements. The description of all geometrical situations in a covariant form which is a-priori independent of approximations of these geometrical features.
7. Covariant contact description for high order approximations including exact representation of the geometry for continua (iso-geometrical approach). Numerical tests show the efficiency for the classical Hertz problem (only a single finite element is sufficient!).
8. Development of the curve-to-curve contact model allowing to consider the complete set of relative motions between curves including a rotational interaction (a novel in the current theory).
9. Application of the curve-to-curve contact algorithm to edge-to-edge contact as well as to beam-to-beam contact. Curved beams possessing C^1-continuity allowing contact (a cable model).
10. A special integration technique based on sub-domain integration is developed for "the segment-to-segment" approach (equivalent to the Mortar method). Numerical tests show the improvement of results for the contact patch-test.
11. Generalization of the classical Coulomb law into complex interface laws in covariant form for arbitrary geometries of the surfaces (e.g. coupled anisotropic friction and adhesion for surfaces). Development of an a-priori stable numerical algorithm for computations.

12. Development of the corresponding constitutive relations together with the corresponding numerical algorithm allowing an anisotropic behavior for the curve-to-curve interaction (various relative adhesion and friction properties).
13. Experimental validation of the proposed anisotropic law for coupled tangential adhesion and friction.

3 Overview of the Developments

Although the specific points of the proposed theory are spread through many publications, they can be summarized under a unified goal. Thus, the current section is giving to the reader the complete structure of specific details which can be found in publications by the authors on the subject [3, 7–16].

The most powerful approach in computational contact mechanics is to work in accordance with the geometry of the surfaces of contacting bodies and construct all computational algorithms in a covariant form. This combination forms the so-called geometrically exact theory of contact interaction.

As is known, *the closest distance* between the surfaces of contacting bodies has become *a natural measure of the contact interaction.* The procedure is introduced via the closest point projection procedure (CPP), solution of which requires the differentiability of the function representing the parameterization of the surfaces of the contacting bodies. *The analysis of the solvability for the CPP procedure* allows then to classify all types of all possible contact pairs given in Section 2. Thus, the consideration of the solvability of the CPP procedure, as presented in [14], forms a basis of the theory. Starting with a consideration of C^2-continuous surfaces, the concept of *the projection domain* is introduced as a domain from which any point can be uniquely projected, and therefore, the contact algorithm can be further constructed. This domain can be constructed for utmost C^1-continuous surfaces. If the surfaces contain edges and vertices then the CPP procedure should be generalized in order to include the projection onto edges and onto vertices. The criteria of uniqueness and existence of these projection routines and corresponding domains is then studied in detail in [14].

3.1 Selection of a Coordinate System

The main idea for the application for contact is then straightforward – the CPP procedure corresponding to a certain geometrical feature gives rise to a special, in general, curvilinear 3D coordinate system. This coordinate system is attached to a geometrical feature and its convective coordinates are directly used for further definition of the contact measures. Thus, all contact pairs listed in Section 2 can be described in the corresponding local coordinate system. The existence requirement for a generalized CPP procedure leads to the transformation rule between types of contact pairs according to which the corresponding coordinate system is taken. Thus, all contact pairs can be uniquely described in most situations.

A surface-to-surface contact pair (see Figure 4) is described via the well known "master-slave" contact algorithm based on the CPP procedure onto the surface. This projection allows to define a coordinate system as follows:

$$\mathbf{r}(\xi^1,\xi^2,\xi^3) = \rho(\xi^1,\xi^2) + \xi^3\mathbf{n}(\xi^1,\xi^2) \tag{1}$$

The vector $\mathbf{r}$ is a vector for the "slave" point, ρ is a parameterization of the "master" surface, $\mathbf{n}$ is a normal to the surface. Eqn. (1) describes, in fact, a coordinate transformation where convective coordinates are used for measure of contact interaction: ξ^3 is a penetration, the increments $\Delta\xi^1, \Delta\xi^2$ of the tangential coordinates ξ^1, ξ^2 are measures for tangent interaction. The algorithm is applied only in the existence domain for the surface CPP procedure.

Consideration of the existence of the CPP procedure for edges allows to define then *a point-to-line* contact algorithm also used for *the line-to-surface* contact pair (see Figure 6). The local coordinate system is constructed as follows:

$$\mathbf{r}(s,r,\varphi) = \rho(s) + r\mathbf{e}(s,\varphi); \quad \mathbf{e} = \nu\cos\varphi + \beta\sin\varphi \tag{2}$$

Here, the vector $\mathbf{r}$ is describing a "slave" point from the surface, $\rho(s)$ is a parameterization of the "master" curve edge; a unit vector describing the shortest distance $\mathbf{e}$ is written via the unit normal ν and the bi-normal β of the curve ρ. The convective coordinates used as measures: r – for normal interaction; s – for tangential interaction; φ – for rotational interaction.

The line-to-surface contact pair, however, can be described dually via the surface-to-surface contact algorithm if we consider a "slave" point on the edge and project it onto the "master" surface, see Figure 5 (*line-to-surface contact as surface-to-surface algorithm*). The contact is described then in the surface coordinate system (1).

The line-to-line contact pair (see Figure 3) requires the projection on both curves, therefore, there is no classical "master" and "slave" and both curves are equivalent. For the description one of the two coordinate systems can be used assigned to the I-th curve:

$$\rho_2(s_1,r,\varphi_1) = \rho_1(s_1) + r\mathbf{e}_1(s_1,\varphi_1); \quad \mathbf{e}_1 = \nu_1\cos\varphi_1 + \beta_1\sin\varphi_1 \quad 1 \leftrightarrows 2. \tag{3}$$

Here, the vector ρ_2 is a vector describing a contact point of the second curve, $\rho_1(s_1)$ is a parameterization of the first curve; a unit vector describing the shortest distance $\mathbf{e}_1$ is written via the unit normal ν_1 and bi-normal β_1 of the first curve. Eqn. (3) describes the motion of the second contact point in the coordinate system attached to the first curve. The description is symmetric with respect to the choice of the curve $1 \leftrightarrows 2$. The convective coordinates are used as measures: r – for normal interaction for both curves; s_I – for tangential interaction and φ_I – for rotational interaction for the I-th curve.

The point-to-point contact pair (see Figure 2) is described then in a coordinate system standard for rigid body rotation problems (e.g. via the Euler angles), however in the contact situation this is a very rare case, and in computations it is rather unprobable unless specially treated, and therefore, because of the unavoidable numerical errors would fall into other contact pair types.

Initially, the computational algorithm is constructed for *non-frictional contact interaction of smooth surfaces, see [7].* Here the description starts in the coordinate system given in Eqn. (1), however, due to the small penetration $\xi^3 \approx 0$ it is mostly falling into a description in the Gaussian surface coordinate system arising from the surface parameterization $\rho(\xi^1, \xi^2)$. All contact parameters such as sliding distance and tangential forces are described then on the tangent plane at $\xi^3 = 0$. The linearization procedure is given in a form of covariant derivatives. This leads to a closed form of the tangent matrix subdivided into main, rotational and curvature parts. The influence of these parts on convergence is studied in numerical examples for the linear and quadratic finite elements.

The approach is easily extended into the problem with a *Coulomb friction*, see [8]. It is shown that for the correct regularization of tangential contact conditions the evolution equation for contact traction should be written in a form of covariant derivatives. The structures of all parts of tangent matrices are obtained due to covariant derivation in a compact tensor closed form. This makes it applicable for any surface approximation. It is shown that the tangent matrix in the sticking case is always symmetric for any kind of approximation. *A classification of parts of the tangent matrix* is given and their influence on convergence with regard to small and large sliding problems is considered. Small sliding problems are introduced as problems where the computation of the sticking-sliding zone is essential, while a sliding path is only of interest for large sliding problems. *An algorithm to transfer history variables* in contact problems overcoming the discontinuity of history variables on element boundaries, see the illustration in Figure 7, is created in a covariant form.

3.2 Applicability to a Majority of Problems – FE Approximation

The *application* of the developed theory in a combination with "solid-shell" *finite elements is presented in [3].* Thus, all known algorithms "Node-to-Segment" (NTS), Mortar like "Segment-to-Segment" (STS) and the algorithm for contact with a surface described by the analytical function "Segment-to-Analytical-Surface" (STAS) are reconsidered within the covariant description. Details of a finite element implementation are considered in the combination with the "solid-shell" elements.

All contact parameters are evaluated at integration points within the segment-to-segment approach, therefore, the method can be seen as a penalty based Mortar approach. In addition, *combinations of various adaptive integration techniques* such as integration with sub-domains with independent application of either Gauss–

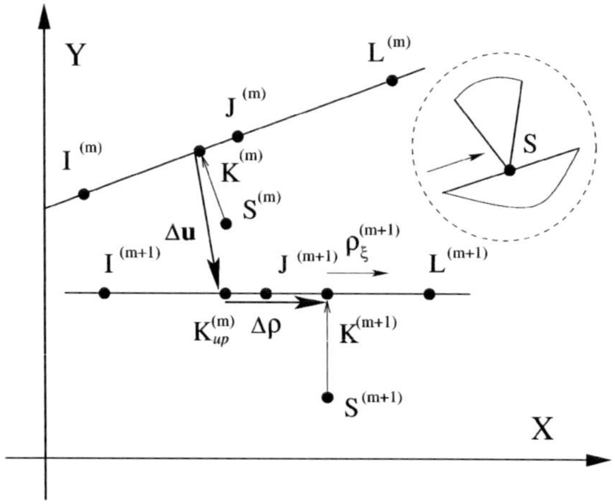

Fig. 7 Transfer of variables between elements in a 2D case.

Legendre, or Gauss–Lobatto quadrature formulas allow satisfying the "contact patch-test" on unstructured distorted meshes for arbitrarily chosen "master" or "slave" segments. The influence of various integration techniques on computed results (especially on the force-displacement curves) is extensively studied in numerical examples. It is shown that the geometrical contact conditions can be satisfied with high tolerance even for linear finite element approximations by the application of adaptive integration techniques.

The necessity of the application of *the history transfer algorithm* (see Figure 7) is shown for deep drawing cases as the closest point is crossing the borders of elements [8]. A closed form solution for the penetration within the STAS approach is given for planes and for quadratic surfaces (cylinder, sphere, torus). A reduced algorithm for the CPP procedure is constructed for surfaces of revolution.

A general geometrical approach to treat contact kinematics in the 2D case either as a reduction of the 3D case, or as a development based on a plane curve geometry is described in [11]. This leads to a more *simple kinematical interpretation of all parts of the tangent matrix*. A straightforward implementation of frictional contact in 2D is proposed. The algorithm to transfer the history variables in contact problems overcoming the discontinuity of history variables on element boundaries and the algorithm to update the history variables in the case of reversible loading are studied and illustrated in detail.

A special development of the covariant approach in combination with *high-order finite element approximations* is given in [15]. Both the penalty, and the Lagrange multipliers method are considered. The Lagrange multipliers are integrated via an integration technique satisfying the discrete Babuska-Brezzi (BB) stability condition. The linearization procedure in the case of an exact geometry description of the contact boundary represented by the blending function method is developed. As a result a contact layer element allowing anisotropic p-refinement is created. A good correlation for the analytically solved Hertz problem is achieved even within a single contact layer element.

3.3 Contact Interface Law – Anisotropic Adhesion-Friction

A systematic generalization of a contact interface law from the Coulomb friction law into the anisotropic region in a covariant form including various known visco-elasto-plastic mechanical models is started in [9]. Thus, *a coupled model including anisotropy for tangential adhesion and for friction* is obtained. This model is formulated via the principle of maximum dissipation in a rate form. Finally, the computational model is derived via the application of the return-mapping scheme to the incremental form. As a result a frictional force is derived in a closed form including both, the adhesion and the friction tensors. The structure of the tensors is derived for various types of anisotropy: a uniform orthotropy of a plane given by the spectral decomposition, a nonuniform orthotropy of a plane inherited with a polar coordinate system and a spiral orthotropy of a cylindrical surface. The update algorithm for history variables is developed for the arbitrary coupled anisotropy. The geometrical interpretation of the return-mapping and the update algorithm are considered via an ellipse on the tangent plane.

The second part for the *anisotropic friction model* in [10] continues the development of the computational algorithm for the coupled anisotropic friction model. The linearization is obtained as a covariant derivation in the local surface coordinate system and, therefore, all tangent matrices possess the simple form as described above for the isotropic friction models. Thus, the main part of the tangent matrix for sliding case is constructed from the following linearized form – on purpose not explaining all quantities in the formulas:

$$\int_s \left((\delta \mathbf{r}_s - \delta \rho) \cdot \frac{\varepsilon_N \mathbf{BFT}^{tr} \otimes \mathbf{n} - |N| \, \mathbf{BFB}}{\sqrt{\mathbf{BFT}^{tr} \cdot \mathbf{FBFT}^{tr}}} (\mathbf{v}_s - \mathbf{v}) \right) ds \tag{4a}$$

$$+ \int_s \left((\delta \mathbf{r}_s - \delta \rho) \cdot \frac{|N| \, \mathbf{BFT}^{tr} \otimes (\mathbf{BFB})^T \mathbf{FBFT}^{tr}}{(\mathbf{BFT}^{tr} \cdot \mathbf{FBFT}^{tr})^{3/2}} (\mathbf{v}_s - \mathbf{v}) \right) ds \tag{4b}$$

The isotropic case is then recovered with the adhesion tensor $\mathbf{B} = -\varepsilon_T \mathbf{E}$ and with the friction tensor $\mathbf{F} = \mu \mathbf{E}$.

The mechanical interpretation as two spring-two slider-mass is discussed. The behavior of contacting bodies for various types of anisotropy is numerically analyzed. The development of the sticking zone for the small displacement case, and the influence of orthotropic properties on a trajectory of the sliding block in the case of large displacements are analyzed for uniform orthotropy of a plane given by the spectral decomposition and for nonuniform orthotropy of a plane inherited with the polar coordinate system. As an interesting result, *geometrically isotropic behavior* of the block has been found: in this case a combination of both, anisotropy for adhesion and anisotropy for friction leads to a trajectory which can be normally observed only for isotropic surfaces. It is shown that *the application of the spiral orthotropy on a cylindrical surface allows to simulate the kinematics of a bolt connection – an angular thread screw* on a relatively coarse mesh.

A *symmetrized algorithm based on the Augmented Lagrangian method* for coupled anisotropic friction is developed in [12]. It is shown that for small sliding problems both normal and tangential tractions should be augmented to enforce the non-penetration resp. sticking conditions. However for large sliding problems the augmentation of only the normal traction leads to satisfactory tolerances for trajectories.

As a key for the practical application, the developed model is *experimentally investigated.* It is shown in [13], that the coupled anisotropic adhesion-friction model can *successfully describe a set of trajectories* of a block on a rubber mat with a periodical wavy surface profile, while the classical anisotropic friction model fails. Special attention is given to an analysis of *the geometrically isotropic behavior*.

3.4 Curve-to-Curve Contact

A geometrically exact description in a covariant form for *a curve-to-curve contact pair* as shown in Figure 3 is developed in [16]. The development begins consistently with the CPP procedure providing a shortest distance between curves as a natural measure of normal contact interaction. The CPP procedure leads to a special local coordinate system in which convective coordinates are used directly as measures of contact interaction between curves: normal, tangential and rotational interactions. The existence and uniqueness of the CPP procedure is studied in detail – projection domains with a-priori unique solution are constructed in this coordinate system for curves with varying geometry.

Several achievements appear to be novel for the curve-to-curve contact description:

1. The consideration of any relative motion separately for each curve is possible.
2. The rotational interactions including corresponding rotational moments between curves can be considered consistently.

The Coulomb friction law for tangential interaction and the Tresca friction law for rotational interaction are considered as examples for constitutive relations between curves. All necessary linearizations for the iterative solution scheme are provided as covariant derivation in the introduced coordinate system for arbitrary large distances between curves. This leads to a closed form of tangent matrices independent of the approximation used for the finite elements. The verification section contains the comparison between beam-to-beam and edge-to-edge finite element models as well as the verification with a famous "equilibrium of Euler elastica problem" computed via a finite difference scheme. Further numerical examples are illustrating the ability to describe various kinematics for curve-to-curve contact situations e.g. partial sticking of a single curve, see [16]. As an interesting result it is possible to model partial tied contact for two deforming beams. Thus, in Figure 8(a) only the upper

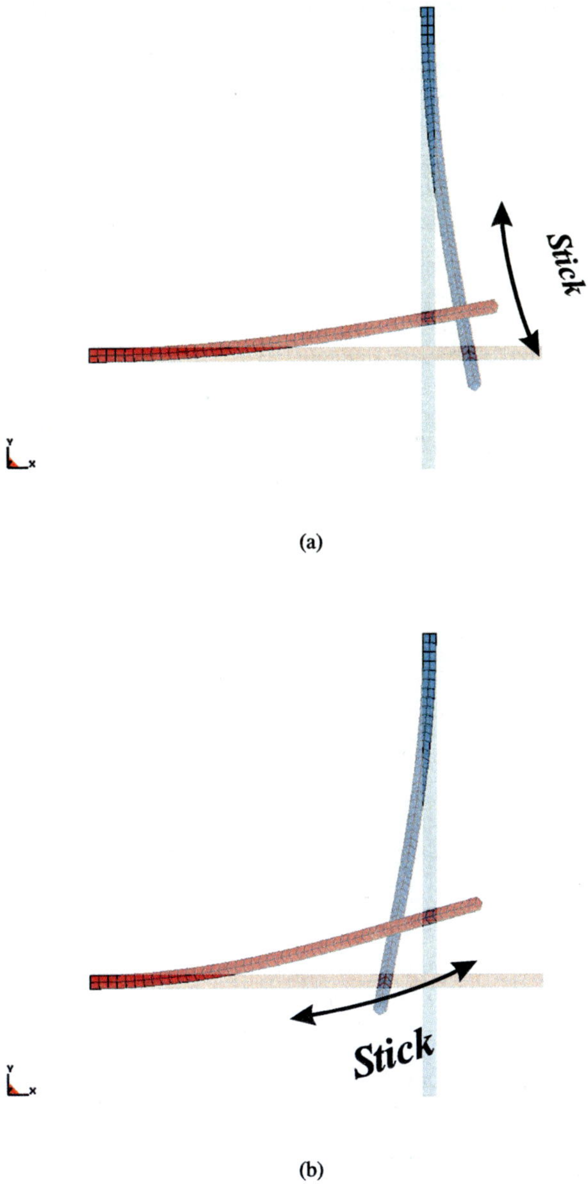

Fig. 8 Partially tied contact: (a) upper beam is sliding, (b) lower beam is sliding.

beam is sliding along the lower beam while sticking in its longitudinal direction and in Figure 8(b) only the lower beam is sliding along the upper beam while sticking in the orthogonal direction.

4 Conclusions

The overview can be summarized as follows:

- The consideration of contact between bodies from a geometrical point of view allows to study systematically all possible contact cases: contact between surfaces, edges, beams and vertices.
- The basis of the theory is the formulation of all parameters in a local coordinate system inherited with a corresponding closest point procedure.
- All known constitutive relations (for elasticity, viscoelasticity, plasticity and viscoplasticity) can be carried into metrics giving the basis for new contact interface laws.

References

1. Alart, P., Heege, A.: Consistent tangent matrices of curved contact operators involving anisotropic friction. Revue Europeenne Des Elements Finis 4, 183–207 (1995)
2. Gurtin, M.E., Weissmueller, J., Larche, F.: A general theory of curved deformable interfaces in solids at equilibrium. Philosophical Magazine A 78, 1093–1109 (1998)
3. Harnau, M., Konyukhov, A., Schweizerhof, K.: Algorithmic aspects in large deformation contact analysis using 'Solid-Shell' elements. Computers and Structures 83, 1804–1823 (2005)
4. Heegaard, J.H., Curnier, A.: Geometric properties of 2D and 3D unilateral large slip contact operators. Computer Methods in Applied Mechanics and Engineering 131, 263–286 (1996)
5. Heege, A., Alart, P.: A frictional contact element for strongly curved contact problems. International Journal for Numerical Methods in Engineering 39, 165–184 (1996)
6. Jones, R.E., Papadopoulos, P.: A geometric interpretation of frictional contact mechanics. Zeitschrift Für Angewandte Mathematik und Physik (ZAMP) 57(6), 1025–1041 (2006)
7. Konyukhov, A., Schweizerhof, K.: Contact formulation via a velocity description allowing efficiency improvements in frictionless contact analysis. Computational Mechanics 33, 165–173 (2004)
8. Konyukhov, A., Schweizerhof, K.: Covariant description for frictional contact problems. Computational Mechanics 35, 190–213 (2005)
9. Konyukhov, A., Schweizerhof, K.: Covariant description of contact interfaces considering anisotropy for adhesion and friction: Part 1. Formulation and analysis of the computational model. Computer Methods in Applied Mechanics and Engineering 196, 103–117 (2006)

10. Konyukhov, A., Schweizerhof, K.: Covariant description of contact interfaces considering anisotropy for adhesion and friction: Part 2. Linearization, finite element implementation and numerical analysis of the model. Computer Methods in Applied Mechanics and Engineering 196, 289–303 (2006)
11. Konyukhov, A., Schweizerhof, K.: A special focus on 2D formulations for contact problems using a covariant description. International Journal for Numerical Methods in Engineering 66, 1432–1465 (2006)
12. Konyukhov, A., Schweizerhof, K.: Symmetrization of various friction models based on an augmented Lagrangian approach. In: Wriggers, P., Nackenhorst, U. (eds.) IUTAM Symposium on Computational Contact Mechanics. IUTAM Bookseries, pp. 97–111. Springer, Heidelberg (2007)
13. Konyukhov, A., Vielsack, P., Schweizerhof, K.: On coupled models of anisotropic contact surfaces and their experimental validation. Wear 264(7/8), 579–588 (2008)
14. Konyukhov, A., Schweizerhof, K.: On the solvability of closest point projection procedures in contact analysis: Analysis and solution strategy for surfaces of arbitrary geometry. Computer Methods in Applied Mechanics and Engineering 197(33/40), 3045–3056 (2008)
15. Konyukhov, A., Schweizerhof, K.: Incorporation of contact for high-order finite elements in covariant form. Computer Methods in Applied Mechanics and Engineering 198, 1213–1223 (2009)
16. Konyukhov, A., Schweizerhof, K.: Geometrically exact covariant approach for contact between curves. Computer Methods in Applied Mechanics and Engineering 199, 2510–2531 (2010)
17. Krstulovic-Opara, L., Wriggers, P., Korelc, J.: A C1-continuous formulation for 3d finite deformation frictional contact. Computational Mechanics 29, 27–42 (2002)
18. Laursen, T.A.: Convected description in large deformation frictional contact problems. International Journal of Solids and Structures 31, 669–681 (1994)
19. Laursen, T.A., Simo, J.C.: A continuum-based finite element formulation for the implicit solution of multibody large deformation frictional contact problems. International Journal for Numerical Methods in Engineering 35, 3451–3485 (1993)
20. Laursen, T.A.: Computational Contact and Impact Mechanics. Fundamentals of Modeling Interfacial Phenomena in Nonlinear Finite Element Analysis. Springer, Berlin (2002)
21. Litewka, P.: Hermite polynomial smoothing in beam-to-beam frictional contact. Computational Mechanics 40, 815–826 (2007)
22. Litewka, P., Wriggers, P.: Contact between 3D beams with rectangular cross-sections. International Journal for Numerical Methods in Engineering 53, 2019–2042 (2002)
23. Litewka, P., Wriggers, P.: Frictional contact between 3D beams. Computational Mechanics 28, 26–39 (2002)
24. Montmitonnet, P., Hasquin, A.: Implementation of an anisotropic friction law in a 3D finite element model of hot rolling. In: Shen, S.-F., Dowson, P. (eds.) Proceedings of NUMIFORM 1995, pp. 301–306 (1995)
25. Parisch, H.: A consistent tangent stiffness matrix for three-dimensional nonlinear contact analysis. International Journal for Numerical Methods in Engineering 28, 1803–1812 (1989)
26. Parisch, H., Luebbing, C.: Ch. Luebbing. A formulation of arbitrarily shaped surface elements for three-dimensional large deformation contact with friction. International Journal for Numerical Methods in Engineering 40, 3359–3383 (1997)

27. Peric, D., Owen, D.R.J.: Computational model for 3-D contact problems with friction based on the penalty method. International Journal for Numerical Methods in Engineering 35, 1289–1309 (1992)
28. Simo, J.C., Laursen, T.A.: An Augmented Lagrangian treatment of contact problems involving friction. Computers and Structures 42, 97–116 (1992)
29. Simo, J.C., Wriggers, P., Schweizerhof, K., Taylor, R.L.: Finite deformation post-buckling analysis involving inelasticity and contact constraints. International Journal for Numerical Methods in Engineering 23, 779–800 (1986)
30. Stadler, M., Holzapfel, G.A., Korelc, J.: C^n continuous modeling of smooth contact surfaces using nurbs and application to 2D problems. International Journal for Numerical Methods in Engineering 57, 2177–2203 (2003)
31. Wriggers, P., Simo, J.C.: A note on tangent stiffness for fully nonlinear contact problems. Communications in Applied Numerical Methods 1, 199–203 (1985)
32. Wriggers, P., Van, V., Stein, E.: Finite element formulation of large deformation impact-contact problem with friction. Computers and Structures 37, 319–331 (1990)
33. Wriggers, P., Zavarise, G.: On contact between three-dimensional beams undergoing large deflection. Communications in Numerical Methods in Engineering 13, 429–438 (1997)
34. Wriggers, P.: Computational Contact Mechanics. John Wiley and Sons, Chichester (2002)
35. Zavarise, G., Wriggers, P.: Contact with friction between beams in 3D space. International Journal for Numerical Methods in Engineering 49, 977–1006 (2000)

Finite Deformation Contact Based on a 3D Dual Mortar and Semi-Smooth Newton Approach

Alexander Popp, Michael W. Gee and Wolfgang A. Wall

Abstract This paper gives a review of the recently proposed dual mortar approach combined with a consistently linearized semi-smooth Newton scheme for 3D finite deformation contact analysis. Some implementation aspects are presented in detail and the most important extensions of the contact model including friction and the treatment of self contact are highlighted. The mortar finite element method, which is applied as discretization scheme, initially yields a mixed formulation with the nodal Lagrange multiplier degrees of freedom as additional primary unknowns. However, by using so-called dual shape functions for Lagrange multiplier interpolation, the global linear system of equations to be solved within each Newton step can be condensed and thus contains only displacement degrees of freedom. All types of nonlinearities, including finite deformations, nonlinear material behavior and contact itself (active set search) are handled within one single iterative solution scheme based on a consistently linearized semi-smooth Newton method. Some very demanding numerical examples are presented to show the high quality of results obtained with this approach as well as to illustrate its superior numerical efficiency and robustness.

1 Introduction

Computational contact mechanics has seen a great thrust of research during the past decade. The vast majority of algorithms developed in the past have been based on the node-to-segment (NTS) approach with the characteristic feature of contact constraints being enforced at specific collocation points. While such methods still enjoy great popularity due to their relatively simple implementation, another class of methods has been advanced through both in-depth mathematical analysis and the development of robust computational schemes: the segment-to-segment approach.

Alexander Popp · Michael W. Gee · Wolfgang A. Wall
Institute for Computational Mechanics, Technische Universität München, Boltzmannstrasse 15, D-85747 Garching, Germany; e-mail: {popp, gee, wall}@lnm.mw.tum.de

G. Zavarise & P. Wriggers (Eds.): Trends in Computational Contact Mechanics, LNACM 58, pp. 57–77.
springerlink.com

Most newly proposed algorithms are based on the mortar method which was originally introduced in the context of domain decomposition techniques [1]. An essential feature of the mortar method is the introduction of an integral (weak) form of continuity conditions across an interface instead of strong, pointwise constraints. When compared with classical NTS methods, there exist many obvious advantages of mortar methods for contact analysis, such as improved robustness in finite deformation and finite sliding situations or guaranteed satisfaction of contact patch tests, for which NTS schemes have been shown to fail under certain circumstances [17]. For further details and state-of-the-art reviews on computational techniques for contact mechanics, we refer to [15,29]. Mortar contact formulations have successfully been applied to the solution of finite deformation contact problems in [21–23, 30]. The mentioned formulations all apply a regularization of contact constraints based on the penalty method or the augmented Lagrangian approach using the Uzawa algorithm. While already clearly demonstrating the accuracy and superior robustness of mortar methods for finite deformation contact analysis, these formulations suffer from one serious drawback. Either the accuracy of the numerical analysis is influenced by an unphysical, user-defined penalty paramater or the computational cost increases due to an additional augmentation loop.

All this obviously motivates the application of direct Lagrange multiplier techniques to the finite deformation contact approach. Some applications based on standard Lagrange multiplier interpolation can be found in [4] and more recently in [9]. Yet, the dual Lagrange multiplier spaces proposed in [27, 28] seem to have the highest potential for efficient algorithms as these alternative spaces allow for a static condensation of the discrete Lagrange multiplier degrees of freedom. Thus, an undesirable increase in global system size, usually typical of direct Lagrange multiplier techniques, is avoided. Mortar methods with dual Lagrange multipliers have first been developed in the context of small deformation contact problems [12, 13], where also the idea of interpreting the active set search as a semi-smooth Newton method (see e.g. [2, 10]) has been adapted. Some first steps towards a finite deformation implementation have been made in [7,8], however still with an incomplete linearization of contact forces and constraints. In the authors' previous work [18, 19], the ideas of dual Lagrange multipliers and a semi-smooth Newton approach for the active set search have been consistently extended to fully nonlinear 2D and 3D frictionless contact problems with linear and quadratic interpolation. The corresponding dual mortar contact formulation with Coulomb friction and its consistent linearization based on semi-smooth Newton ideas have been proposed recently [6].

Our current contribution presents and highlights some key aspects of the work presented in [6, 18, 19] in more detail. Section 2 provides a rough overview of the general continuum mechanics problem setup for 3D finite deformation contact. The mortar approach and dual Lagrange multiplier interpolation as the two main ingredients of the finite element discretization are described in Section 3. In Sections 4 and 5, the semi-smooth Newton type active set strategy, its consistent linearization and the resulting solution algorithm are outlined. Section 6 highlights some very important aspects of implementation, including the evaluation of mortar integrals in 3D, higher-order interpolation, linear and angular momentum conservation and efficient

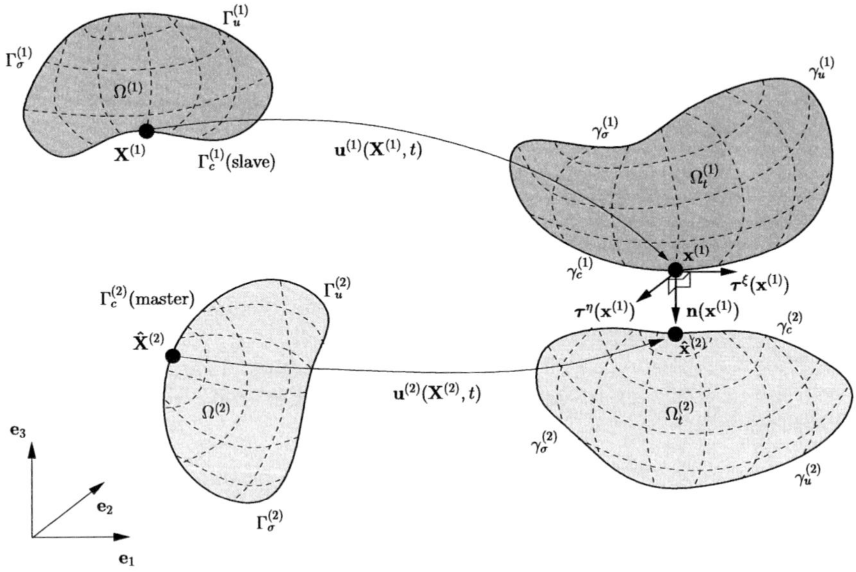

Fig. 1 Notation for the two body finite deformation contact problem in 3D.

parallel contact search algorithms. Moreover, we allude to the treatment of self contact and friction. The finite deformation dual mortar approach and its capabilities are evaluated by means of three numerical examples in Section 7 and some conclusions are drawn in Section 8.

2 General Formulation of the Finite Deformation Contact Problem

The basic problem definition has been described in detail in [19], thus we only give a very short overview here. Figure 1 shows the reference and current configurations of two elastic bodies undergoing a finite deformation process and introduces some notation. The surfaces $\partial\Omega_t^{(i)}$, $i = 1,2$ are divided into three disjoint boundary sets, namely the common Dirichlet and Neumann boundaries $\gamma_u^{(i)}$ and $\gamma_\sigma^{(i)}$ as well as the potential contact surfaces $\gamma_c^{(i)}$. Although a mortar discretization will be applied later, we still retain the customary nomenclature of slave and master contact surfaces.

Displacement vectors $\mathbf{u}^{(i)} = \mathbf{x}^{(i)} - \mathbf{X}^{(i)}$ describe the motion of the deformable bodies from reference configuration $\mathbf{X}^{(i)}$ to current configuration $\mathbf{x}^{(i)}$. Material nonlinearity is taken into account by assuming a compressible Neo-Hookean material behavior (see e.g. [11]) based on the second Piola–Kirchhoff stress tensor $\mathbf{S}$ and the Green–Lagrange strain tensor $\mathbf{E} = \frac{1}{2}\left(\mathbf{F}^T\mathbf{F} - \mathbf{I}\right)$, where $\mathbf{F}$ is the deformation gradient. Based on these definitions, the well-known boundary value problem (BVP) of finite deformation elasticity and the inital boundary value problem (IBVP) of finite deformation elastodynamics can be formulated. This paper focuses on contact interaction itself, which is typically described by a gap function $g(\mathbf{X},t)$ in the current configuration:

$$g(\mathbf{X},t) = -\mathbf{n}\big(\mathbf{x}^{(1)}(\mathbf{X}^{(1)},t)\big) \cdot \big[\mathbf{x}^{(1)}(\mathbf{X}^{(1)},t) - \hat{\mathbf{x}}^{(2)}(\hat{\mathbf{X}}^{(2)},t)\big], \tag{1}$$

where $\mathbf{n}$ represents the current outward unit normal on the slave surface $\gamma_c^{(1)}$ in $\mathbf{x}^{(1)}$, $\hat{\mathbf{x}}^{(2)}$ denotes the projection of $\mathbf{x}^{(1)}$ onto the master surface $\gamma_c^{(2)}$ along $\mathbf{n}$ and $\hat{\mathbf{X}}^{(2)}$ is the corresponding point in the reference configuration, see Figure 1. Our definition of the discrete version of contact normal $\mathbf{n}$ will be motivated briefly in Section 6.1. Together with the two tangent vectors $\boldsymbol{\tau}^{\xi}$ and $\boldsymbol{\tau}^{\eta}$, $\mathbf{n}$ builds an orthonormal basis in $\mathbf{x}^{(1)}$. The classical Karush–Kuhn–Tucker (KKT) conditions of normal contact and the frictionless sliding conditions then read as follows:

$$g(\mathbf{X},t) \geq 0, \quad p_n \leq 0, \quad p_n g(\mathbf{X},t) = 0, \tag{2}$$

$$t_\tau^{\xi} = t_\tau^{\eta} = 0, \tag{3}$$

where the normal and tangential components of the slave contact traction $\mathbf{t}_c^{(1)}$ are denoted as p_n, t_τ^{ξ} and t_τ^{η}. A corresponding formulation of Coulomb's law for frictional contact is omitted here, but can be found in our recent work [6]. For deriving a weak formulation the solution space $\mathscr{U}^{(i)}$ and the weighting space $\mathscr{V}^{(i)}$ are defined as

$$\mathscr{U}^{(i)} = \left\{ u_j^{(i)} \in H^1(\Omega^{(i)}),\ j = 1,2,3 \mid u_j^{(i)} = \hat{u}_j^{(i)} \text{ on } \Gamma_u^{(i)} \right\}, \tag{4}$$

$$\mathscr{V}^{(i)} = \left\{ \delta u_j^{(i)} \in H^1(\Omega^{(i)}),\ j = 1,2,3 \mid \delta u_j^{(i)} = 0 \text{ on } \Gamma_u^{(i)} \right\}. \tag{5}$$

Here, $H^1(\Omega^{(i)})$ denotes the usual Sobolev space of functions with square integrable values and first derivatives, respectively. The method of weighted residuals then yields: Find $u_j^{(i)} \in \mathscr{U}^{(i)}$ such that

$$\delta\Pi(\mathbf{u},\delta\mathbf{u}) = \delta\Pi_{\text{int,ext}}(\mathbf{u},\delta\mathbf{u}) + \int_{\gamma_c^{(1)}} \boldsymbol{\lambda} \cdot (\delta\mathbf{u}^{(1)} - \delta\hat{\mathbf{u}}^{(2)})\,\mathrm{d}\gamma = 0$$

$$\forall\, \delta u_j^{(i)} \in \mathscr{V}^{(i)}, \quad j = 1,2,3, \tag{6}$$

where the negative contact traction on the slave side of the interface has been replaced by a Lagrange multiplier vector $\boldsymbol{\lambda} = -\mathbf{t}_c^{(1)}$. By arbitrary choice, the Lagrange multipliers $\lambda_j \in \mathscr{M}$, $j = 1,2,3$, have been introduced on the slave side, where $\mathscr{M}$ is defined as dual space of the trace space $\mathscr{W}^{(1)}$ of $\mathscr{U}^{(1)}$ restricted to $\Gamma_c^{(1)}$. Under the assumption that $\Gamma_c^{(1)}$ is compactly embedded in $\partial\Omega^{(1)} \setminus \Gamma_u^{(1)}$, we obtain $\mathscr{W}^{(1)} = H^{1/2}(\Gamma_c^{(1)})$ and $\mathscr{M} = H^{-1/2}(\Gamma_c^{(1)})$, see [28]. Corresponding test functions $\delta\lambda_j$ serve as weighting functions for the non-penetration constraint in (2):

$$\int_{\gamma_c^{(1)}} \delta\lambda_n \, g(\mathbf{X},t)\,\mathrm{d}\gamma \geq 0 \qquad \forall\, \delta\lambda_n \in \mathscr{M}. \tag{7}$$

The remaining strong contact conditions in (2) and (3) are then rewritten as

$$\lambda_n \geq 0, \quad \lambda_n g(\mathbf{X},t) = 0, \quad \lambda_\tau^\xi = \lambda_\tau^\eta = 0. \tag{8}$$

Altogether, equations (6)–(8) establish a mixed variational formulation with the solution $u_j^{(i)} \in \mathscr{U}^{(i)}$ and $\lambda_j \in \mathscr{M}$. An overview of mortar finite element discretization using dual Lagrange multipliers follows in the upcoming paragraph. Some of the contact constraints are still formulated as inequality conditions, which necessitates the application of a suitable active set strategy. The primal-dual active set strategy (PDASS) and our reformulation as a semi-smooth Newton scheme including consistent linearization will be presented in Section 4.

3 Dual Mortar Finite Element Discretization

Spatial discretization of the contact virtual work term in (6) requires a discretization of slave and master surface, which is directly connected to the underlying structural discretization based on their trace space relationship. In this paper, mostly first-order Lagrangian finite elements are considered, thus contact surfaces may consist of 3-node triangular and 4-node quadrilateral elements. Exemplarily, we give the general definition of slave and master displacements, which reads as follows:

$$\mathbf{d}^{(1)h}|_{\Gamma_c^{(1)h}} = \sum_{k=1}^{n_{sl}} N_k^{(1)}(\boldsymbol{\xi}^{(1)})\,\mathbf{d}_k^{(1)}, \quad \mathbf{d}^{(2)h}|_{\Gamma_c^{(2)h}} = \sum_{l=1}^{n_m} N_l^{(2)}(\boldsymbol{\xi}^{(2)})\,\mathbf{d}_l^{(2)}. \tag{9}$$

Spatially discretized quantities are labeled with a superscript h and the total number of slave and master nodes is n_{sl} and n_m, respectively. Nodal displacements are given by $\mathbf{d}_k^{(1)}$, $\mathbf{d}_l^{(2)}$ and shape functions $N_k^{(1)}$, $N_l^{(2)}$ are defined with respect to the finite element parameter space $\boldsymbol{\xi}^{(i)} = (\xi^{(i)}, \eta^{(i)})$. Lagrange multiplier interpolation is then based on dual shape functions Φ_j, as pioneered by Wohlmuth [27], with the same polynomial degree as used for surface geometry interpolation:

$$\boldsymbol{\lambda}^h = \sum_{j=1}^{n_{sl}} \Phi_j(\boldsymbol{\xi}^{(1)})\,\mathbf{z}_j, \tag{10}$$

with discrete nodal Lagrange multipliers $\mathbf{z}_j$. The construction of dual shape functions is based on a biorthogonality relation as introduced in [27], [28] and [13]:

$$\int_{\gamma_c^{(1)h}} \Phi_j(\boldsymbol{\xi}^{(1)})\,N_k^{(1)}(\boldsymbol{\xi}^{(1)})\,\mathrm{d}\gamma = \delta_{jk} \int_{\gamma_c^{(1)h}} N_k^{(1)}(\boldsymbol{\xi}^{(1)})\,\mathrm{d}\gamma, \tag{11}$$

where δ_{jk} is the Kronecker delta. This approach will later allow for static condensation of the discrete multipliers $\mathbf{z}_j$. Note, that in the context of finite deformations dual shape functions become deformation-dependent themselves. For an overview and exemplary local calculations of element-specific dual shape functions in 3D contact analysis, we refer to [7,19]. Nodal blocks of the two mortar integral matrices $\mathbf{D} \in \mathbb{R}^{3n_{sl} \times 3n_{sl}}$ and $\mathbf{M} \in \mathbb{R}^{3n_{sl} \times 3n_m}$ are then defined as

$$\mathbf{D}[j,j] = D_{jj}\mathbf{I}_3 = \int_{\gamma_c^{(1)h}} N_j^{(1)} \mathrm{d}\gamma \, \mathbf{I}_3, \tag{12}$$

$$\mathbf{M}[j,l] = M_{jl}\mathbf{I}_3 = \int_{\gamma_c^{(1)h}} \Phi_j N_l^{(2)} \mathrm{d}\gamma \, \mathbf{I}_3, \tag{13}$$

with $j = 1,\ldots,n_{sl}$, $l = 1,\ldots,n_m$ and with the identity $\mathbf{I}_3 \in \mathbb{R}^{3\times 3}$. Owing to the biorthogonality relation (11), $\mathbf{D}$ reduces to a diagonal matrix. This yields as algebraic notation of the discretized contact virtual work

$$\delta\Pi_c^h = (\delta\mathbf{d}^{(1)})^T \mathbf{D}^T \mathbf{z} - (\delta\mathbf{d}^{(2)})^T \mathbf{M}^T \mathbf{z}, \tag{14}$$

where all discrete nodal values of Lagrange multipliers and nodal test function values are assembled into global vectors $\mathbf{z}$, $\delta\mathbf{d}^{(1)}$ and $\delta\mathbf{d}^{(2)}$, respectively. To make upcoming algebraic representations clearer, the set of all finite element nodes is now split into three subsets: a subset $\mathscr{S}$ containing all n_{sl} potential slave side contact nodes, a subset $\mathscr{M}$ of all n_m potential master side contact nodes and the set of all remaining nodes $\mathscr{N}$. The global nodal displacement vector can be sorted accordingly, yielding $\mathbf{d} = (\mathbf{d}_{\mathscr{N}}, \mathbf{d}_{\mathscr{M}}, \mathbf{d}_{\mathscr{S}})^T$. Then, the vector of discrete contact forces is

$$\mathbf{f}_c = [\mathbf{0} \quad -\mathbf{M} \quad \mathbf{D}]^T \mathbf{z}. \tag{15}$$

The contact forces extend the fully discretized force residual resulting from standard finite element discretization of internal and external virtual work in (6). This yields the total nonlinear force residual

$$\mathbf{r} := \mathbf{f}_{\text{int}}(\mathbf{d}) - \mathbf{f}_{\text{ext}} + \mathbf{f}_c(\mathbf{d},\mathbf{z}) = 0. \tag{16}$$

A discrete version of the weak non-penetration condition is obtained by inserting the Lagrange multiplier interpolation (10) into (7)

$$\int_{\gamma_c^{(1)}} \delta\lambda_n g \,\mathrm{d}\gamma \approx \sum_{j=1}^{n_{sl}} (\delta z_n)_j \int_{\gamma_c^{(1)h}} \Phi_j \hat{g} \,\mathrm{d}\gamma := \sum_{j=1}^{n_{sl}} (\delta z_n)_j \tilde{g}_j \geq 0. \tag{17}$$

Here, $\hat{g}$ is the discrete version of the gap function defined in (1), and for each slave node $j \in \mathscr{S}$ a discrete normal weighted gap $\tilde{g}_j$ has been introduced. Discretization of the remaining contact conditions yields as discrete formulation of the KKT conditions and the frictionless sliding conditions:

$$\tilde{g}_j \geq 0, \quad (z_n)_j \geq 0, \quad (z_n)_j \tilde{g}_j = 0, \tag{18}$$

$$(z_\tau^\xi)_j = (z_\tau^\eta)_j = 0. \tag{19}$$

It is worth mentioning that although a mortar discretization has been employed, the weighted constraints at discrete nodal points are enforced independently.

4 Semi-Smooth Newton Method and Consistent Linearization

The basic idea of any active set strategy is to find the correct subset of all slave nodes which are currently in contact with the master surface. The set $\mathscr{A} \subseteq \mathscr{S}$ is called the active set and definition of the inactive set $\mathscr{I} = \mathscr{S} \setminus \mathscr{A}$ is straightforward. In our recent papers [18], [19] the active set search has been successfully interpreted as a semi-smooth Newton method for both 2D and 3D finite deformation contact analysis. The main advantage of this new approach is the fact that all sources of nonlinearities, i.e. finite deformations, nonlinear material behavior and contact itself, can be efficiently treated within one single iterative scheme. There is no need for a nested approach with the outer loop solving for the correct active set as presented in [8, 9]. Here, only a very short overview will be given starting with the reformulation of the discrete KKT conditions in (18) within a so-called nonlinear complementarity function C_j for each slave node $j \in \mathscr{S}$:

$$C_j(\mathbf{z}_j, \mathbf{d}) = (z_n)_j - \max(0, (z_n)_j - c_n \tilde{g}_j) = 0, \quad c_n > 0. \tag{20}$$

It can be easily shown that (20) is equivalent to the set of KKT conditions and that this equivalence holds for arbitrary positive values of the purely algorithmic complementarity parameter c_n. While C_j is a continuous function, it is non-smooth and has no uniquely defined derivative at positions $(z_n)_j - c_n \tilde{g}_j = 0$. Yet, it is well-known from mathematical literature on constrained optimization [10, 24] and from applications in small deformation contact analysis [12] that the max-function is semi-smooth and therefore a Newton method can still be applied. The complete nonlinear problem reads as follows:

$$\begin{aligned} \mathbf{r} = \mathbf{f}_{\text{int}}(\mathbf{d}) - \mathbf{f}_{\text{ext}} + \mathbf{f}_c(\mathbf{d}, \mathbf{z}) &= \mathbf{0}, \\ C_j(\mathbf{z}_j, \mathbf{d}) &= 0 \qquad \forall\, j \in \mathscr{S}, \\ (z_\tau^\xi)_j = (z_\tau^\eta)_j &= 0 \qquad \forall\, j \in \mathscr{S}. \end{aligned} \tag{21}$$

Deriving a consistent Newton method for the iterative solution of the nonlinear contact problem under consideration relies on the full linearization of all deformation-dependent quantities, such as nodal normal and tangential vectors or mortar integral matrices. This linearization process has been presented in great detail for both 2D and 3D case in the authors' previous work [18, 19].

5 Solution Algorithm

The resulting semi-smooth Newton algorithm to be solved within one time or load increment is summarized next. All types of nonlinearities including the search for the correct active set are resolved within one Newton iteration, with the active and incative sets $\mathscr{A}$ and $\mathscr{I}$ being updated after each semi-smooth Newton step:

Algorithm 1

1. Set $i = 0$ and initialize the solution $(\mathbf{d}_0, \mathbf{z}_0)$.
2. Initialization: $\mathscr{A}_0 \cup \mathscr{I}_0 = \mathscr{S}$ and $\mathscr{A}_0 \cap \mathscr{I}_0 = \emptyset$.
3. Find the primal-dual pair $(\Delta\mathbf{d}_i, \mathbf{z}_{i+1})$ by solving

$$\Delta\mathbf{r}|_i = -\mathbf{r}|_i, \tag{22}$$

$$\mathbf{z}_j|_{i+1} = 0 \qquad \forall\, j \in \mathscr{I}_i, \tag{23}$$

$$\Delta\tilde{g}_j|_i = -\tilde{g}_j|_i \qquad \forall\, j \in \mathscr{A}_i, \tag{24}$$

$$\Delta\boldsymbol{\tau}_j^{\xi} \cdot \mathbf{z}_j|_i + \boldsymbol{\tau}_j^{\xi} \cdot \mathbf{z}_j|_{i+1} = 0 \qquad \forall\, j \in \mathscr{A}_i, \tag{25}$$

$$\Delta\boldsymbol{\tau}_j^{\eta} \cdot \mathbf{z}_j|_i + \boldsymbol{\tau}_j^{\eta} \cdot \mathbf{z}_j|_{i+1} = 0 \qquad \forall\, j \in \mathscr{A}_i. \tag{26}$$

4. Update $\mathbf{d}_{i+1} = \mathbf{d}_i + \Delta\mathbf{d}_i$.
5. Set $\mathscr{A}_{i+1}$ and $\mathscr{I}_{i+1}$ to

$$\mathscr{A}_{i+1} := \left\{ j \in \mathscr{S} \,|\, (z_n)_j|_{i+1} - c_n\tilde{g}_j|_{i+1} > 0 \right\},$$

$$\mathscr{I}_{i+1} := \left\{ j \in \mathscr{S} \,|\, (z_n)_j|_{i+1} - c_n\tilde{g}_j|_{i+1} \leq 0 \right\}. \tag{27}$$

6. If $\mathscr{A}_{i+1} = \mathscr{A}_i$, $\mathscr{I}_{i+1} = \mathscr{I}_i$ and $\|\mathbf{r}_{\text{tot}}\| \leq \varepsilon_r$, then stop, else set $i := i + 1$ and go to step (3).

Here, ε_r represents an absolute Newton convergence tolerance for the $\mathscr{L}_2$-norm of the total residual vector $\mathbf{r}_{\text{tot}}$, which comprises the force residual $\mathbf{r}$ and the residual of the contact constraints (23)–(26). Numerous tests reveal that even for large step sizes and fine contacting meshes the correct active set is found after a few Newton steps. Once the sets remain constant, quadratic convergence is obtained due to the underlying consistent linearization. The interested reader is referred to [18, 19] for the matrix notation of the final linear system (22)–(26). Owing to the dual Lagrange multiplier shape functions, the mortar matrix $\mathbf{D}$ becomes diagonal, which makes its inversion trivial. Thus, the discrete multiplier values are eliminated by condensation and the resulting linear system of equations is unsymmetric but positive definite and contains only displacement degrees of freedom. The undesirable increase in global system size, typical for direct Lagrange multiplier methods, is avoided. Here, we restrict the presentation to a schematic algebraic form

$$\mathbf{L}_{dd}|_i \Delta\mathbf{d}_i = -\tilde{\mathbf{r}}_{\text{tot}}|_i, \tag{28}$$

which illustrates the modified residual vector $\tilde{\mathbf{r}}_{\text{tot}}$ emanating from $\mathbf{r}_{\text{tot}}$ by condensation of the discrete Lagrange multiplier degrees of freedom. Similarly, we obtain a modified effective tangent stiffness matrix $\mathbf{L}_{dd}$ including contact stiffness terms and the condensed constraint contributions. Is is worth mentioning that the presented method uses neither a regularization (usually introduced via an unphysical and problem-specific penalty parameter) nor a costly augmentation scheme. The dual Lagrange multipliers allow for the contact constraints to be enforced exactly.

6 Implementation Details and Extensions

In this section, some important aspects concerning the implementation of the algorithms described above and some possible model extensions are outlined. This includes the evaluation of mortar integrals in 3D, higher-order interpolation, linear and angular momentum conservation, efficient parallel contact search algorithms as well as the treatment of self contact and friction. In each subsection we provide an overview on how the respective algorithmic detail is accurately implemented and how the respective additional model complexity is handled.

6.1 Evaluation of Mortar Integrals

Evaluation of the discrete mortar integral terms and discrete weighted gaps requires so-called contact segments suitable for numerical integration. These contact segments must be defined such that shape function integrands in (12), (13) and (17) are continuous on these surface subsets. The most common choice for 3D mortar algorithms available in the literature is a simplified coupling algorithm, in which integration is not performed on the slave surface itself but on its geometrical approximation using piecewise flat segments. This coupling algorithm has been proposed in [20, 22, 23] for standard Lagrange multiplier shape functions and a review including all details of the consistent linearization in the context of a dual mortar scheme can also be found in [19]. For contact normal discretization, the so-called continuous field of slave normals based on nodal averaging has first been suggested in [30]. The numerical examples in [7, 18, 19, 30] demonstrate that this approach provides the desired robustness, while it is of course not unique. At each slave node $j \in \mathscr{S}$ an averaged nodal unit normal is defined based on averaging of the unit normal vector $\mathbf{n}_{j,e}$ at node j on the adjacent slave element $e = 1,\ldots,n^j_{\text{ele}}$, where n^j_{ele} represents the total number of slave surface contact elements adjacent to slave node j. The definition reads as follows:

$$\mathbf{n}_j = \frac{\sum_{e=1}^{n_{\text{ele}}^j} \mathbf{n}_{j,e}}{\| \sum_{e=1}^{n_{\text{ele}}^j} \mathbf{n}_{j,e} \|} . \tag{29}$$

6.2 Quadratic Interpolation

Treatment of quadratic interpolation in the context of a 3D dual mortar method needs two basic ingredients: an extension of the segmentation and integration algorithm as well as the definition of suitable dual Lagrange multiplier spaces. As outlined previously, the usual method for linear interpolation is based on the projection of flattened surface elements. This approach has been directly extended to the quadratic case in [21] and has also been employed in [19]. The algorithm basically subdivides each quadratic surface element into a certain number of linearly interpolated segments. However, by establishing geometric mappings from parent element space to sub-segment space and vice versa, one is still able to properly evaluate the higher-order shape function products in (12) and (13). Is is important to point out that this approach certainly introduces an approximation of the integration domain, which no longer reflects the underlying quadratic surfaces correctly. Yet, as explained in [21], this is already the case in the first-order integration algorithm, where warped slave elements are flattened to ease mortar projection.

The second important aspect here is the definition of a suitable dual Lagrange multiplier space for 3D quadratic interpolation. While a definition based on the biorthogonality condition (11) is straightforward for 9-node quadrilateral surface elements, an extension to the remaining cases (i.e. 8-node quadrilateral and 6-node triangular surface elements) is subject to current research. Basically, this is due to the non-positiveness of shape function integrals $\int N_j^{(1)} \mathrm{d}\gamma$, which prohibits the proper use of (11) for *quad8* and *tri6* surfaces and thus the definition of dual shape functions. Alternatives have already been studied for the small deformation tied contact case in [14]. It has been shown there that optimal rates of convergence of the mortar method can also be achieved when reducing the polynomial order of the Lagrange multiplier interpolation by one compared to the displacements.

6.3 Conservation of Linear and Angular Momentum

The proposed method *exactly* conserves linear momentum if master and slave side (i.e. the mortar integral matrices **D** and **M**) are integrated based on the same numerical integration scheme. This can be proven considering the sum of the standard finite element basis functions being 1 on both slave and master surface, see [19].

As mentioned by other authors, e.g. in [22], conservation of angular momentum and thus rotational invariance of the semi-discrete system is challenging in the context of mortar contact formulations. This is due to the fact that in general the variation of the mortar integrals is neglected when deriving the discrete contact virtual work expression, see [22, 23]. Yet, as has been demonstrated recently in [9] for the 2D case, these additional terms must be considered in order to assure *exact* angular momentum conservation. Therefore, modifications of the presented approach have to be made so that the variation of the mortar integrals **D** and **M** is taken into account in the discrete contact virtual work in cases where exact angular momentum conservation is considered important. Linearization of this extended formulation includes second derivatives of deformation-dependent quantities in **D** and **M**.

6.4 Efficient Parallel Contact Search

The solution of finite deformation contact problems is difficult to be managed due to the strong nonlinearities involved. A brute force approach for contact search may add another very time-consuming part, at least to the simulation of very large contact problems. Therefore, efficient parallel search algorithms are needed. Here, we employ a recently developed approach [31] based on so-called discretized-orientation-polytopes (k-DOPs) as bounding volumes. Compared to the commonly employed axis-aligned bounding boxes (AABBs), the k-DOPs allow for a much tighter and thus more efficient geometrical representation of the contact surfaces. Slave and master contact surface are organized within hierarchical binary tree structures so that very fast search and tree update procedures can be applied. The approach given in [31] for the single-processor case has been extended to fit into a parallel finite element simulation framework based on overlapping domain decomposition. This allows an efficient parallel search, even for very large contact problems.

6.5 Treatment of Self Contact

The approach described in Section 6.4 can also be extended to self contact, but this requires some fundamental algorithmic adjustments. A very efficient method for self contact search has been presented in [32] and has also been included into the simulation framework presented here. The search algorithm again is based on a bounding volume hierarchy organized as a binary tree. It is however crucial for self contact simulations to set up the bounding volume hierarchy in a *bottom-up* way based on mesh connectivity (e.g. using a dual graph). Moreover, a curvature criterion may be used to accelerate the searching procedure. Finally, dynamic assignment of slave and master sub-surfaces, which are unknown *a priori* for self contact contact problems, becomes necessary.

6.6 Treatment of Friction

An extension of the described mortar contact algorithms to the 2D frictional case has recently been proposed by the authors [6]. All steps outlined in Sections 2–5, including definition of dual shape functions and mortar integrals as well as semi-smooth Newton treatment of the non-penetration constraint, remain unchanged. Only the frictionless sliding conditions introduced in (3) are replaced by frictional sliding conditions, e.g. using Coulomb's law. Based on an objective formulation of friction kinematic variables [30] the additional constraints again are recast into non-smooth nonlinear complementarity functions, as presented for the non-penetration constraint in (20). By applying full linearization, the frictional constraints are consistently integrated into the semi-smooth Newton scheme outlined in Section 4. All nonlinearities, including contact (i.e. search for the correct active set), friction (i.e. search for the current stick and slip regions), finite deformations and nonlinear material behavior are then treated within one single iterative scheme. For further details and numerical examples, the interested reader is referred to [6].

7 Examples

We present three numerical examples to illustrate the capabilities of the proposed approach. All simulations are based on a parallel implementation of the contact algorithms described above in our in-house multiphysics research code BACI [26]. A compressible Neo-Hookean constitutive law determined by Young's modulus E and Poisson's ratio ν is employed. Convergence of a Newton iterative scheme is measured in terms of the total residual norm $\|\tilde{\mathbf{r}}_{\text{tot}}\|$ with a relative convergence tolerance of 10^{-12}. The complementarity parameter described in Section 4 is set to $c_n = 1$, which guarantees for all problem setups considered that the correct active set is found within only a few semi-smooth Newton steps.

7.1 Hertzian Contact

In contrast to the 2D case, where Hertzian contact of an elastic half-cylinder and a rigid planar surface is one of the most frequently investigated problems, fully 3D Hertzian contact of an elastic hemisphere and a rigid planar surface is rarely analyzed. In this section, the proposed mortar contact formulation is validated with such a 3D Hertzian contact example based on a small deformation assumption. Without finite deformations being present, the semi-smooth Newton scheme described in Section 4 needs to solve for the contact nonlinearity (i.e. finding the correct active set) only. Finite deformation and finite strain situations will be presented in the following subsections. Here, the focus lies on the accuracy of the proposed method, which is analyzed by comparing numerical results for the Lagrange multipliers (con-

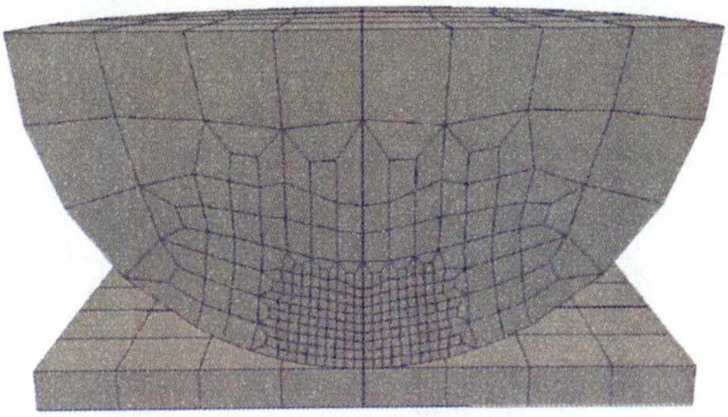

Fig. 2 3D Hertzian contact of an elastic hemisphere – geometry and finite element mesh (the model is cut in the middle to show the finite element mesh in the potential contact zone).

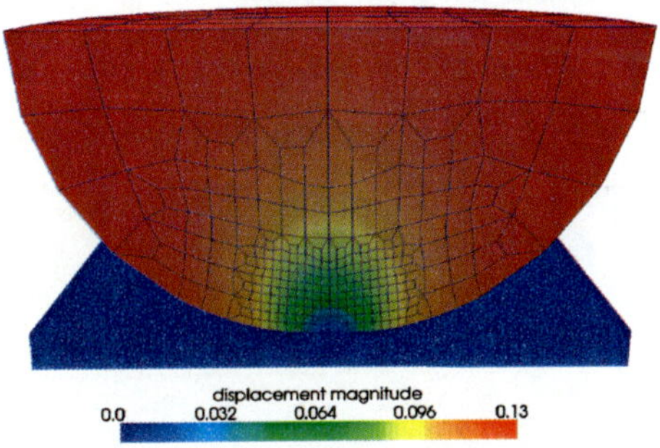

Fig. 3 3D Hertzian contact of an elastic hemisphere – deformed geometry and displacements.

tact traction) and analytical solutions known from literature. The two characteristic quantities of the Hertzian contact traction distribution are the maximum pressure $p_{c,\max}$ and the contact zone radius a, see [25]:

$$p_{c,\max} = 0.388\sqrt[3]{4p\pi E^2}, \quad a = 1.109\sqrt[3]{\frac{pR^3\pi}{2E}}. \tag{30}$$

Geometry and finite element mesh for the 3D Hertzian contact example are illustrated in Figure 2. A constant pressure $p = 0.2$ is applied onto the top surface of the hemisphere with radius $R = 8$. Material data is given as $E = 200$ and $\nu = 0.3$, which altogether yields $p_{c,\max} = 18.0$ and $a = 1.03$ as analytical reference solution. Figure 3 shows the deformed configuration and the computed displacement magnitude. It can be seen very clearly from both Figures 2 and 3 that the mesh, consisting of 8-node hexahedral elements, has been locally refined around the pole of the hemisphere. Yet, the mesh can still be considered quite coarse, with only five elements along a radial line of the actual contact zone.

In order to evaluate the accuracy of our dual mortar contact approach, contact traction results need to be visualized, which is done in Figure 4 from three different perspectives. The maximum pressure $p_{c,\max}$, represented in the simulation by the nodal dual Lagrange multiplier value at the pole of the hemisphere, differs from

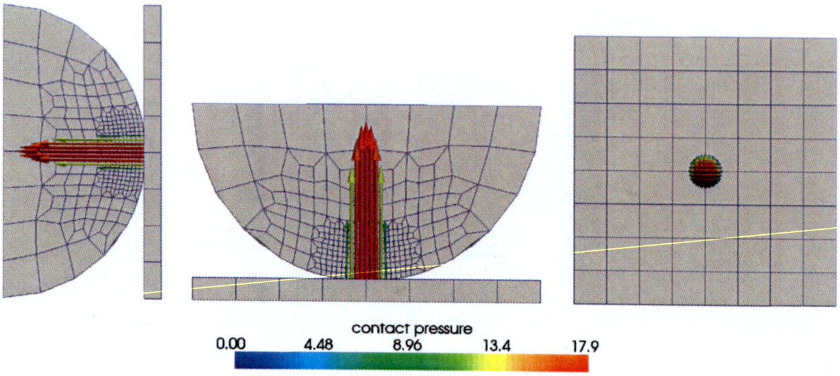

Fig. 4 3D Hertzian contact of an elastic hemisphere – numerical contact traction results in three different perspectives (pairwise orthogonal projections).

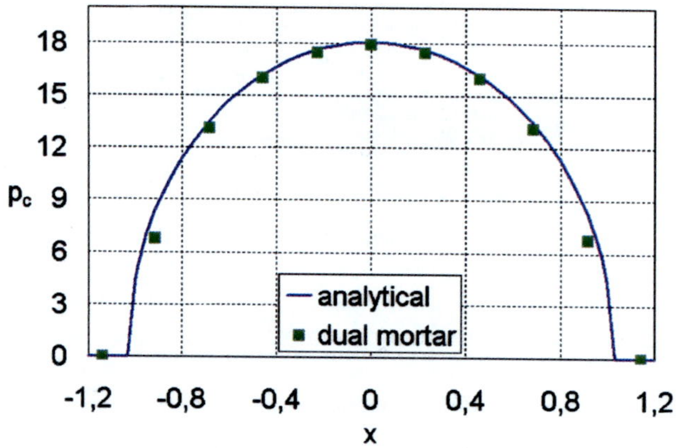

Fig. 5 3D Hertzian contact of an elastic hemisphere – numerical contact traction results in a cutting plane and comparison with axisymmetric analytical solution.

the reference solution by less than 0.5%. The accuracy of the computed numerical solution can be seen even more clearly in Figure 5, where the analytical normal contact traction distribution and the nodal dual Lagrange multiplier solution are compared in the cutting plane already visualized in Figures 2 and 3.

The effect of very small discrepancies appearing in the outermost part of the contact zone is well-known. It further diminishes when mesh refinement or higher-order interpolation is applied. The interested reader is referred to [5] for a detailed investigation of this phenomenon in the context of a 2D penalty approach.

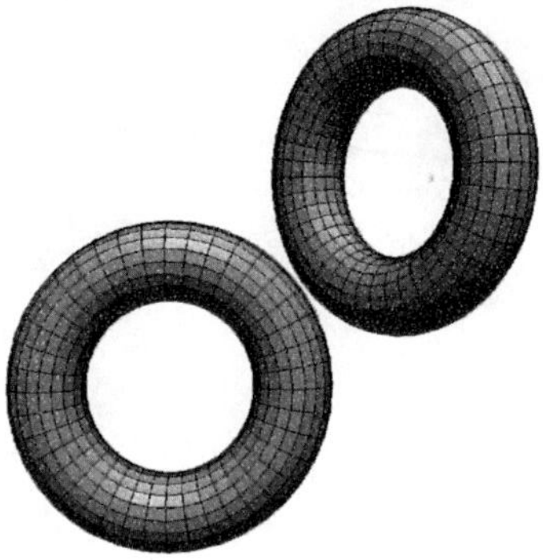

Fig. 6 Two torus impact – geometry and finite element mesh.

Table 1 Convergence behavior of the proposed semi-smooth Newton scheme in terms of the total residual norm for two representative time steps $\Delta t = 0.05$ starting from (a) $t = 4.5$ and (b) $t = 5.5$.

Newton step	(a) $t = 4.5$	(b) $t = 5.5$
1	1.09e+06 (*)	1.04e+06 (*)
2	2.16e+04 (*)	3.60e+04 (*)
3	4.05e+00	1.38e+01 (*)
4	1.53e−04	2.11e−03
5	2.46e−08	2.77e−08

(*) = change in active contact set

7.2 Two Torus Impact

In this example, a finite deformation impact problem of two toruses ($E = 2250$, $\nu = 0.3$) is investigated. Both geometry and loading conditions are based on a quite similar analysis presented in [31] to evaluate contact search strategies. Figure 6 illustrates the initial configuration and the finite element mesh with 1600 hexahedral elements in total. The major and minor radius of the two hollow toruses is 76 and 24, repectively and the wall thickness is 4.5. The upper torus is rotated around the vertical axis by 45 degrees. We apply transient structural dynamics here using a Generalized-α time integration scheme [3] with the initial density of the contacting bodies chosen as $\rho = 0.1$.

During a total of 200 time steps (step size $\Delta t = 0.05$) the lower torus is first accelerated towards the upper torus and then a very general oblique impact situation with large structural deformations occurs. In addition to that, the proposed method needs to resolve continuous changes of the active contact set due to the large amount of (frictionless) sliding. Figure 7 shows six characteristic deformed configurations associated with the simulation stages described above.

Table 1 illustrates convergence of the presented fully linearized semi-smooth Newton scheme in terms of the total residual norm for two representative time steps both including large deformations and considerable changes of the active contact

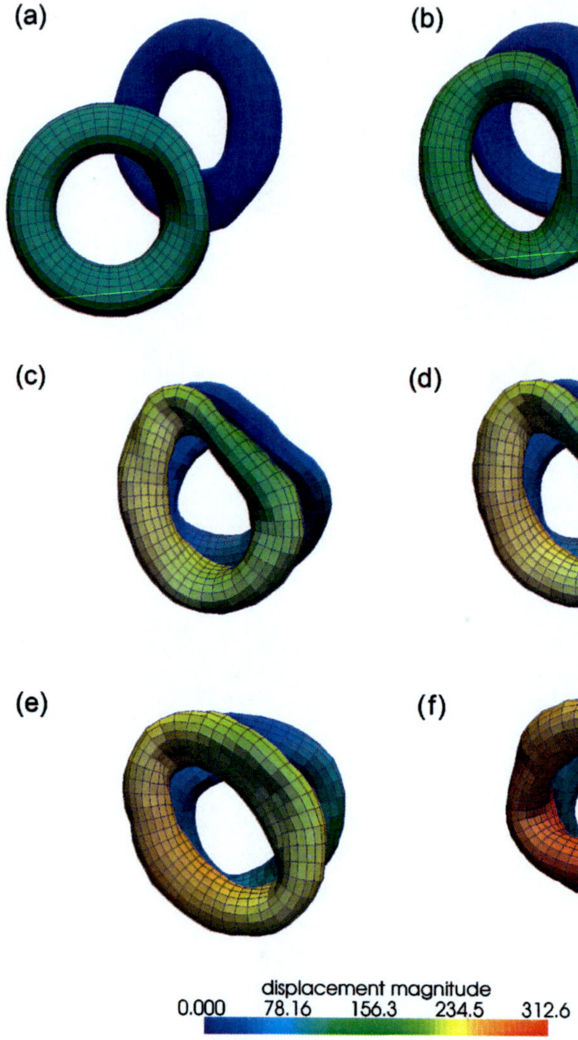

Fig. 7 Two torus impact – displacement magnitude at characteristic stages of deformation: (a) $t = 3$; (b) $t = 5$; (c) $t = 7$; (d) $t = 7.5$; (e) $t = 8$; (f) $t = 10$.

set. The results demonstrate that the semi-smooth Newton method features excellent convergence in this example. The integration of all types of nonlinearities into a semi-smooth Newton scheme avoids tremendous computational cost as compared with a fixed-point type approach for the active set (see e.g. [8, 9]). Moreover, it is obvious that consistent linearization of all nonlinear quantities, including contact forces, normal and tangential vectors, is crucial to avoid unacceptable deterioration of convergence. We omit here the discussion of respective results for other

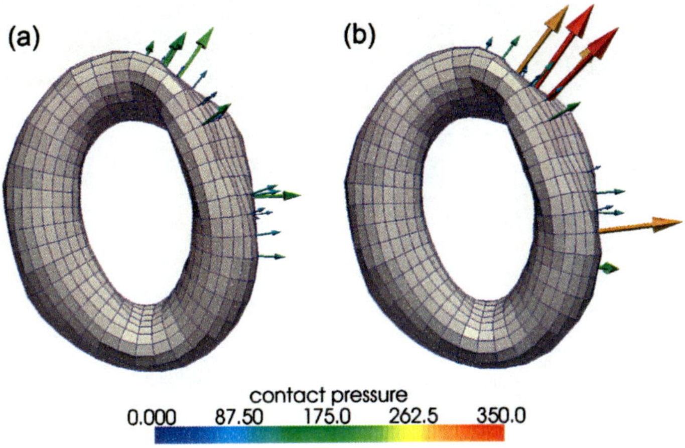

Fig. 8 Two torus impact – exemplary visualization of the deformed lower torus and of the computed contact traction results at (a) $t = 5.5$ and (b) $t = 5.55$.

approaches available in the literature, such as the fixed-point type approach for the active set or an incomplete linearization of contact terms (see e.g. [7]), but instead refer to our previous work [18, 19] for in-depth comparisons.

To illustrate the strong nonlinearities involved in this problem setup even more clearly, Figure 8 shows the deformed lower torus at the beginning and at the end of the time step analyzed in column (b) of Table 1 (i.e. at $t = 5.5$ and at $t = 5.55$). The normal contact traction distribution, represented by the nodal Lagrange multiplier solution, is visualized with arrows and confirms that despite significant changes of the contact region, all nonlinearities are resolved efficiently and without deterioration of convergence within one single semi-smooth Newton iteration.

7.3 Sphere in Sphere

In this example, we consider a rigid sphere ($R = 0.6$) being pressed into an elastic hollow sphere ($R_i = 0.7, R_o = 2, E = 1000, \nu = 0$) by incrementally applying a prescribed displacement $v = 1.125$ in vertical direction, see Figure 9. A total of 5102 finite elements (8-node hexahedrals) are used and the outer sphere is chosen as slave surface. The total displacement is applied in 100 steps. Similar investigations can be found in the literature for both 2D and 3D, see e.g. [4, 20, 23]. Some characteristic stages of deformation are given in Figure 9. Based on the finite deformation dual mortar formulation and the consistently linearized semi-smooth Newton method presented in the previous sections, we obtain optimal Newton convergence and robust results, even though very large strains occur. Of course, if one is interested in accurate stress results within the highly deformed contact zone, an adaptive mesh re-

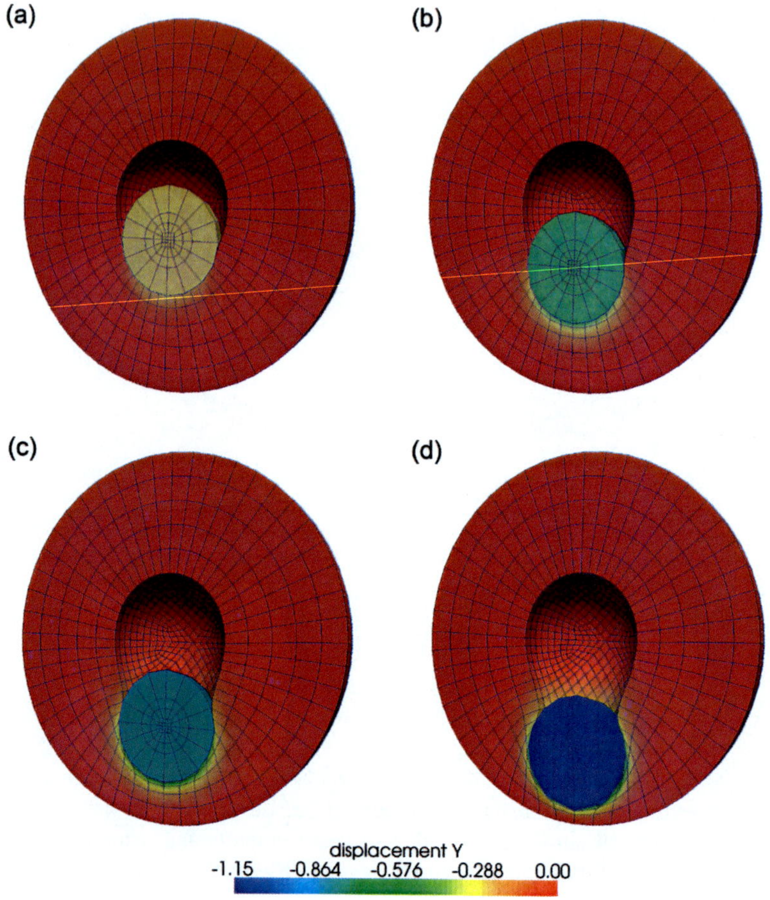

Fig. 9 Sphere in sphere – four characteristic stages of deformation: (a) step 25; (b) step 50; (c) step 75; (d) step 100 (corresponding to a prescribed displacement $v = 1.125$).

finement strategy becomes necessary. Figure 10 shows the total contact force versus the prescribed displacements v in vertical direction.

8 Conclusions

The dual mortar method recently extended by the authors to 3D finite deformation contact analysis has been reviewed and some important implementation aspects as well as model extensions have been highlighted. Consistent linearization of contact forces and constraints together with an interpretation of the active set search as a semi-smooth Newton scheme yields a very efficient solution algorithm. The accur-

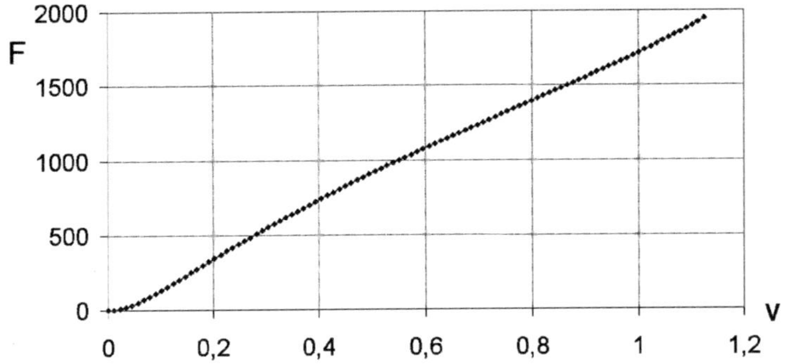

Fig. 10 Sphere in sphere – contact force F vs. prescribed displacements v.

acy of the presented method and its superior numerical robustness and efficiency have been demonstrated with several examples including finite strain situations. Future work will focus on the integration of the presented mortar contact approach into an XFEM-based fluid-structure interaction framework [16]. It is anticipated that the ability to simulate finite deformation contact of deformable solids in fluid flows will allow to investigate a wide range of engineering and biomechanics problems, which so far have hardly been accessible to numerical analysis.

Acknowledgements The support of the first author (A.P.) by the TUM Graduate School of Technische Universität München is gratefully acknowledged. The work in this paper was partly funded by the German Federal Ministry of Economics and Technology through project 20T0608A in cooperation with Rolls Royce Deutschland under contract T004.008.000.

References

1. Bernardi, C., Maday, Y., Patera, A.T.: A new nonconforming approach to domain decomposition: the mortar element method. In: Brezis, H., Lions, J.L. (eds.) Nonlinear Partial Differential Equations and Their Applications, pp. 13–51. Pitman/Wiley, London, New Yark (1994)
2. Christensen, P.W., Klarbring, A., Pang, J.S., Strömberg, N.: Formulation and comparison of algorithms for frictional contact problems. International Journal for Numerical Methods in Engineering 42(1), 145–173 (1998)
3. Chung, J., Hulbert, G.M.: A time integration algorithm for structural dynamics with improved numerical dissipation: The generalized-alpha method. Journal of Applied Mechanics 60, 371–375 (1993)
4. Fischer, K.A., Wriggers, P.: Frictionless 2D contact formulations for finite deformations based on the mortar method. Computational Mechanics 36(3), 226–244 (2005)
5. Franke, D., Düster, A., Nübel, V., Rank, E.: A comparison of the h-, p-, hp-, and rp-version of the FEM for the solution of the 2D Hertzian contact problem. Computational Mechanics 45(5), 513–522 (2010)

6. Gitterle, M., Popp, A., Gee, M.W., Wall, W.A.: Finite deformation frictional mortar contact using a semi-smooth Newton method with consistent linearization. International Journal for Numerical Methods in Engineering 84(5), 543–571 (2010)
7. Hartmann, S.: Kontaktanalyse dünnwandiger Strukturen bei grossen Deformationen. PhD Thesis, Institut für Baustatik und Baudynamik, Universität Stuttgart (2007)
8. Hartmann, S., Brunssen, S., Ramm, E., Wohlmuth, B.: Unilateral non-linear dynamic contact of thin-walled structures using a primal-dual active set strategy. International Journal for Numerical Methods in Engineering 70(8), 883–912 (2007)
9. Hesch, C., Betsch, P.: A mortar method for energy-momentum conserving schemes in frictionless dynamic contact problems. International Journal for Numerical Methods in Engineering 77(10), 1468–1500 (2009)
10. Hintermüller, M., Ito, K., Kunisch, K.: The primal-dual active set strategy as a semismooth Newton method. SIAM Journal on Optimization 13(3), 865–888 (2002)
11. Holzapfel, F.: Nonlinear Solid Mechanics – A Continuum Approach for Engineering. John Wiley & Sons, Chichester (2000)
12. Hüeber, S., Stadler, G., Wohlmuth, B.I.: A primal-dual active set algorithm for three-dimensional contact problems with Coulomb friction. SIAM Journal on Scientific Computing 30(2), 572–596 (2008)
13. Hüeber, S., Wohlmuth, B.I.: A primal-dual active set strategy for non-linear multibody contact problems. Computer Methods in Applied Mechanics and Engineering 194(27-29), 3147–3166 (2005)
14. Lamichhane, B.P., Stevenson, R.P., Wohlmuth, B.I.: Higher order mortar finite element methods in 3D with dual Lagrange multiplier bases. Numerische Mathematik 102(1), 93–121 (2005)
15. Laursen, T.A.: Computational Contact and Impact Mechanics. Springer-Verlag, Heidelberg (2002)
16. Mayer, U., Popp, A., Gerstenberger, A., Wall, W.: 3D fluid-structure-contact interaction based on a combined XFEM FSI and dual mortar contact approach. Computational Mechanics 46(1), 53–67 (2010)
17. Papadopoulos, P., Taylor, R.L.: A mixed formulation for the finite element solution of contact problems. Computer Methods in Applied Mechanics and Engineering 94(3), 373–389 (1992)
18. Popp, A., Gee, M.W., Wall, W.A.: A finite deformation mortar contact formulation using a primal-dual active set strategy. International Journal for Numerical Methods in Engineering 79(11), 1354–1391 (2009)
19. Popp, A., Gitterle, M., Gee, M.W., Wall, W.A.: A dual mortar approach for 3D finite deformation contact with consistent linearization. International Journal for Numerical Methods in Engineering 83(11), 1428–1465 (2010)
20. Puso, M.A.: A 3D mortar method for solid mechanics. International Journal for Numerical Methods in Engineering 59(3), 315–336 (2004)
21. Puso, M.A., Laursen, T.A., Solberg, J.: A segment-to-segment mortar contact method for quadratic elements and large deformations. Computer Methods in Applied Mechanics and Engineering 197(6/8), 555–566 (2008)
22. Puso, M.A., Laursen, T.A.: A mortar segment-to-segment contact method for large deformation solid mechanics. Computer Methods in Applied Mechanics and Engineering 193(6-8), 601–629 (2004)

23. Puso, M.A., Laursen, T.A.: A mortar segment-to-segment frictional contact method for large deformations. Computer Methods in Applied Mechanics and Engineering 193(45/47), 4891–4913 (2004)
24. Qi, L., Sun, J.: A nonsmooth version of Newton's method. Mathematical Programming 58(1), 353–367 (1993)
25. Timoshenko, S.P., Goodier, J.N.: Theory of Elasticity. McGraw-Hill, New York (1970)
26. Wall, W.A., Gee, M.W.: Baci – A multiphysics simulation environment. Technical Report, Technische Universität München (2009)
27. Wohlmuth, B.I.: A mortar finite element method using dual spaces for the Lagrange multiplier. SIAM Journal on Numerical Analysis 38(3), 989–1012 (2000)
28. Wohlmuth, B.I.: Discretization Methods and Iterative Solvers Based on Domain Decomposition. Springer, Berlin (2001)
29. Wriggers, P.: Computational Contact Mechanics. John Wiley & Sons, Chichester (2002)
30. Yang, B., Laursen, T.A., Meng, X.: Two dimensional mortar contact methods for large deformation frictional sliding. International Journal for Numerical Methods in Engineering 62(9), 1183–1225 (2005)
31. Yang, B., Laursen, T.: A contact searching algorithm including bounding volume trees applied to finite sliding mortar formulations. Computational Mechanics 41(2), 189–205 (2008)
32. Yang, B., Laursen, T.A.: A large deformation mortar formulation of self contact with finite sliding. Computer Methods in Applied Mechanics and Engineering 197(6-8), 756–772 (2008)

The Contact Patch Test for Linear Contact Pressure Distributions in 2D Frictionless Contact

G. Zavarise and L. De Lorenzis

Abstract It is well known that the classical one-pass node-to-segment algorithms for the enforcement of contact constraints fail the contact patch test. The situation is particularly critical when using the penalty method. In a recent study, the authors proposed a modified one-pass node-to-segment algorithm able to pass the contact patch test also in conjunction with the penalty method. However, in a general situation the pressure distribution transferred across a contact surface is non-uniform. Hence, even for a contact element which passes the contact patch test under a uniform distribution of the contact pressures, the transfer of a non-uniform state of stress across the contact surface may give rise to disturbances related to the discretization. Such disturbances introduce local solution errors and ultimately affect the accuracy of the analysis. This paper, following on from the previous study, develops an enhanced node-to-segment formulation able to pass a modified version of the contact patch test whereby a linear distribution of pressures has to be transmitted across the contact surface.

1 Introduction

Several investigations have shown that the classical one-pass node-to-segment (NTS) algorithm for the enforcement of contact constraints fails the contact patch test proposed in [3] (Figure 1a) [1–3]. This implies that solution errors may be introduced at the contacting surfaces, and these errors do not necessarily decrease with mesh refinement. The previous research has mainly focused on the Lagrange multiplier method, to exactly enforce the contact geometry conditions. The situation is even worse with the penalty method, due to its inherent approximation which yields a solution affected by a non-zero penetration.

G. Zavarise · L. De Lorenzis
Department of Innovation Engineering, University of Salento, Ed. La Stecca, via per Monteroni, 73100 Lecce, Italy; email{giorgio.zavarise, laura.delorenzis}@unisalento.it

G. Zavarise & P. Wriggers (Eds.): Trends in Computational Contact Mechanics, LNACM 58, pp. 79–100.
springerlink.com

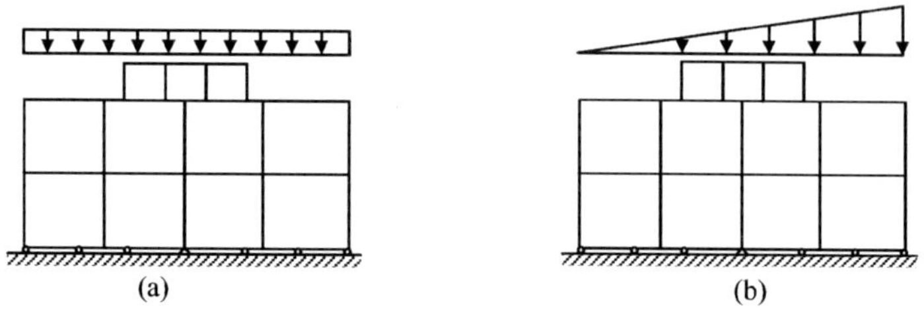

Fig. 1 Patch test for constant and linear contact pressure distributions.

In a recent study, the authors analyzed and improved the contact patch test behavior of the one-pass NTS algorithm used in conjunction with the penalty method [7]. For this purpose, several sequential modifications of the basic formulation were considered, which yielded incremental improvements in results of the contact patch test. The final proposed formulation was a modified one-pass NTS algorithm able to correctly reproduce the transfer of a constant contact pressure with a constant proportional penetration.

In a general situation, the pressure distribution transferred across a contact surface is non-uniform. Hence, even for a contact element which passes the "uniform contact patch test" (Figure 1a), the transfer of a non-uniform state of stress across the contact surface may give rise to disturbances related to the discretization. Such disturbances introduce local solution errors and affect the accuracy of the analysis.

This paper, following on from the study mentioned earlier [7], develops an enhanced NTS formulation able to pass the "linear contact patch test" depicted in Figure 1b. When the continuum is discretized with linear elements, a contact element passing the linear contact patch test achieves the highest possible level of accuracy. Hence, non-penetration is here enforced with the penalty method and the treatment is limited to the case of linear continuum elements.

2 NTS Formulation

The penalty method consists of a local approximate enforcement of the geometrical non-penetration condition, obtained by minimizing a suitable modification of the potential

$$\bar{\Pi} = \Pi + \cup_A \frac{1}{2} \varepsilon_N g_N^2 \tag{1}$$

where Π and $\bar{\Pi}$ are, respectively, the unmodified (contactless) and the modified potential functionals of the problem; the summation is extended to all the slave nodes where the non-penetration condition has been violated; g_N is the measure of the penetration (see Figure 2); ε_N is the penalty parameter. This approach corresponds to locating linear discrete springs, of zero initial length and stiffness ε_N, at the "active" slave nodes.

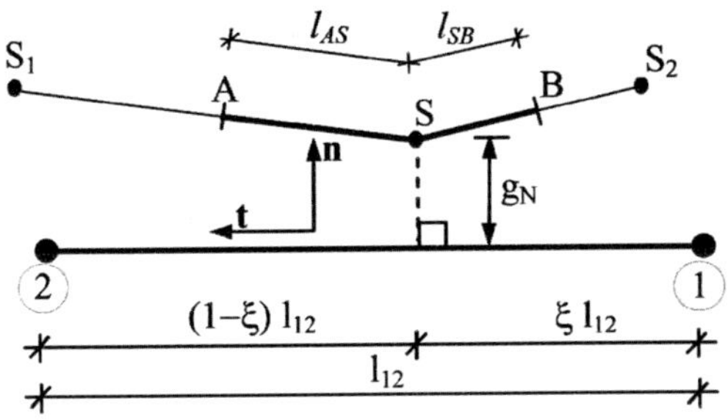

Fig. 2 Area of competence of a slave node S and NTS geometry.

Prior to any other consideration, a specific issue of the penalty method in the classical NTS algorithm has to be solved. The use of discrete springs at the slave nodes prevents the contact patch test (uniform or linear) from being passed, due to the non-uniform contact areas and the constant penalty parameter associated with the various slave nodes. This problem can be solved by considering the stiffness of each nodal spring as the result of an integration over the "area of competence" of the slave node. This area can be roughly defined as the sum of the half-lengths of the two segments adjacent to the slave node [8]. Details about this geometry are provided in Figure 2, where S is the slave node, 1 and 2 are the end nodes of the master segment, **t** and **n** are the tangent and normal unit vectors, l_{12} is the master segment length and ξ is the normalized projection of the slave node onto the master segment, $0 \leq \xi \leq 1$ [5]. The discrete spring stiffness located at each slave node is then the resultant of the distributed stiffness of the springs located along the area of competence, i.e. $\varepsilon_N = \hat{\varepsilon}_N(l_{AS} + l_{SB})$, where $\hat{\varepsilon}_N$ is the stiffness per unit length of the bed of springs, and $l_S = l_{AS} + l_{SB}$ is the area of competence of the slave node (Figure 2). Such equivalence is computed for each slave node. The input of the analysis is the stiffness per unit length, $\hat{\varepsilon}_N$. This algorithm is named penalty method with Area Regularization (AR), or NTS-AR.

3 The Basis for an Enhanced NTS Formulation

In order to discuss the patch test performance of the NTS algorithm, we have first to look not at a single slave node, but rather at the two end nodes of a slave segment. Then, two main categories of cases can be introduced. The first one includes cases where the projection of each slave segment is contained within a single segment on the master surface. These are indicated as "normal" cases (Figure 3a). The second category includes all the remaining cases, where the projection involves more than one segment on the master surface. These cases are indicated in the following as "pathologic" cases (Figure 3b).

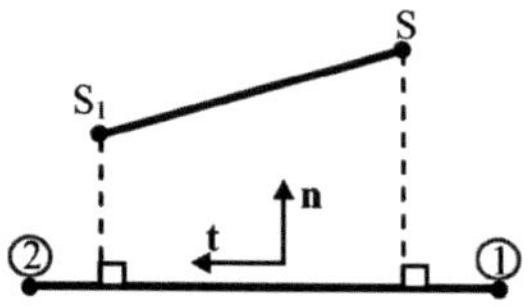

(a) "Normal" case

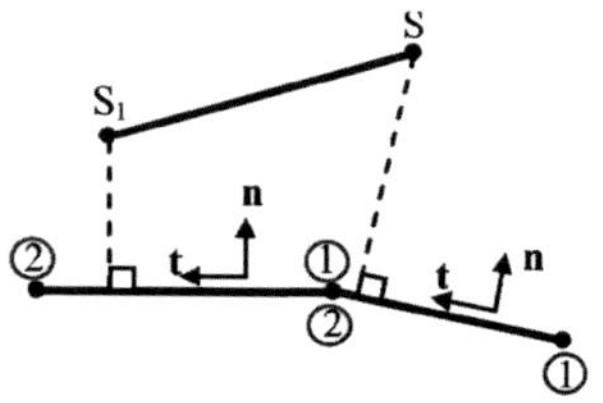

(b) "Pathologic" case

Fig. 3 Projection cases.

The basic idea underlying the proposed enhancements to the NTS formulation is the following: in order for the uniform (linear) contact patch test to be passed, there has to be equivalence of forces and moments between any uniform (linear) distribution of contact pressures acting on the slave segment and the concentrated forces transmitted to the corresponding master nodes by using the NTS algorithm. Such equivalence has to hold *locally*, i.e. for each slave segment (or for the area of competence of each slave node). Once the aforementioned equivalence holds, with the additional use of area regularization when the penalty method is adopted, the contact patch test is passed.

In "normal" cases, this equivalence is automatically satisfied for each slave segment, both for a uniform and for a linear contact pressure distribution. Therefore, in these cases, the NTS-AR algorithm passes both the uniform and the linear contact patch test. Conversely, in presence of "pathologic" cases, this equivalence does no longer hold for each slave segment, while it continues to hold at the global level. As a result, the patch test is not passed by the NTS-AR algorithm, and additional enhancements are needed. To deal with these cases with an NTS strategy, we have to look again at a single slave node. In this context, two distinct ideal steps are involved in the transfer of a uniform (linear) contact pressure between two discretized contacting bodies: the *first step* is the slave nodal force computation; the *second step* is the slave nodal force transmission to the master nodes.

In order for the uniform (linear) patch test to be passed, equivalence of forces and moments between the concentrated forces and a uniform (linear) contact pressure must hold at each contact element during both of the above phases.

4 The Contact Element Passing the Uniform Contact Patch Test

The virtual slave node technique (Virtual-to-segment or VTS), proposed in [6], consists in changing the integration scheme usually adopted in NTS contact elements. An arbitrary number of points is specified inside each contact element, and each one is treated by the NTS algorithm as a (virtual) slave node. In [7] it was shown that this technique can be employed to satisfy equivalence of forces and moments between a uniform contact pressure and the contact forces transferred at each slave

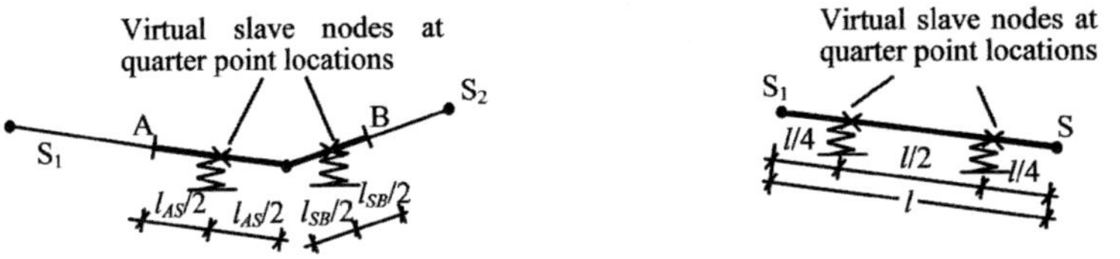

(a) Slave node scheme

(b) Slave segment scheme

Fig. 4 Modified virtual slave node technique.

node. For this purpose, it is sufficient to place one virtual slave node at the centroid of each of the two half-segments adjacent to the generic slave node S, see Figure 4a. A modified virtual slave node technique results, whereby the NTS-AR strategy is applied to the virtual slave nodes, and these are located at the quarter points of each slave segment (Figure 4b, [7]). This technique is such that Momentum Equilibrium (ME) is locally satisfied, hence it is termed VTS-ME.

The VTS-ME algorithm passes the contact patch test provided that the area of competence of each virtual slave node, when projected to the master surface, falls within a single segment on the master surface. If the previous condition is not satisfied, then the concentrated contact forces situated at the virtual slave nodes are incorrectly transformed into forces at the master nodes.

In order to solve this problem, care must be taken while transferring the contact forces acting at the virtual slave nodes to the elements of the master surface. The detailed procedure is reported in the original paper [7], where it was used to devise a contact element that passed the uniform patch test. The corresponding algorithm was indicated as VTS-PPT.

5 A New Contact Element Passing the Linear Contact Patch Test

The VTS-PPT algorithm is based on a piecewise constant contact pressure interpolation across the contacting surface. As a result, it cannot pass the linear patch test. In the following, suitable modifications to the NTS-AR algorithm are conducted. The attention will be directly focused on "pathologic" cases. In such cases, two types of errors occur during the slave nodal force computation and the slave nodal force transmission. Both problems, solved in the earlier work for the uniform case, will be solved henceforth for the linear case.

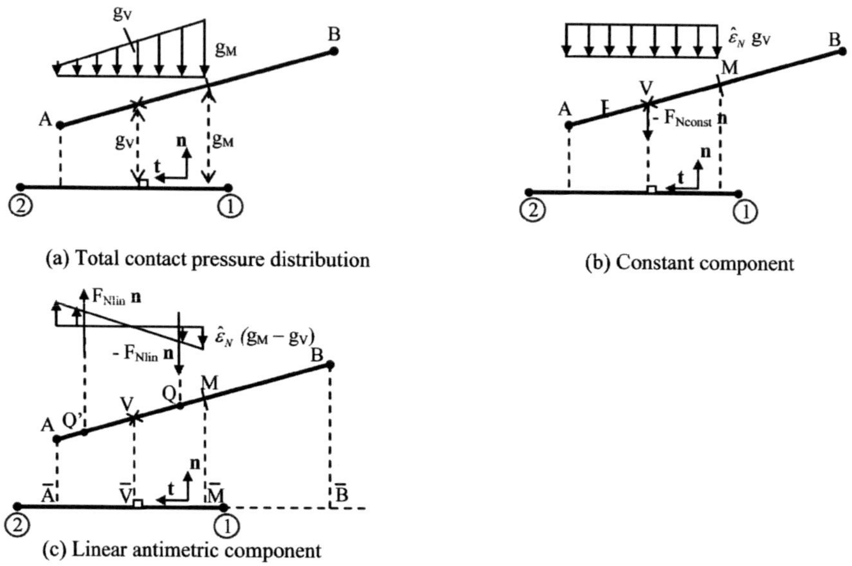

(a) Total contact pressure distribution

(b) Constant component

(c) Linear antimetric component

Fig. 5 Slave nodal force computation.

5.1 Step 1: Slave Nodal Force Computation (VTS-MEL)

The VTS-ME algorithm is no longer successful in case of a linear contact pressure, as in this case the quarter points of a slave segment no longer coincide with the centroids of the contact pressure distributions acting on the two halves of the slave segment. To solve this problem there are different possible strategies. In order to keep the formulation as simple as possible, the strategy adopted as follows still adopts two virtual slave nodes at the quarter points of each slave segment. The next consideration is that any linear distribution of contact pressures along the area of competence of each virtual slave node (Figure 5a) can be thought of as the superposition of a constant average component (Figure 5b) and a linear antimetric distribution (Figure 5c). The centroid of the constant component is located at the virtual slave node, therefore it is correctly transformed into slave nodal forces. For the linear antimetric distribution, the computation of the slave nodal forces has to be conducted appropriately. This distribution can be treated as two opposite concentrated forces located at the centroids of the two triangles. The entire formulation is reported in a later section.

The resulting algorithm, which guarantees Moment Equilibrium during step 1 also for a Linear contact pressure distribution, will be indicated as VTS-MEL. In cases where the area of competence of a virtual slave node is projected entirely within a single master segment, the VTS-MEL algorithm conducts to the exact computation not only of the slave but also of the master nodal forces. Therefore, in such cases this algorithm passes the linear patch test.

Case Test A

In case test A, shown in Figure 6a, two sets of springs at non-matching locations are pressed into each other with a linear pressure distribution. The upper surface is the slave and the lower is the master one. The use of 1D springs instead of 2D blocks is due to the fact that, under a linear applied pressure distribution, a bending stress regime would arise in 2D blocks. The modeling of such regime by linear quadrilateral elements is affected by the shear effect, and the resulting inaccuracies prevent a clear examination of the role of the contact algorithm in the transfer of stresses across the contact interface.

Figure 6b shows the expected correct values of the contact forces, obtained by moving to the nodes on both surfaces the externally applied linear pressure distribution. Figure 6c shows how the upper set of forces at the slave nodes is transmitted to the master nodes with the classical NTS algorithm. It is obvious that the resulting distribution is incorrect, i.e. it is not compatible with a linear pressure over the master surface. Hence, in this case the penalty method, either with or without AR, cannot pass the patch test. This is shown in Figure 7a, which plots the reaction forces in the lower springs. A large discrepancy is visible between the exact reactions and those predicted by the NTS-AR algorithm, the latter coinciding with those depicted in Figure 6c.

For the discretization in Figure 6a, the area of competence of each virtual slave node (located at the quarter points of each slave segment) falls within one master segment. Therefore, the VTS-ME algorithm would pass the uniform patch test (see case test B in [7]). However, it does not pass the linear one, as shown in Figure 7. Nevertheless, it is evident that a significant improvement is achieved with respect to the NTS-AR algorithm. Finally, the VTS-MEL algorithm carries out a proper transformation of the linear contact pressures into slave and master nodal forces. As a result, the linear patch test is passed (see Figures 6d and 7).

5.2 *Step 2: Slave Nodal Force Transmission (VTS-PPTL)*

In the most general discretization case, the projection of the area of competence of a virtual slave node onto the master surface involves more than one master segment (Figure 8). As a result, the VTS-MEL algorithm yields the exact computation of the slave but not of the master nodal forces. In order to solve this problem, the following steps have to be followed for each virtual slave node:

1. projection of the area of competence over the master surface;
2. transformation of the concentrated contact force acting at the virtual slave node into a linear contact pressure acting over this projected area;
3. transformation of the linear contact pressure acting over the projected area into equivalent concentrated contact forces acting at the end nodes of all the elements on the master surface involved by the projected area.

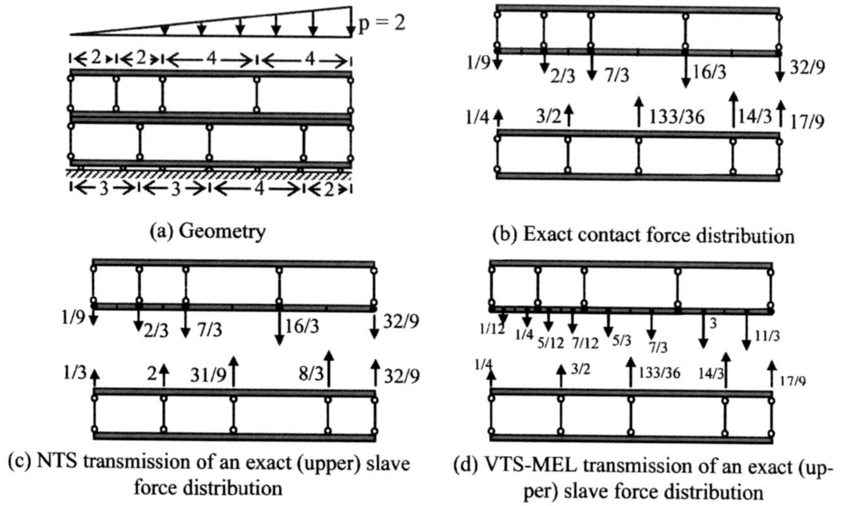

(a) Geometry

(b) Exact contact force distribution

(c) NTS transmission of an exact (upper) slave force distribution

(d) VTS-MEL transmission of an exact (upper) slave force distribution

Fig. 6 Case test A.

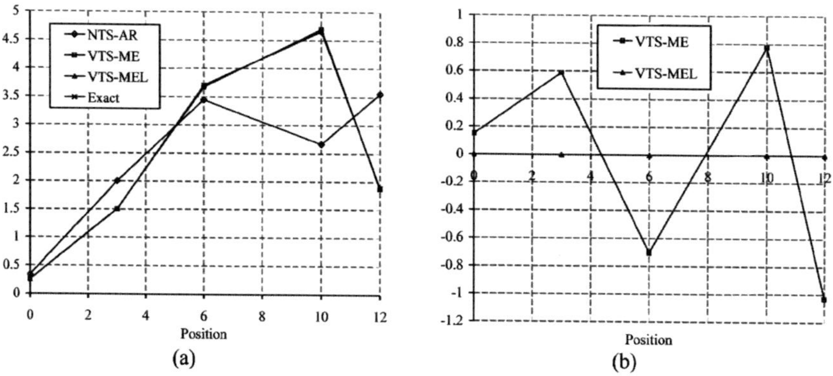

(a) (b)

Fig. 7 Case test A – contact force distributions.

Following the above strategy, a contact element Passing the Patch Test also for Linear contact pressures (PPTL) can be devised. The corresponding algorithm will be indicated as VTS-PPTL. Its detailed description is reported in Section 7. Once again, the linear pressure distribution (Figure 8a) is considered as the sum of a constant component (Figure 8b) and a linear antimetric component (Figure 8c). Both have to be appropriately transformed into forces at the master nodes.

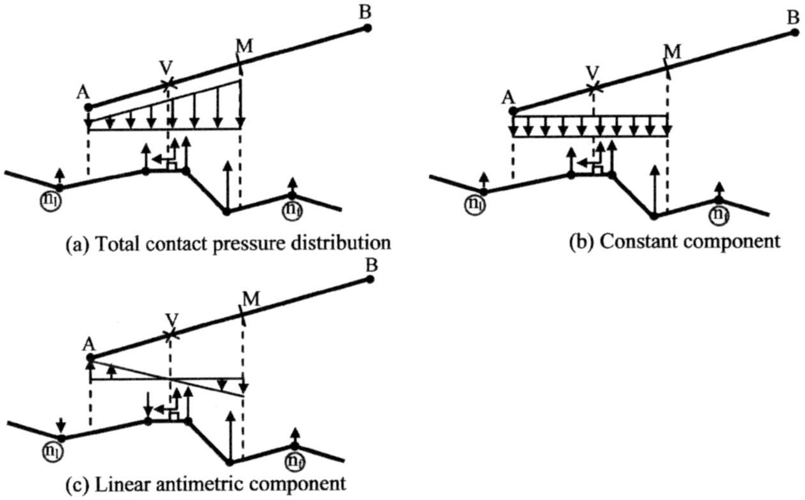

Fig. 8 Contact force transmission to more than one segment on the master surface.

Case Test B

The geometry of case test B is depicted in Figure 9a. The small difference in the size of the last two master segments prevents the projection of the area of competence inside a unique master segment. Figure 9b reports the expected correct values of the contact forces, whereas Figure 9c shows the contact forces obtained by assuming the correct distribution at the slave nodes, and computing the distribution of forces at the master nodes resulting from application of the NTS algorithm. As in case test A, the latter distribution is obviously incorrect (Figure 10a), hence the patch test cannot be passed with the standard formulation.

Unlike for case test A, using the VTS-MEL technique does not completely solve the problem (Figure 10), although it improves results with respect to the NTS-AR algorithm. The reason is that, for this geometry, the area of competence of one of the virtual slave nodes intercepts two segments on the master surface. Hence the transfer of contact forces to the master nodes is incorrect. Also the VTS-PPT algorithm, which is able to pass the uniform patch test, is unsuccessful for the linear patch test (Figure 10). The problem is solved by using the strategy summarized above (and described in detail in Section 6), as shown in Figures 9d and 10.

6 The VTS-MEL Algorithm

As outlined earlier, the VTS-MEL algorithm is obtained from the standard one-pass NTS algorithm with the following modifications:

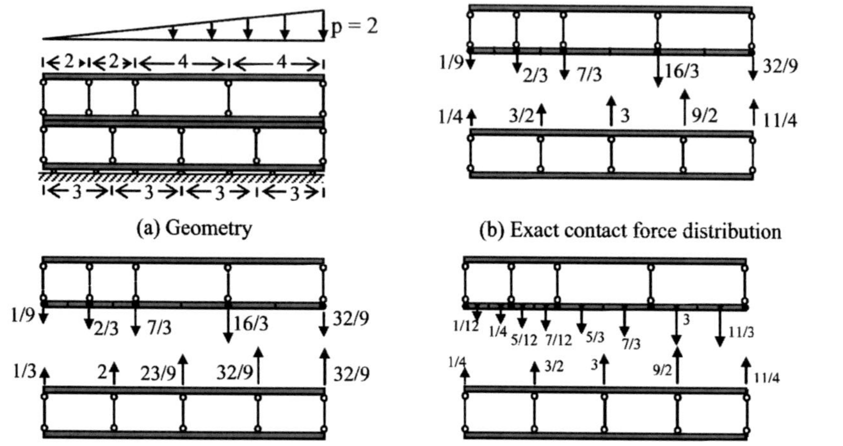

(a) Geometry

(b) Exact contact force distribution

(c) NTS transmission of an exact (upper) slave force distribution

(d) VTS-PPTL transmission of an exact (upper) slave force distribution

Fig. 9 Case test B.

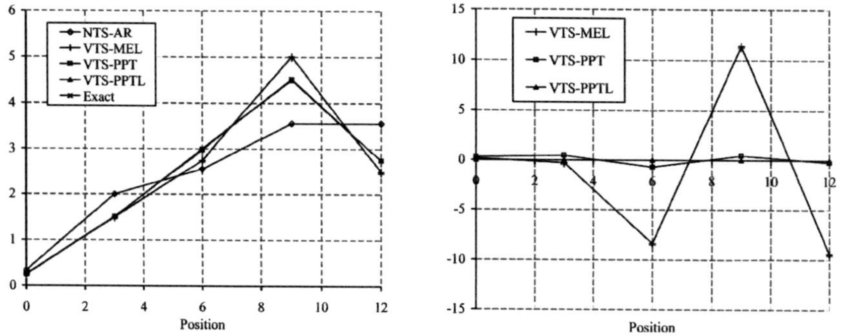

(a)

(b)

Fig. 10 Case test B – contact force distributions.

1. coupling of two slave nodes to consider the slave segment;
2. adoption of the virtual slave node technique [6];
3. modification of the technique by replacing the two virtual slave nodes, originally located at the Gauss points, with two nodes located at the quarter points of each slave segment;
4. proper transformation of a linear contact pressure distribution into slave and master nodal forces.

Steps 1 to 3 are quite straightforward and have been discussed earlier. Step 4 is detailed in the following.

6.1 Computation of the Residual Vector

In the NTS algorithm using the virtual slave node technique, the residual vector has the following expression [6, 7] (see also Figure 5):

$$[R] = \begin{bmatrix} \mathbf{R}_A \\ \mathbf{R}_B \\ \mathbf{R}_1 \\ \mathbf{R}_2 \end{bmatrix} = F_N \begin{bmatrix} \frac{1-c_{ip}}{2}\mathbf{n} \\ \frac{1+c_{ip}}{2}\mathbf{n} \\ -(1-\xi)\mathbf{n} \\ -\xi\mathbf{n} \end{bmatrix} \tag{2}$$

where $\mathbf{R}_A$, $\mathbf{R}_B$, $\mathbf{R}_1$ and $\mathbf{R}_2$ are, respectively, the contact forces acting at the two end nodes of the slave segment containing the virtual slave node and at the two end nodes of the master segment, F_N is the total contact force transferred from the virtual slave node to the master surface, ξ is the normalized projection of the virtual slave node onto the master segment, $0 \leq \xi \leq 1$, and c_{ip} is the normalized coordinate of the virtual slave node, $-1 \leq c_{ip} \leq 1$. Hence the first four degrees of freedom are those of the two end nodes of the slave segment containing the virtual slave node, and the subsequent ones are those of the two end nodes of the master segment. In the VTS-ME algorithm, the virtual slave nodes are placed at the quarter points of the slave segment, then $c_{ip} = \pm 1/2$ and the corresponding weights are equal to unity. In this case, the normal force is

$$F_N = \varepsilon_N g_V = \frac{\hat{\varepsilon}_N l_{AB}}{2} g_V \tag{3}$$

where g_V is the normal gap evaluated at the virtual slave node.

In case of a linear contact pressure distribution, the virtual slave nodes located at the quarter points do not coincide with the centroids of the pressure to be transmitted. Therefore, the above approach does not satisfy local equilibrium of moments. However, any linear distribution can be considered as the sum of two components, as shown in Figure 5. The first component is a constant one (Figure 5b), whose resultant is

$$F_{N\,\text{const}} = \varepsilon_N g_V = \frac{\hat{\varepsilon}_N l_{AB}}{2} g_V \tag{4}$$

For this component the VTS-ME computation is correct. Hence the residual contribution stemming from the constant component is obtained from equation (2) as

$$[R]_{\text{const}} = \frac{\hat{\varepsilon}_N l_{AB}}{2} g_V \begin{bmatrix} \frac{1-c_{ip}}{2}\mathbf{n} \\ \frac{1+c_{ip}}{2}\mathbf{n} \\ -(1-\xi)\mathbf{n} \\ -\xi\mathbf{n} \end{bmatrix} \tag{5}$$

Let us now focus on the second, linear antimetric component (Figure 5c). This is equivalent to two opposite forces

$$F_{N\,\text{lin}} = \frac{1}{2}\hat{\varepsilon}_N(g_M - g_V)l_{\bar{V}\bar{M}} = \frac{1}{8}\hat{\varepsilon}_N(g_M - g_V)l_{\bar{A}\bar{B}} \tag{6}$$

where g_M is the normal gap evaluated at the midpoint of the slave segment (see Figure 5a), and $l_{\bar{V}\bar{M}}$ and $l_{\bar{A}\bar{B}}$ are the lengths of the projections of segments VM and AB, respectively, along the direction of the master segment. The second equality stems from the quarter-point location of the virtual nodes, which gives

$$l_{\bar{V}\bar{M}} = \frac{l_{\bar{A}\bar{B}}}{4} \tag{7}$$

The virtual variation of the contact contribution to the potential associated with the linear component, $\delta W_{c\,\text{lin}}$, can be computed by considering the virtual work of the two opposite forces $F_{N\,\text{lin}}$ for the normal gaps of their respective application points, Q and Q′ (see Figure 5c)

$$\delta W_{c\,\text{lin}} = F_{N\,\text{lin}}\delta g_Q + F_{N\,\text{lin}}\delta g_{Q'} = 2F_{N\,\text{lin}}\delta g_Q = 2F_{N\,\text{lin}}\frac{2}{3}(\delta g_M - \delta g_V) \tag{8}$$

where g_Q and $g_{Q'}$ are the normal gaps evaluated at points Q and Q′, respectively, and the symbol δ denotes virtual variation. Equation (8) considers that both forces give the same contribution to $\delta W_{c\,\text{lin}}$, and that, being from simple geometry

$$g_Q = \frac{2}{3}(g_M - g_V) \tag{9}$$

(see Figure 5c), it is also

$$\delta g_Q = \frac{2}{3}(\delta g_M - \delta g_V) \tag{10}$$

By substituting equation (6) into equation (8), the following result is obtained:

$$\delta W_{c\,\text{lin}} = \frac{\hat{\varepsilon}_N l_{\bar{A}\bar{B}}}{6}(g_M - g_V)(\delta g_M - \delta g_V) \tag{11}$$

From the analysis of the simple NTS geometry (see Figure 2), it follows that

$$\delta g_N = [\delta_S]^T[N_S] \tag{12}$$

with

$$[\delta_S] = \begin{bmatrix} \delta \mathbf{x}_S \\ \delta \mathbf{x}_1 \\ \delta \mathbf{x}_2 \end{bmatrix}, \quad [N_S] = \begin{bmatrix} \mathbf{n} \\ -(1-\xi)\mathbf{n} \\ -\xi\mathbf{n} \end{bmatrix} \tag{13}$$

where $\mathbf{x}_S$, $\mathbf{x}_1$ and $\mathbf{x}_2$ are the position vectors of nodes S, 1 and 2, respectively. As already evident from their earlier definitions, g_N and ξ are both related to the slave node. Equation (12) is well known from the classical NTS formulation. Being based on purely geometric considerations, they continue to apply if the slave node is substituted by any other point. In particular, for V and M we can write

$$\delta g_V = [\delta_V]^T [N_V], \quad \delta g_M = [\delta_M]^T [N_M] \tag{14}$$

where

$$[\delta_V] = \begin{bmatrix} \delta \mathbf{x}_V \\ \delta \mathbf{x}_1 \\ \delta \mathbf{x}_2 \end{bmatrix}, \quad [N_V] = \begin{bmatrix} \mathbf{n} \\ -(1-\xi_V)\mathbf{n} \\ -\xi_V \mathbf{n} \end{bmatrix} \tag{15}$$

$$[\delta_M] = \begin{bmatrix} \delta \mathbf{x}_M \\ \delta \mathbf{x}_1 \\ \delta \mathbf{x}_2 \end{bmatrix}, \quad [N_M] = \begin{bmatrix} \mathbf{n} \\ -(1-\xi_M)\mathbf{n} \\ -\xi_M \mathbf{n} \end{bmatrix} \tag{16}$$

In equations (15) and (16), $\mathbf{x}_V$ and $\mathbf{x}_M$ are the position vectors of points V and M, respectively. Moreover, ξ_V and ξ_M are the normalized projections of V and M onto the master segment, respectively. For the moment, we are considering $0 \leq \xi_V$, $\xi_M \leq 1$.

From simple geometric considerations, we obtain

$$\mathbf{x}_M = \frac{1}{2}(\mathbf{x}_A + \mathbf{x}_B), \quad \mathbf{x}_V = \frac{1-c_{ip}}{2}\mathbf{x}_A + \frac{1+c_{ip}}{2}\mathbf{x}_B \tag{17}$$

where $\mathbf{x}_A$ and $\mathbf{x}_B$ are the position vectors of the endpoints A and B of the slave segment, respectively. Therefore, equation (14) can be rewritten as follows:

$$\delta g_V = [\delta]^T [N_{Ve}], \quad \delta g_M = [\delta]^T [N_{Me}] \tag{18}$$

where

$$[\delta] = \begin{bmatrix} \delta \mathbf{x}_A \\ \delta \mathbf{x}_B \\ \delta \mathbf{x}_1 \\ \delta \mathbf{x}_2 \end{bmatrix}, \quad [N_{Ve}] = \begin{bmatrix} \frac{1-c_{ip}}{2}\mathbf{n} \\ \frac{1+c_{ip}}{2}\mathbf{n} \\ -(1-\xi_V)\mathbf{n} \\ -\xi_V \mathbf{n} \end{bmatrix}, \quad [N_{Me}] = \begin{bmatrix} \frac{1}{2}\mathbf{n} \\ \frac{1}{2}\mathbf{n} \\ -(1-\xi_M)\mathbf{n} \\ -\xi_M \mathbf{n} \end{bmatrix} \tag{19}$$

By substitution of equation (18) into equation (11), the following expression is obtained:

$$\begin{aligned} \delta W_{c\,\mathrm{lin}} &= \frac{\hat{\varepsilon}_N l_{\bar{A}\bar{B}}}{6}(g_M - g_V)[\delta]^T([N_{Me}] - [N_{Ve}]) \\ &= \frac{\hat{\varepsilon}_N l_{\bar{A}\bar{B}}}{6}(g_M - g_V)[\delta]^T \begin{bmatrix} \frac{c_{ip}}{2}\mathbf{n} \\ -\frac{c_{ip}}{2}\mathbf{n} \\ -(\xi_V - \xi_M)\mathbf{n} \\ (\xi_V - \xi_M)\mathbf{n} \end{bmatrix} \end{aligned} \tag{20}$$

Finally, the expression of the residual can be easily obtained by casting $\delta W_{c\,\text{lin}}$ in the following form:

$$\delta W_{c\,\text{lin}} = [\delta]^T [R]_{\text{lin}} \tag{21}$$

hence

$$[R]_{\text{lin}} = \frac{\hat{\varepsilon}_N l_{\bar{A}\bar{B}}}{6}(g_M - g_V) \begin{bmatrix} \frac{c_{ip}}{2}\mathbf{n} \\ -\frac{c_{ip}}{2}\mathbf{n} \\ -(\xi_V - \xi_M)\mathbf{n} \\ (\xi_V - \xi_M)\mathbf{n} \end{bmatrix} \tag{22}$$

Finally, the complete contact contribution to the residual vector is

$$[R] = [R]_{\text{const}} + [R]_{\text{lin}} = \frac{\hat{\varepsilon}_N l_{AB}}{2} g_V \begin{bmatrix} \frac{1-c_{ip}}{2}\mathbf{n} \\ \frac{1+c_{ip}}{2}\mathbf{n} \\ -(1-\xi)\mathbf{n} \\ -\xi\mathbf{n} \end{bmatrix} + \frac{\hat{\varepsilon} l_{\bar{A}\bar{B}}}{6}(g_M - g_V) \begin{bmatrix} \frac{c_{ip}}{2}\mathbf{n} \\ -\frac{c_{ip}}{2}\mathbf{n} \\ -(\xi_V - \xi_M)\mathbf{n} \\ (\xi_V - \xi_M)\mathbf{n} \end{bmatrix} \tag{23}$$

6.2 Consistent Linearization

In order to obtain the consistent tangent stiffness matrix, the virtual variation of the contact contribution to the potential needs to be linearized. Also in this case, two components will be separately considered.

Using the virtual slave node technique and locating the virtual slave nodes at the quarter points (VTS-ME algorithm), the tangent stiffness matrix is

$$[K_t] = \frac{\hat{\varepsilon}_N l_{AB}}{2} \left\{ [N_{Se}][N_{Se}]^T - \frac{g_N^2}{l_{12}^2}[N_{0e}][N_{0e}]^T - \frac{g_N}{l_{12}}\left([T_{Se}][N_{0e}]^T + [N_{0e}][T_{Se}]^T\right) \right\} \tag{24}$$

where

$$[N_{Se}] = \begin{bmatrix} \frac{1-c_{ip}}{2}\mathbf{n} \\ \frac{1+c_{ip}}{2}\mathbf{n} \\ -(1-\xi)\mathbf{n} \\ -\xi\mathbf{n} \end{bmatrix}, \quad [T_{Se}] = \begin{bmatrix} \frac{1-c_{ip}}{2}\mathbf{t} \\ \frac{1+c_{ip}}{2}\mathbf{t} \\ -(1-\xi)\mathbf{t} \\ -\xi\mathbf{t} \end{bmatrix}, \quad [N_{0e}] = \begin{bmatrix} \mathbf{0} \\ \mathbf{0} \\ -\mathbf{n} \\ \mathbf{n} \end{bmatrix} \tag{25}$$

The above vectors involve the degrees of freedom of the two end nodes of the slave segment, and those of the two end nodes of the master segment.

Repeating the procedure used earlier for the residual vector, the tangent stiffness matrix can be considered the sum of two contributions, related to the constant and to the linear antimetric components of the contact pressure distribution. The first contribution can be directly obtained from equation (24) by substituting g_N with g_V, as follows:

$$[K_t]_{\text{const}} = \frac{\hat{\varepsilon}_N l_{AB}}{2}\left\{[N_{Se}][N_{Se}]^T - \frac{g_V^2}{l_{12}^2}[N_{0e}][N_{0e}]^T - \frac{g_V}{l_{12}}([T_{Se}][N_{0e}]^T + [N_{0e}][T_{Se}]^T)\right\} \tag{26}$$

In order to derive the second contribution, the expression of $\delta W_{c\ \text{lin}}$ given by equation (11) needs to be linearized as follows:

$$\begin{aligned}\Delta\delta W_{c\ \text{lin}} &= \frac{\hat{\varepsilon}_N l_{\bar{A}\bar{B}}}{6}\{(\Delta g_M - \Delta g_V)(\delta g_M - \delta g_V) + (g_M - g_V)(\Delta\delta g_M - \Delta\delta g_V)\} \\ &+ \frac{\hat{\varepsilon}_N}{6}(g_M - g_V)(\delta g_M - \delta g_V)\Delta l_{\bar{A}\bar{B}}\end{aligned} \tag{27}$$

where the symbol Δ denotes linearization. Due to the symmetry between virtual variation and linearization, equation (18) yields also

$$\Delta g_V = [N_{Ve}]^T[\Delta], \quad \Delta g_M = [N_{Me}]^T[\Delta] \tag{28}$$

where

$$[\Delta] = \begin{bmatrix}\Delta\mathbf{x}_A \\ \Delta\mathbf{x}_B \\ \Delta\mathbf{x}_1 \\ \Delta\mathbf{x}_2\end{bmatrix} \tag{29}$$

Once again, well known results for the simple NTS geometry give [4, 7]

$$\Delta\delta g_N = -\frac{1}{l_{12}}[\delta_S]^T\left\{[T_S][N_0]^T + [N_0][T_S]^T + \frac{g_N}{l_{12}}[N_0][N_0]^T\right\}[\Delta_S] \tag{30}$$

where

$$[\Delta_S] = \begin{bmatrix}\Delta\mathbf{x}_S \\ \Delta\mathbf{x}_1 \\ \Delta\mathbf{x}_2\end{bmatrix} \tag{31}$$

The above equation, stemming from purely geometric arguments, can be rewritten for V and M as follows:

$$\Delta\delta g_V = -\frac{1}{l_{12}}[\delta_V]^T\left\{[T_V][N_0]^T + [N_0][T_V]^T + \frac{g_V}{l_{12}}[N_0][N_0]^T\right\}[\Delta_V] \tag{32}$$

$$\Delta\delta g_M = -\frac{1}{l_{12}}[\delta_M]^T\left\{[T_M][N_0]^T + [N_0][T_M]^T + \frac{g_M}{l_{12}}[N_0][N_0]^T\right\}[\Delta_M] \tag{33}$$

where

$$[\Delta_V] = \begin{bmatrix}\Delta\mathbf{x}_V \\ \Delta\mathbf{x}_1 \\ \Delta\mathbf{x}_2\end{bmatrix}, \quad [T_V] = \begin{bmatrix}\mathbf{t} \\ -(1-\xi_V)\mathbf{t} \\ -\xi_V\mathbf{t}\end{bmatrix} \tag{34}$$

$$[\Delta_M] = \begin{bmatrix} \Delta \mathbf{x}_M \\ \Delta \mathbf{x}_1 \\ \Delta \mathbf{x}_2 \end{bmatrix}, \quad [T_M] = \begin{bmatrix} \mathbf{t} \\ -(1-\xi_M)\mathbf{t} \\ -\xi_M \mathbf{t} \end{bmatrix} \tag{35}$$

Considering again equation (17), equations (32) and (33) can be rewritten, respectively, as

$$\Delta\delta g_V = -\frac{1}{l_{12}}[\delta]^T \left\{ [T_{Ve}][N_{0e}]^T + [N_{0e}][T_{Ve}]^T + \frac{g_V}{l_{12}}[N_{0e}][N_{0e}]^T \right\} [\Delta] \tag{36}$$

$$\Delta\delta g_M = -\frac{1}{l_{12}}[\delta]^T \left\{ [T_{Me}][N_{0e}]^T + [N_{0e}][T_{Me}]^T + \frac{g_M}{l_{12}}[N_{0e}][N_{0e}]^T \right\} [\Delta] \tag{37}$$

where

$$[T_{Ve}] = \begin{bmatrix} \frac{1-c_{ip}}{2}\mathbf{t} \\ \frac{1+c_{ip}}{2}\mathbf{t} \\ -(1-\xi_V)\mathbf{t} \\ -\xi_V \mathbf{t} \end{bmatrix}, \quad [T_{Me}] = \begin{bmatrix} \frac{1}{2}\mathbf{t} \\ \frac{1}{2}\mathbf{t} \\ -(1-\xi_M)\mathbf{t} \\ -\xi_M \mathbf{t} \end{bmatrix} \tag{38}$$

Finally, geometrical considerations yield easily

$$l_{\bar{A}\bar{B}} = (\mathbf{x}_A - \mathbf{x}_B) \bullet \mathbf{t} \tag{39}$$

Hence

$$\Delta l_{\bar{A}\bar{B}} = (\Delta\mathbf{x}_A - \Delta\mathbf{x}_B) \bullet \mathbf{t} + (\mathbf{x}_A - \mathbf{x}_B) \bullet \Delta\mathbf{t} \tag{40}$$

From well known geometric results [4, 7] we have

$$\Delta\mathbf{t} = (\Delta\mathbf{x}_2 - \Delta\mathbf{x}_1) \bullet \mathbf{n}\mathbf{n}\frac{1}{l_{12}} \tag{41}$$

Using (41), equation (40) becomes

$$\Delta l_{\bar{A}\bar{B}} = [M]^T [\Delta] \tag{42}$$

where

$$[M] = \begin{bmatrix} \mathbf{t} \\ -\mathbf{t} \\ -\frac{(\mathbf{x}_A - \mathbf{x}_B)\bullet\mathbf{n}}{l_{12}}\mathbf{n} \\ \frac{(\mathbf{x}_A - \mathbf{x}_B)\bullet\mathbf{n}}{l_{12}}\mathbf{n} \end{bmatrix} \tag{43}$$

Substituting equations (28), (32), (33) and (42) into equation (27), and casting $\Delta\delta W_{c\ \mathrm{lin}}$ in the form

$$\Delta\delta W_{c\ \mathrm{lin}} = [\delta]^T [K_t]_{\mathrm{lin}} [\Delta] \tag{44}$$

the following expression is obtained:

$$[K_t]_{\text{lin}} = \frac{\hat{\varepsilon}_N l_{\bar{A}\bar{B}}}{6} \left\{ [N_{Me}][N_{Me}]^T + [N_{Ve}][N_{Ve}]^T - [N_{Ve}][N_{Me}]^T - [N_{Me}][N_{Ve}]^T \right.$$

$$- \frac{(g_M - g_V)}{l_{12}} \left([T_{Me}][N_{0e}]^T + [N_{0e}][T_{Me}]^T + \frac{g_M}{l_{12}} [N_{0e}][N_{0e}]^T \right)$$

$$\left. + \frac{(g_M - g_V)}{l_{12}} \left([T_{Ve}][N_{0e}]^T + [N_{0e}][T_{Ve}]^T + \frac{g_V}{l_{12}} [N_{0e}][N_{0e}]^T \right) \right\}$$

$$+ \frac{\hat{\varepsilon}_N}{6} (g_M - g_V)([N_{Me}][M]^T - [N_{Ve}][M]^T) \tag{45}$$

Finally, we obtain

$$[K_t] = [K_t]_{\text{const}} + [K_t]_{\text{lin}} \tag{46}$$

7 The VTS-PPTL Algorithm

The VTS-PPTL algorithm is described as follows. It is obtained from the standard one-pass NTS algorithm with the following modifications:

1. coupling of two slave nodes to consider the slave segment;
2. adoption of the virtual slave node technique [6];
3. modification of the technique by placing the two virtual slave nodes at the quarter points of each slave segment;
4. for each virtual slave node, identification of the projection of its area of competence over the master surface;
5. transformation of the concentrated contact force acting at the virtual slave node into its equivalent *linear* contact pressure acting over this master projected area;
6. transformation of the *linear* contact pressure acting over the projected area into equivalent concentrated contact forces acting at the end nodes of all the involved elements on the master surface.

Steps 1 to 3 are quite straightforward and have been discussed earlier. Step 4 has been described in [7]. More details are then provided only on steps 5 and 6.

7.1 Contact Contribution to the Residual Vector

The residual vector terms in equation (24) coincide with the contact forces transferred to the end nodes of the slave and of the master segments. Such forces are collectively equivalent to the linear contact pressure distribution in Figure 5a. However, the forces transferred to the master nodes, and thus the expression of the residual vector in equation (23), are both correct only if the projection of AM (or MB) along the direction of the normal **n** to the master segment falls within the master segment itself, such as in Figure 5a.

If the above condition is not satisfied, the contact forces transferred to the end nodes of all the segments of the master surface involved by the projection can be computed by dividing the procedure in sub-problems, as shown in Figures 11 to 13. In particular, the figures illustrate more in detail the computation of the forces on the master nodes in the three possible cases where, respectively, the segment on the master surface is the first, an intermediate one, or the last among those involved by the projection. In this example it is assumed that V belongs to AM, but the same procedure can be applied to any virtual slave node.

The basic version of the proposed algorithm, which was just illustrated, results into a modification of the classical residual vector which now involves more than one segment on the master surface. This procedure is conceptually equivalent to the one implemented in the VTS-PPT algorithm, but is here outlined for clarity. The residual given by equation (23) can be rewritten as follows:

$$[R] = \begin{bmatrix} \mathbf{R}_A \\ \mathbf{R}_B \\ \mathbf{R}_1 \\ \mathbf{R}_2 \end{bmatrix} = \begin{bmatrix} \mathbf{R}_V \\ \mathbf{R}_M \end{bmatrix} \tag{47}$$

where $\mathbf{R}_V$ and $\mathbf{R}_M$ indicate synthetically the contributions of the nodes of the slave segment (i.e. the segment containing the virtual slave node) and of the master segment. This residual is now replaced by an enhanced one of the following form:

$$[R] = [\mathbf{R}_V \;\; \mathbf{R}_M \;\; \mathbf{R}_f \;\; \mathbf{R}_i \;\; \mathbf{R}_l]^T \tag{48}$$

where $\mathbf{R}_f$, $\mathbf{R}_i$ and $\mathbf{R}_l$ are, respectively, the contributions associated to the nodes of the first segment, of the intermediate segment(s) and of the last segment on the master surface involved by the projection of the area of competence of the virtual slave node. Note that the master segment may coincide with the first, with the last, or with one of the intermediate segments on the master surface involved by the projection of the area of competence of the virtual slave node. Therefore, $\mathbf{R}_M$ may coincide with $\mathbf{R}_f$, with $\mathbf{R}_l$, or with one of the $\mathbf{R}_i$ terms, respectively. As shown by equation (48), the dimension of the residual vector depends on the number of elements on the master surface involved by the projection of the area of competence of the virtual slave node.

As done earlier, the generic linear contact pressure acting on the area of competence of a virtual slave node (Figure 8a) is considered as the sum of two components: a constant distribution (Figure 8b) and a linear antimetric distribution (Figure 8c). The residual vector can thus be written as

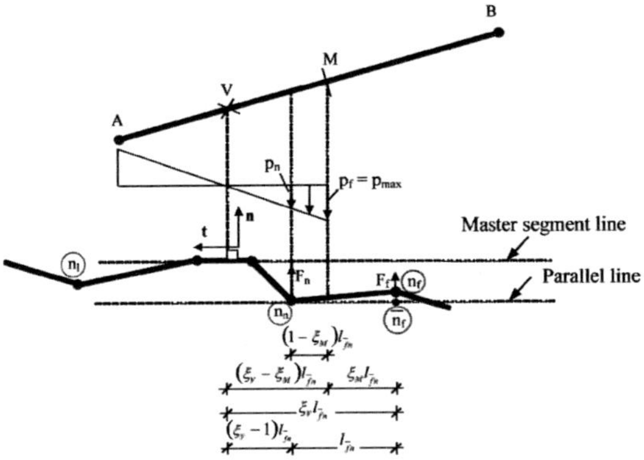

Fig. 11 Contact force transmission to more than one segment on the master surface: first segment subcase.

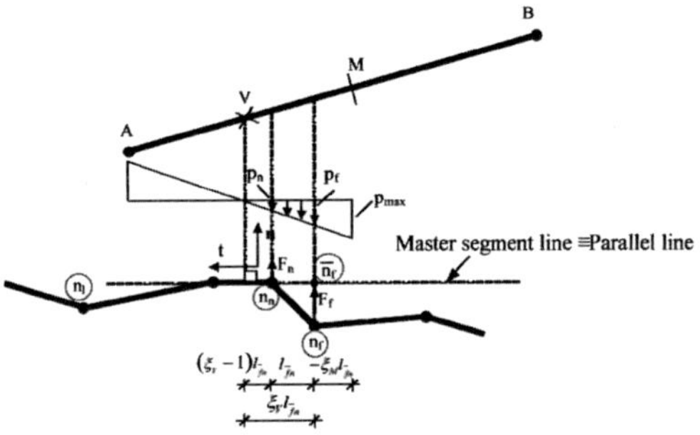

Fig. 12 Contact force transmission to more than one segment on the master surface: intermediate segment subcase.

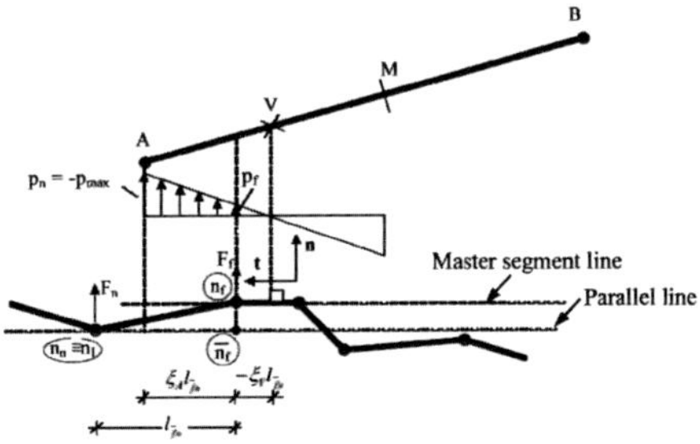

Fig. 13 Contact force transmission to more than one segment on the master surface: last segment subcase.

$$[R] = [R]_{\text{const}} + [R]_{\text{lin}} = \begin{bmatrix} \mathbf{R}_V \\ \mathbf{R}_M \\ \mathbf{R}_f \\ \mathbf{R}_i \\ \mathbf{R}_l \end{bmatrix}_{\text{const}} + \begin{bmatrix} \mathbf{R}_V \\ \mathbf{R}_M \\ \mathbf{R}_f \\ \mathbf{R}_i \\ \mathbf{R}_l \end{bmatrix}_{\text{lin}} \quad (49)$$

The computation of the component stemming from the constant distribution, $[R]_{\text{const}}$, is described in detail in [7]. The following subsections illustrate the computation of the various terms in the component stemming from the linear antimetric distribution, $[R]_{\text{lin}}$.

7.2 Contact Contribution to the Tangent Stiffness Matrix

It should be noticed that the enhanced residual vector is not associated with any modification in the tangent stiffness matrix. In fact, starting from the classical penalty contribution to the potential, only the two end nodes of the master segment play a role in determining the penetration. The tangent stiffness matrix for the standard virtual slave node technique has the following form:

$$[K_t] = \begin{bmatrix} \mathbf{K}_{VV} & \mathbf{K}_{VM} \\ \mathbf{K}_{MV} & \mathbf{K}_{MM} \end{bmatrix} \quad (50)$$

where the subscripts V and M in the submatrices refer to the nodes of the slave segment (i.e. the segment containing the virtual slave node) and of the master segment, respectively. In the proposed algorithm, the stiffness matrix becomes as follows:

$$[K_t] = \begin{bmatrix} \mathbf{K}_{VV} & \mathbf{K}_{VM} & \mathbf{0} & \mathbf{0} & \mathbf{0} \\ \mathbf{K}_{MV} & \mathbf{K}_{MM} & \mathbf{0} & \mathbf{0} & \mathbf{0} \\ \mathbf{0} & \mathbf{0} & \mathbf{0} & \mathbf{0} & \mathbf{0} \\ \mathbf{0} & \mathbf{0} & \mathbf{0} & \mathbf{0} & \mathbf{0} \\ \mathbf{0} & \mathbf{0} & \mathbf{0} & \mathbf{0} & \mathbf{0} \end{bmatrix} \quad (51)$$

As shown by equation (51), the degrees of freedom involved by the non-zero terms in the tangent stiffness matrix remain those of the virtual slave node (in turn interpolated from those of the two end nodes of the slave segment), and those of the two nodes of the corresponding master segment. Hence the terms of the tangent stiffness matrix are computed in the same way as in the standard virtual slave node technique, which corresponds to a small modification of the classical NTS algorithm. The final set of algebraic equations is

$$[K_t][u] = -[R] \quad (52)$$

where $[K_t]$ and $[R]$ are respectively given by equations (51) and (48), and

$$[u] = [\mathbf{u}_V \;\; \mathbf{u}_M \;\; \mathbf{u}_f \;\; \mathbf{u}_i \;\; \mathbf{u}_l]^T \quad (53)$$

is the vector of the unknown displacements. Once again, subscripts V and M refer to the nodes of the slave and of the master segments, respectively. Subscripts f, i and l refer to the nodes of the first segment, of the intermediate segment(s) and of the last segment on the master surface involved by the projection of the area of competence of the virtual slave node.

This inconsistency between the tangent stiffness matrix and the residual vector implies that the rate of convergence of the non-linear problem solved by the Newton–Raphson method is no longer quadratic. However, as shown in [7] for the VTS-PPT algorithm, the consistent tangent stiffness matrix could be easily obtained through the linearization of the correct residual vector. This aspect will be presented in a forthcoming paper.

8 Conclusions

This study has illustrated several sequential modifications of the basic NTS formulation for 2D frictionless contact, used in conjunction with the penalty method. These modifications yield incremental improvements in results of the linear contact patch test. In particular, the last proposed formulation (VTS-PPTL) is a modified one-pass NTS algorithm which is able to correctly transfer a linear contact pressure distribution from the slave to the master surface, hence it passes the linear contact patch test. The differences between the proposed algorithm and the standard NTS one are:

(i) the use of the modified virtual slave node technique, with virtual slave nodes located at the quarter points of each slave segment and a proper computation of the slave nodal forces;
(ii) the use of a specific procedure to correctly compute the contact forces at the master nodes.

References

1. Crisfield, M.A.: Revisiting the contact patch test. International Journal for Numerical Methods in Engineering 44, 1334–1355 (2000)
2. El-Abbasi, N., Bathe, K.J.: Stability and patch test performance of contact discretizations and a new solution algorithm. Computers and Structures 79, 1473–1486 (2001)
3. Taylor, R.L., Papadopoulos, P.: On a patch test for contact problems in two dimensions. In: Wriggers, P., Wanger, W. (eds.) Computational Methods in Nonlinear Mechanics, pp. 690–702. Springer, Berlin (1991)
4. Wriggers, P.: Computational Contact Mechanics. Springer, Heidelberg (2006)
5. Zavarise, G.: Problemi termomeccanici di contatto – Aspetti fisici e computazionali. Tesi di Dottorato in Meccanica delle Strutture, Istituto di Scienza e Tecnica delle Costruzioni, Padova, Italy (1991)

6. Zavarise, G., Boso, D., Schrefler, B.A.: A contact formulation for electrical and mechanical resistance. In: Martins, J.A.C., Monteiro Marques, M.D.P. (eds.) Proceedings of CMIS, III Contact Mechanics International Symposium, Praja de Consolacao, Portugal, pp. 211–218 (2001)
7. Zavarise, G., De Lorenzis, L.: A modified node-to-segment algorithm passing the contact patch test. International Journal for Numerical Methods in Engineering 79, 379–416 (2009)
8. Zavarise, G., Wriggers, P., Stein, E., Schrefler, B.A.: Real contact mechanisms and finite element formulation – A coupled thermomechanical approach. International Journal for Numerical Methods in Engineering 35, 767–785 (1992)

Finite Deformation Thermomechanical Contact Homogenization Framework

İlker Temizer and Peter Wriggers

Abstract A finite deformation homogenization framework is developed to predict the macroscopic thermal response of contact interfaces between rough surface topographies. The overall homogenization framework transfers macroscopic contact variables such as surfacial stretch, pressure and heat flux as boundary conditions on a test sample within a micromechanical interface testing procedure. An analysis of the thermal dissipation within the test sample reveals a thermodynamically consistent identification for the macroscopic thermal contact conductance parameter that enables the solution of a homogenized thermomechanical contact boundary value problem based on standard computational approaches. The homogenized contact response effectively predicts a temperature jump across the macroscale contact interface.

1 Introduction

Rough contact interface topographies are potential sources of large thermal energy dissipations within a variety of modern engineering applications ranging from micro-electromechanical systems and microprocessors to general microelectronics and electronic packaging. Therefore, the development of methods for alleviating the sources of dissipation has accompanied the progress in these fields. The predominant technology that has appeared over the years for this purpose is a class of materials commonly referred to as *thermal interface materials* (TIMs). The fundamental functional contribution of TIMs is to provide a high degree of conformity among the contacting surfaces by filling the gaps that would otherwise be present due to roughness. The reader is referred to [10] for a recent review of TIMs, its potential applications and current open issues.

İlker Temizer · Peter Wriggers
Institute of Continuum Mechanics, Leibniz University of Hannover, Appelstr. 11, 30169 Hannover, Germany; e-mail: {temizer, wriggers}@ikm.uni-hannover.de

G. Zavarise & P. Wriggers (Eds.): Trends in Computational Contact Mechanics, LNACM 58, pp. 101–119.
springerlink.com

From classical solder-based materials and coatings to modern polymeric substances such as gels, adhesives and elastomers, a detailed understanding of the thermomechanical response of TIMs, and that of the raw contact interface in general, under a prescribed confinement pressure and heat flux requires a consideration of the microstructure of the contact interface and a modeling of the interaction between the surfaces at the microscopic level. From an engineering point of view, the overall macroscopic reflection of the microscale physics at the interface is conveniently characterized within a continuum thermomechanics setting by a *thermal contact conductance* parameter k_c. As a particular example that pertains to the present framework where one body is brought into contact with a surface at a prescribed temperature θ_c, the knowledge of a constitutive model for k_c predicts a temperature jump ϑ_c for a given normal heat flux h across the contact interface via $h = -k_c\,\vartheta_c$. On the other hand, the framework to be developed predicts the thermal dissipation $\mathscr{D}_c$ per unit area of the contact interface associated with the above formulation to be (see Equation (22)) $\mathscr{D}_c = -h\ln(1+\vartheta_c/\theta_c) \geq 0$. Clearly, $\lim_{\vartheta_c \to 0} \mathscr{D}_c = 0$ so that the thermal contact dissipation disappears only for a perfectly conducting interface ($k_c \to \infty$) where the temperature jump identically vanishes. Hence, the prediction of $\mathscr{D}_c$ as the primary engineering quantity of interest is entirely associated with the determination of k_c.

A fairly complete review of the classical literature on methods of predicting k_c based on the micromechanics of metallic contact interfaces as well as experimental aspects is presented in [9]. Employing linearized thermomechanical models with a possibility of inelastic deformation for the asperities at the contact interface, complemented with statistical considerations regarding the asperity distribution within the plane of contact and across multiple length scales, the standard approach to thermal contact conductance characterization is predominantly analytical. See also [19] for a critical overview and [7] for an explicit incorporation of length scale effects in asperity deformation models. Specific applications to TIMs have recently been considered in [4, 11].

Analytical approaches provide, in many cases, qualitatively and quantitatively very good predictions of the macroscopic experimental observations [1, 5]. However, they are not directly capable of incorporating realistic contact topographies that require a three-dimensional analysis into the framework due to analytical limitations. Computational strategies that aim towards this purpose have been presented in, e.g., [13, 21] although simplifications regarding the thermomechanical coupling or the actual three-dimensional nature of the solution fields are still embedded in these approaches. A fully three-dimensional linear thermoelasticity strategy incorporating computer scanned surface topographies has recently been pursued in [16] within a finite element method framework, however based on an identification for the temperature jump ϑ_c that will be found as inconsistent within a finite deformation theory framework.

Within the limitations of continuum thermodynamics, the state of the art for thermal contact conductance prediction highlights a need for robust and generalized computational analysis frameworks which (i) are derived from a multiscale approach that consistently incorporates the effect of fine-scale roughness features in the analysis of a macroscale thermomechanical contact problem, (ii) are in the finite deformation regime for a more realistic modeling of polymeric TIMs and the large deformation inelastic response of the asperities for metallic interfaces, and (iii) are able to consider arbitrarily rough surface topographies. The present work is a first step towards the realization of such a framework. In particular, a *contact homogenization* technique will be developed within the finite deformation regime that naturally allows a possibility of considering fully nonlinear bulk responses for contacting bodies with rough boundary topographies in a computational setting.

The fundamental homogenization approach is based on the construction of a micromechanical contact interface testing procedure that delivers common notions of macroscopic contact pressure and heat flux as employed in engineering, in addition to a thermodynamically consistent identification of the thermal contact conductance parameter. The qualitative implications and capabilities of the developed technique as well as its limitations will be highlighted in detail. Overall, the present work is aimed to assist in the efforts towards the establishment of robust computational multiscale contact homogenization techniques that are needed for the analysis of similar microheterogeneous interfaces with potential application to technological materials of interest such as TIMs.

For further details and possible extensions of the presented approach, the interested reader is referred to [22, 23].

2 Thermomechanical Contact Formulation

Consider a thermomechanical problem posed on a macrostructure with reference configuration $\mathscr{R}_o$ that subsequently comes into contact, in its deformed configuration $\mathscr{R}$, with a surface $\mathscr{S}$ that will be assumed smooth and rigid. The problem is heterogeneous on the microscopic scale of the contact interface due to a highly nonuniform boundary topography that is associated with surface roughness on $\partial\mathscr{R}_o$. The material comprising the bulk of the macrostructure except for a thin boundary layer is denoted symbolically $\mathscr{M}$ whereas the thin boundary layer that encompasses the roughness features is denoted $\mathscr{B}$. Due to the heterogeneous boundary layer $\mathscr{B}$, the topography of the contact interface $\partial\mathscr{R}^c$ in the deformed configuration is highly oscillatory. An accurate representation of the solution fields $\{\mathbf{x}, \theta\}$ on $\partial\mathscr{R}^c$ and hence throughout $\mathscr{R}$ requires a prohibitively fine numerical resolution on the potential contact interface, rendering the solution of this original boundary value problem (BVP) very challenging in its original form. Therefore, an approximate solution is sought instead, where a homogenization methodology is employed to construct a homogeneous *effective boundary layer* $\mathscr{B}^*$. The effective boundary layer is homogeneous in the sense that the roughness features are smeared out in order to obtain a microscopically smooth surface. Consequently, the *effective contact interaction* (ECI) of the new boundary $\mathscr{B}^*$ with $\mathscr{S}$ is expected to be constitutively of a different nature than, but macroscopically representative of, the interaction between $\mathscr{B}$ and $\mathscr{S}$. A *homogenized thermomechanical BVP* on $\mathscr{R}_o$ can now be structured by replacing $\mathscr{B}$ with $\mathscr{B}^*$ in the original heterogeneous BVP and employing the same boundary conditions with the ECI on $\partial\mathscr{R}^c$, which delivers a homogenized solution $\{\mathbf{x}^*, \theta^*\}$. The homogenized problem is subject to the numerical discretization concerns of standard contact mechanics problems and hence is considerably easier to solve compared with the original heterogenous one.

Clearly, the major aspect of the homogenized BVP formulation is the characterization of the ECI. The construction of the methodology with which the ECI is to be determined relies on the geometrical and constitutive properties of the smooth rigid surface $\mathscr{S}$ as well as on the nature of its interaction with the heterogeneous boundary layer $\mathscr{B}$. The following remarks highlight some of the fundamental assumptions underlying subsequent developments:

1. *$\mathscr{S}$ is rigid*: In the linear elasticity regime, the frictionless contact analysis of two deformable surfaces can be equivalently represented by a rigid surface interacting with a single deformable one that is assigned equivalent elastic properties and roughness features [8]. This approach is not applicable to the frictional thermomechanical contact of two rough surfaces in the finite deformation regime. Nevertheless, it motivates the assignment of deformability and roughness to one of the surfaces in order to investigate the representative features of the macroscale behavior within a numerically less expensive setting.
2. *$\mathscr{S}$ is smooth*: The smoothness assumption allows the homogenization procedure to be concentrated exclusively on the thermal interaction of the deformable body with $\mathscr{S}$, cf. [18] or [15].
3. *$\mathscr{S}$ is a heat bath*: Similar to the rigidness assumption for the mechanical response, the thermal response of $\mathscr{S}$ is idealized as that of a heat bath at temperature θ_c. Consequently, the thermomechanical contact BVP is isolated to the analysis of the deformable body alone.
4. *$\mathscr{S}$ and $\mathscr{B}$ interact through spot conduction only:* Convective heat transfer through an interstitial gas or radiative heat transfer between the surfaces can significantly influence the thermal interaction across the macroscale contact interface [9, 14, 20]. The former effect can be omitted by postulating the contact interface to be free of any external species. The latter effect may become negligible for small temperature differences between $\partial\mathscr{R}$ and $\mathscr{S}$. In general, both effects are relevant for microscopically *smooth* surfaces as well for finite temperature differences. Hence, in order to concentrate on roughness effects, it is assumed that heat transfer occurs only through the contact spots on the microscale so that the bodies do not exchange energy when they are not in contact, which is a common assumption in the literature [2].

In order to formulate thermal contact under the stated assumptions, the continuity of the temperature across the contact interface is postulated at the microscopic scale, i.e. $\vartheta_c = 0$ on $\partial\mathscr{R}^c$ where $\vartheta_c := \theta - \theta_c$. This is equivalent to the statement that *the contact spot interfaces between the rough microscale deformable surface and $\mathscr{S}$ are perfectly conforming at the next hierarchically lower scale*. Macroscale continuity of the temperature across increasingly conforming surfaces is also implied by the present homogenization framework. Hence, attention is effectively focused on a single scale of heterogeneity and additional sources of microscale dissipation such as ballistic resistance is ignored [10]. The thermal contact formulation employed in this work then enforces $\vartheta_c = 0$ via a penalty formulation that regularizes the contact heat flux integral in the weak form via

$$\int_{\partial \mathscr{R}^c} \delta\theta \, h \, \mathrm{d}a = \int_{\partial \mathscr{R}^c} \delta\theta \, [-\varepsilon_\theta \, \vartheta_c] \, \mathrm{d}a, \tag{1}$$

where ε_θ is an ideally infinitely large penalty parameter. Clearly, if the value of ε_θ is taken as a finite value that depends on contact variables such as the contact pressure and heat flux then it serves the same purpose as the thermal contact conductance parameter with which a finite temperature jump is induced across the contact interface. The determination of the ECI under the stated assumptions, i.e. the constitutive characterization for the macroscopic contact heat flux $\overline{h}$ with respect to the macroscopic temperature jump $\overline{\vartheta}_c$, is the main goal of the next section. It is noted that in the present work, all contact constraint integrals are evaluated exactly via numerical integration as in the mortar method, i.e. the contact constraints are enforced at the integration points of the contact elements. See also [17] for general aspects of numerical thermomechanical contact analysis formulations.

3 Thermal Contact Homogenization Methodology

3.1 Interface Testing Procedure

In order to characterize the coupled macroscale thermal contact response of the heterogeneous boundary layer $\mathscr{B}$ and the surface $\mathscr{S}$, a cut from $\mathscr{B}$ is employed as a micromechanical contact test sample that is subjected to a two-phase deformation that defines an interface testing procedure (Figure 1). Within this procedure, the intrinsic interface response is isolated by (i) remaining close to mechanical and thermal equilibrium states, so that all rate terms disappear in the formulation, and (ii) additionally ignoring body forces and heat supplies to preclude external effects. Consequently, using standard continuum mechanics notation (see, e.g., [6]), the thermomechanical response of the sample is extracted through the solution of the following BVP at all stages of the deformation:

Weak Form of the Interface Test BVP Between $\mathscr{C}_o$ and $\mathscr{S}$:
Determine $\mathbf{x}(\mathbf{X})$ and $\theta(\mathbf{X})$ so that the linear momentum balance

$$-\int_{\mathscr{C}_o} \delta\mathbf{F} \cdot \mathbf{P} \, \mathrm{d}V + \int_{\partial \mathscr{C}_o^t} \delta\mathbf{x} \cdot \mathbf{p} \, \mathrm{d}A + \int_{\partial \mathscr{C}^{r,c}} \delta\mathbf{x} \cdot \mathbf{t} \, \mathrm{d}a = 0$$

and the energy balance

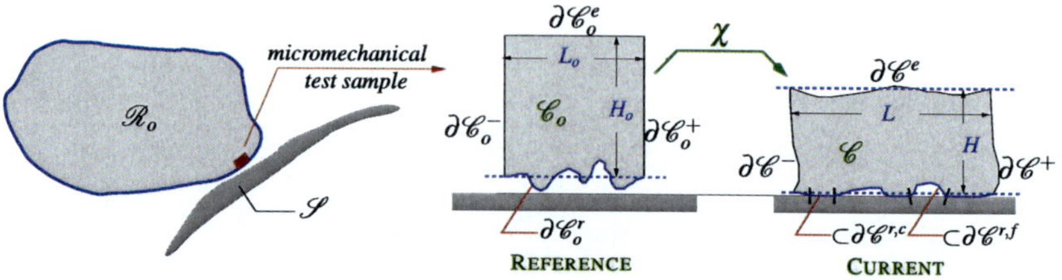

Fig. 1 The micromechanical contact test is summarized: the reference undeformed configuration of the test sample is mapped to the current deformed configuration via the map $\chi = \chi_N \circ \chi_T$ (see Equation (2)).

$$\int_{\mathscr{C}_o} \delta \mathbf{g}_o \cdot \mathbf{q}_o \, dV + \int_{\partial\mathscr{C}_o^h} \delta\theta \, h_o \, dA + \int_{\partial\mathscr{C}^{r,c}} \delta\theta \, h \, da = 0$$

are satisfied for all $\delta\mathbf{x}$ subject to $\mathbf{x} = \hat{\mathbf{x}}$ with $\delta\mathbf{x} = \mathbf{0}$ on $\partial\mathscr{C}_o^x$ and for all $\delta\theta$ subject to $\theta = \hat{\theta}$ with $\delta\theta = 0$ on $\partial\mathscr{C}_o^\theta$, complemented by the constitutive models for $\mathbf{P}$ and $\mathbf{q}_o$ and a contact formulation on $\partial\mathscr{C}^{r,c}$.

Denoting the reference undeformed configuration of the sample by $\mathscr{C}_o$, the testing procedure maps $\mathscr{C}_o$ to the current deformed configuration $\mathscr{C}$ via $\mathbf{x} = \chi(\mathbf{X})$. This motion is now split to two phases that define χ through the composite function

$$\chi = \chi_N \circ \chi_T \tag{2}$$

such that $\mathbf{x}_\bullet = \chi_T(\mathbf{X})$ and $\mathbf{x} = \chi_N(\mathbf{x}_\bullet)$. The motion χ_T will be referred to as the *tangential map* and χ_N as the *normal map*. The position vector $\mathbf{x}_\bullet$ is associated with a global intermediate deformed configuration $\mathscr{C}_\bullet$ that corresponds to the end of the phase defined by the tangential map: $\mathscr{C}_o \xrightarrow{\chi_T} \mathscr{C}_\bullet \xrightarrow{\chi_N} \mathscr{C}$. Hence, the deformation gradient can be globally decomposed multiplicatively as

$$\mathbf{F} = \mathbf{F}_N \mathbf{F}_T \tag{3}$$

where $\mathbf{F}_T := \partial\mathbf{x}_\bullet / \partial\mathbf{X}$ and $\mathbf{F}_N := \partial\mathbf{x} / \partial\mathbf{x}_\bullet$.

3.2 Macro-to-Micro Transition

An explicit assignment to the motions χ_T and χ_N follows from the monitoring of the macroscale homogenized BVP at a point of interest $\mathscr{P}$ on the macroscale domain boundary $\partial\mathscr{R}_o$ ($\partial\mathscr{R}$) with normal $\overline{\mathbf{n}}_o$ ($\overline{\mathbf{n}}$). To this end, the projection operators $\mathbf{D}_N := \overline{\mathbf{n}}_o \otimes \overline{\mathbf{n}}_o$ and $\mathbf{D}_T := \mathbf{I} - \mathbf{D}_N$ allow an additive decomposition of the macroscopic deformation gradient $\overline{\mathbf{F}}$ within the boundary layer in the vicinity of $\mathscr{P}$ via

$$\overline{\mathbf{F}} = \overline{\mathbf{F}}\mathbf{D}_N + \overline{\mathbf{F}}\mathbf{D}_T. \tag{4}$$

Similarly, the macroscopic temperature gradient $\overline{\mathbf{g}}$ can be additively decomposed as

$$\overline{\mathbf{g}} = \mathbf{d}_N\overline{\mathbf{g}} + \mathbf{d}_T\overline{\mathbf{g}}, \tag{5}$$

where $\mathbf{d}_N := \overline{\mathbf{n}} \otimes \overline{\mathbf{n}}$ and $\mathbf{d}_T := \mathbf{I} - \mathbf{d}_N$. In these decompositions, normal gradients $\overline{\mathbf{F}}\mathbf{D}_N$ and $\mathbf{d}_N\overline{\mathbf{g}}$ are independent of the distributions of the macroscale variables $\overline{\mathbf{x}}$ and $\overline{\theta}$ on the boundary while the surfacial gradients $\overline{\mathbf{F}}\mathbf{D}_T$ and $\mathbf{d}_T\overline{\mathbf{g}}$ are determined solely by these distributions.

In order to characterize the macroscopic thermal contact response of the test sample, variables extracted from the macroscale solution fields on the boundary surface of the homogenized BVP will be transferred to the microscale as boundary conditions within the testing procedure. In particular, one observes:

- The macroscopic contact pressure $\overline{p} = -\overline{\mathbf{t}} \cdot \overline{\mathbf{n}}$ and the macroscopic contact heat flux $\overline{h} = -\overline{\mathbf{q}} \cdot \overline{\mathbf{n}}$ naturally accompany the formulation of the macroscale contact problem.
- $\overline{\mathbf{F}}\mathbf{D}_T$ is the macroscale variable associated with the surfacial deformation which effectively causes the surface to appear rougher or smoother in the intermediate configuration than in the reference configuration.
- The sought constitutive response is to be idealized as being intrinsically embedded in the contact surface of the homogenized BVP. Macroscale variables $\overline{\mathbf{F}}\mathbf{D}_N$ and $\mathbf{d}_N\overline{\mathbf{g}}$ which cannot be determined from a surfacial distribution are dropped from the overall framework.
- The surfacial temperature gradient $\mathbf{d}_T\overline{\mathbf{g}}$ does not additionally contribute to the interface energy exchange in the thermal contact formulation and hence is ruled out as a transfer variable.

These choices are consistent with the formulation of a multiscale contact homogenization framework within a coupled micro-macro approach as attempted in [15] where one extracts the unknown constitutive response at the integration point of a macroscale contact element from an embedded microstructural contact test. Such a multiscale approach will not be pursued in the present work. Instead, $\{\theta_c, \overline{\mathbf{F}}\mathbf{D}_T, \overline{p}, \overline{h}\}$ will be regarded as variables that parameterize the macroscale response in the numerical investigations.

3.3 Boundary Conditions

The tangential map χ_T is now chosen to prescribe $\overline{\mathbf{F}}\mathbf{D}_T$ whereas χ_N will enforce $\overline{p}$ and $\overline{h}$. Hence, within the motion split (2) a purely surfacial deformation is followed by the application of the contact pressure and heat flux. However, in an actual macroscale BVP, surfacial deformation is dependent on the contact phase as well. Consequently, the proposed motion split where the full surfacial deformation is applied before the contact phase is guaranteed to be an exact representation of the macroscale conditions only for a path-independent behavior such as frictionless thermoelastic contact. For frictional interaction or bulk mechanical dissipation effects, the proposed split is interpreted as a tool aimed at investigating the macroscale thermomechanical contact response. Details on the methods of imposing the motion maps are delineated below where the associated boundary conditions (BCs) are partitioned into three main categories according to their geometrical domain. Only mixed (periodic) type BCs are considered due to periodic contact microstructures employed in the numerical investigations. The effect of the application of these BCs on the sample dimensions is depicted in Figure 1.

For convenience, orthonormal basis vector sets $\{\mathbf{V}_1, \mathbf{V}_2, \overline{\mathbf{n}}_o\}$ and $\{\mathbf{v}_1, \mathbf{v}_2, \overline{\mathbf{n}}\}$ are defined at the macroscale point $\mathscr{P}$ such that $\overline{\mathbf{n}}_o = \mathbf{V}_1 \times \mathbf{V}_2$ and $\overline{\mathbf{n}} = \mathbf{v}_1 \times \mathbf{v}_2$. However, $\overline{\mathbf{F}}\mathbf{D}_T$ involves a rigid body motion associated with the mapping of $\overline{\mathbf{n}}_o$ to $\overline{\mathbf{n}}$. In order to be consistent with the construction of the sample configuration where $\mathscr{C}_o$ was chosen such that the normal to $\partial\mathscr{C}_o^e$ is $-\overline{\mathbf{n}}$, it is necessary to remove this rotation in a preliminary step. For this purpose, the complementary tensor

$$\overline{\mathbf{F}}^c = \overline{F}^c_{\alpha\beta}\, \mathbf{v}_\alpha \otimes \mathbf{v}_\beta + \overline{\mathbf{n}} \otimes \overline{\mathbf{n}} \tag{6}$$

will be employed in the enforcement of the boundary conditions such that

$$\overline{F}^c_{\alpha\beta} := \overline{\mathbf{F}}\mathbf{D}_T \cdot (\mathbf{v}_\alpha \otimes \mathbf{V}_\beta) \equiv \overline{\mathbf{F}} \cdot (\mathbf{v}_\alpha \otimes \mathbf{V}_\beta). \tag{7}$$

Clearly, $\overline{\mathbf{F}}^c \mathbf{d}_T = \overline{F}^c_{\alpha\beta}\, \mathbf{v}_\alpha \otimes \mathbf{v}_\beta$.

3.3.1 Rough Interface Surface

The lower boundary of the test sample is the rough surface $\partial\mathscr{C}_o^r$ where contact BCs are imposed on a deformed configuration as described in Section 2. More specifically, contact is not initiated by the tangential map χ_T and hence the traction and normal heat flux vanishes in view of the pure spot conduction assumption of Section 2:

$$\textit{During } \chi_T\text{:} \qquad \mathbf{t} = \mathbf{0} \quad \text{and} \quad h = 0 \quad \text{on} \quad \partial\mathscr{C}^r. \tag{8}$$

During the normal loading phase, $\partial\mathscr{C}^r$ comes into contact with $\mathscr{S}$ on $\partial\mathscr{C}^{r,c} \subset \partial\mathscr{C}^r$ where contact conditions are enforced as summarized in Section 2. The remaining portion of $\partial\mathscr{C}^r$ is traction and flux free:

$$\textit{During } \chi_N\text{:} \qquad \mathbf{t} = \mathbf{0} \quad \text{and} \quad h = 0 \quad \text{on} \quad \partial\mathscr{C}^{r,f} = \partial\mathscr{C}^r \setminus \partial\mathscr{C}^{r,c}. \tag{9}$$

3.3.2 Interior Lateral Surface

The method of enforcing the boundary conditions on the lateral surface $\partial\mathscr{C}_o^i := \partial\mathscr{C}_o^- \cup \partial\mathscr{C}_o^+$ ($\partial\mathscr{C}_o^i \xrightarrow{\chi_T} \partial\mathscr{C}_\bullet^{\,i} \xrightarrow{\chi_N} \partial\mathscr{C}^i$) is imposed by the assumed periodic characteristics of the boundary layer $\mathscr{B}$. Let $\partial\mathscr{C}_o^-$ and $\partial\mathscr{C}_o^+$ denote partitions of opposing lateral surfaces. They are coupled to each other through the mixed BCs that enforce the periodicity of the deformation and the temperature and the anti-periodicity of the tractions and the normal heat fluxes:

$$\begin{aligned}
&\textit{During } \chi_T: \quad && \mathbf{x}^+ - \mathbf{x}^- = \overline{\mathbf{F}}^c (\mathbf{x}_o^+ - \mathbf{x}_o^-) && \text{and} \quad \mathbf{t}_o^+ = -\mathbf{t}_o^-, \\
& && \theta^+ - \theta^- = 0 && \text{and} \quad h_o^+ = -h_o^-, \\
&\textit{During } \chi_N: \quad && \mathbf{x}^+ - \mathbf{x}^- = \mathbf{x}_\bullet{}^+ - \mathbf{x}_\bullet{}^- && \text{and} \quad \mathbf{t}_o^+ = -\mathbf{t}_o^-, \\
& && \theta^+ - \theta^- = 0 && \text{and} \quad h_o^+ = -h_o^-.
\end{aligned} \tag{10}$$

3.3.3 Exterior Observable Test Surface

While the lower surface of the test sample is associated with the potential contact interface and the lateral surface is regarded as interior, the upper surface $\partial\mathscr{C}_o^e$ is the exterior surface of the test sample where all macroscopic observables may be monitored. Here, one enforces

$$\begin{aligned}
&\textit{During } \chi_T: \quad && \mathbf{x}\cdot\overline{\mathbf{n}} = \overline{\mathbf{F}}^c\,\mathbf{x}_o\cdot\overline{\mathbf{n}}, && \mathbf{t}\cdot\mathbf{v}_\alpha = 0 && \text{on} \quad \partial\mathscr{C}_o^e, \\
&\textit{During } \chi_N: \quad && \mathbf{t}_o = \overline{p}\,\det\overline{\mathbf{F}}^c\,\overline{\mathbf{n}}, && h_o = -\overline{h}\,\det\overline{\mathbf{F}}^c && \text{on} \quad \partial\mathscr{C}_o^e.
\end{aligned} \tag{11}$$

Additionally, $\mathbf{x}\cdot\mathbf{v}_\alpha = \overline{\mathbf{F}}^c\,\mathbf{x}_o\cdot\mathbf{v}_\alpha$ is enforced at a single point in order to constrain rigid body translation in the plane of contact at all phases of the deformation and $\theta = \theta_T$ is enforced at a single point during χ_T which induces a unique uniform temperature distribution (θ_T) at the end of the tangential phase. Also, the sample needs to be first brought into contact with the surface in order to apply pressure and heat flux BCs during the normal phase. This preliminary step is displacement controlled. Note that $\mathbf{t}\cdot\mathbf{v}_\alpha = 0$ is satisfied automatically on $\partial\mathscr{C}^e$ during χ_N.

3.4 Averaging Theorems

The solution fields on the external observable test surface $\partial\mathscr{C}_o^e$ can be employed to recover the macroscopic mechanical and thermal contact variables of interest after a surface averaging procedure

$$\langle \mathscr{Q} \rangle_\Omega := \frac{1}{|\Omega|}\int_\Omega \mathscr{Q}\,\mathrm{d}\Omega$$

Identification of Macroscopic Thermomechanical Variables:

$$\overline{\mathbf{F}}^c\mathbf{d}_T = \langle \mathbf{F}^p \rangle_{\overline{\partial\mathscr{C}_o^e}}, \quad \overline{p} = \frac{1}{|\mathscr{A}|}\int_{\partial\mathscr{C}^e} \mathbf{t}\cdot\overline{\mathbf{n}}\,da, \quad \overline{h} = -\frac{1}{|\mathscr{A}|}\int_{\partial\mathscr{C}^e} h\,da. \tag{12}$$

Here, $|\mathscr{A}|$ represents the nominal contact area and $\overline{\partial \mathscr{C}_o^e}$ is the image of $\partial \mathscr{C}_o^e$ projected onto the rigid surface $\mathscr{S}$ with $\mathbf{F}^p$ as the deformation gradient within the projected image.

It is noted that the absence of body forces and accelerations within the thermo-mechanical testing procedure is crucial since otherwise the identification of $\overline{\mathbf{F}}^c \mathbf{d}_T$, $\overline{p}$ and $\overline{h}$ would entail the evaluation of integrals over the volume of the test sample which is not accessible. Additionally, the special construction of the BCs effectively eliminate the presence of surface integrals over the inaccessible internal surface $\partial \mathscr{C}_o^i$. Therefore, these identifications can alternatively be regarded as a restriction that needs to be satisfied by the thermal and mechanical BCs employed during the contact interface testing procedure in order to facilitate the measurement of the macroscopic contact variables.

3.5 Micro-to-Macro Transition

The *macro-to-micro* transition transfers macroscale variables of interest as BCs on the micromechanical test sample. The *micro-to-macro* transition scheme homogenizes the response of the test sample under the prescribed BCs and effectively induces a temperature jump $\overline{\vartheta}_c \neq 0$ across the contact interface of the macroscale homogenized problem. In order to identify this temperature jump, a thermodynamically consistent procedure is introduced based on the thermal dissipation

$$\mathscr{D}_o := -\frac{\mathbf{q}_o \cdot \mathbf{g}_o}{\theta} \geq 0$$

per unit volume of the undeformed test sample. Specifically, employing the *logarithmic temperature* $\ln \theta$ where θ must explicitly be chosen as the thermodynamic absolute temperature and noting that $\text{Grad}[\ln \theta] = \mathbf{g}_o / \theta$, the total thermal dissipation Φ will be monitored:

$$\Phi := \int_{\mathscr{C}_o} \mathscr{D}_o \, \mathrm{d}V = -\int_{\mathscr{C}_o} \mathbf{q}_o \cdot \text{Grad}[\ln \theta] \, \mathrm{d}V. \tag{13}$$

3.5.1 Total Thermal Dissipation in the Test Sample

The absence of heat supply and rate effects implies

$$\Phi \equiv \int_{\partial \mathscr{C}_o} h_o \ln \theta \, \mathrm{d}A + \int_{\mathscr{C}_o} \text{Div}[\mathbf{q}_o] \ln \theta \, \mathrm{d}V = \int_{\partial \mathscr{C}_o} h_o \ln \theta \, \mathrm{d}A. \tag{14}$$

Decomposing the boundary to the specific domains of each BC category, and recalling that θ_c is a constant, one can further specialize this result via

$$\begin{aligned}
\Phi &= \int_{\partial\mathscr{C}_o^i} h_o \ln\theta \; dA + \int_{\partial\mathscr{C}_o^e} h_o \ln\theta \; dA + \int_{\partial\mathscr{C}_o^r} h_o \ln\theta \; dA \\
&= \int_{\partial\mathscr{C}_o^i} h_o \ln\theta \; dA - \int_{\partial\mathscr{C}_o^e} \overline{h} \det \overline{\mathbf{F}}^c \ln\theta \; dA + \int_{\partial\mathscr{C}^{r,c}} h \ln\theta_c \; da \\
&= \int_{\partial\mathscr{C}_o^i} h_o \ln\theta \; dA - \overline{h} \det \overline{\mathbf{F}}^c |\partial\mathscr{C}_o^e| \langle \ln\theta \rangle_{\partial\mathscr{C}_o^e} + \ln\theta_c \int_{\partial\mathscr{C}^{r,c}} h \; da.
\end{aligned} \tag{15}$$

For periodic BCs on $\partial\mathscr{C}_o^i$, the periodicity of θ ensures the periodicity of $\ln\theta$ and together with the anti-periodicity of h_o this integral is identically zero. On the other hand, due to energy balance $\int_{\partial\mathscr{C}^{r,c}} h \; da = \overline{h}|\mathscr{A}|$. Upon defining $|\partial\mathscr{C}_o^e| = |\partial\mathscr{C}_o^e| =: |\mathscr{A}_o|$ and rearranging using $|\mathscr{A}| = \det \overline{\mathbf{F}}^c |\mathscr{A}_o|$, one obtains the identity

$$\Phi = \overline{h}|\mathscr{A}|\left(\ln\theta_c - \langle \ln\theta \rangle_{\partial\mathscr{C}_o^e}\right). \tag{16}$$

If $\overline{h} > 0$ then heat flows into the test sample through the contact interface and $\theta < \theta_c$ on $\partial\mathscr{C}_o^e$. Similarly, $\overline{h} < 0$ induces $\theta > \theta_c$ on $\partial\mathscr{C}_o^e$ so that in both cases the total thermal dissipation in the sample is guaranteed to be non-negative. The significance of Equation (16) is that the total thermal dissipation for a given total heat flux $\overline{h}|\mathscr{A}|$ through the sample and a contact surface temperature θ_c is measurable through the knowledge of the average temperature over the observable test surface only.

3.5.2 Accompanying and Homogenized Bar Problems

For a homogeneous test sample, i.e. where $\partial\mathscr{C}_o^r$ is smooth, the imposed BCs induce solution fields that vary only in the direction away from the contact interface. For identical BCs on the heterogeneous and homogeneous samples, the solution to the homogeneous problem induces a dissipation

$$\Phi_{ABP} = \overline{h}|\mathscr{A}| \left(\ln\theta_c - \ln\theta_{ABP}\right), \tag{17}$$

where θ_{ABP} is the constant temperature on $\partial\mathscr{C}_o^e$. Clearly, $\Phi_{ABP} \neq \Phi$. The subscript "ABP" is used to distinguish the solution on this problem, which stands for the *accompanying bar problem.* Note that the lateral dimension L_o (see Figure 1) of the test sample for the ABP is irrelevant due to uniform solution fields in this direction. However, the sample height H_o will be chosen so as to match the height of a heterogeneous sample.

For heterogeneous test samples with a large height H_o, the solution fields sufficiently away from the contact interface are qualitatively similar to that obtained from the ABP. Hence, on $\partial\mathscr{C}^e$ one measures $\mathbf{t} \approx \overline{p}\overline{\mathbf{n}}$ and $\theta \approx$ *constant.* Alternatively stated, the non-uniform variation in the solution fields in the vicinity of the contact interface occurs over a length distance that is comparatively much smaller than the sample height. The situation can be idealized by confining the non-uniform solution fields to an infinitesimally thin region near the contact surface over which a temperature jump $\overline{\vartheta}_c$ occurs. This observation motivates the construction of a *homo-*

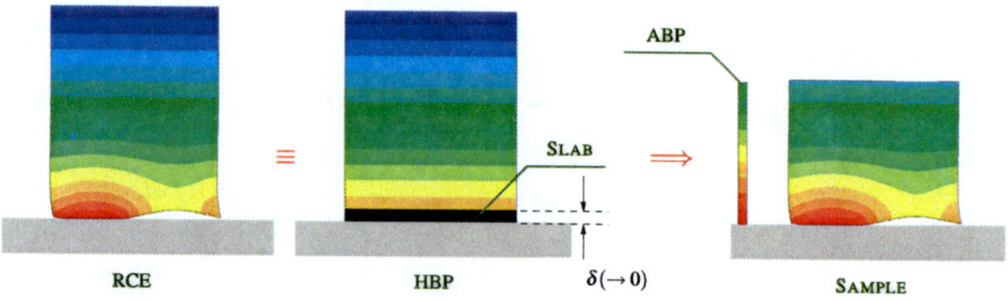

Fig. 2 The HBP idealization of the original heterogeneous solution on an RCE is depicted, which forms the motivation of the homogenization scheme by the introduction of the ABP for arbitrary sample dimensions.

genized bar problem (HBP) where a heterogeneous test sample *of arbitrary height* is replaced by a homogeneous deformable bar with a thin resistive slab replacing the rough surface as depicted in Figure 2. Similar to the ABP, the lateral dimension of the homogenized bar is insignificant.

Clearly, the idealization of the HBP approaches the actual response of the test sample as the sample height H_o increases. However, independent of the sample height, the dissipation in the original test sample and the HBP are enforced to be equivalent for thermodynamical consistency. Consequently, a comparison of the original sample response and the HBP solution yields

$$\begin{aligned} \overline{h}|\mathscr{A}|\left(\ln\theta_c - \langle\ln\theta\rangle_{\partial\mathscr{C}_o^e}\right) &= \Phi \equiv \Phi_{HBP} = \overline{h}|\mathscr{A}|(\ln\theta_c - \ln\theta_{HBP}) \\ &= \underbrace{\overline{h}|\mathscr{A}|(\ln\theta_c - \ln(\theta_c + \overline{\vartheta}_c))}_{=:\Phi^s_{HBP}} + \underbrace{\overline{h}|\mathscr{A}|(\ln(\theta_c + \overline{\vartheta}_c) - \ln\theta_{HBP})}_{=:\Phi^d_{HBP}} \end{aligned} \tag{18}$$

where $\theta_{HBP} = \exp(\langle\ln\theta\rangle_{\partial\mathscr{C}_o^e})$ is the constant temperature on $\partial\mathscr{C}_o^e$ for the HBP. Here, Φ^s_{HBP} and Φ^d_{HBP} represent the total thermal dissipations through the thin slab and the deformable portions of the HBP, respectively, with $\theta_c + \overline{\vartheta}_c$ as the interface temperature between the two.

The identification of $\overline{\vartheta}_c$, which is arbitrary at this point, now follows from postulating $\Phi^d_{HBP} = \Phi_{ABP}$. Alternatively stated, it is postulated that *the additional thermal dissipation introduced into a homogeneous problem through a heterogeneous boundary layer is entirely contained within the dissipation of the idealized thin slab*:

$$\Phi^s_{HBP} = \Phi - \Phi_{ABP}, \tag{19}$$

or

$$\ln(1 + \overline{\vartheta}_c/\theta_c) = \langle\ln\theta\rangle_{\partial\mathscr{C}_o^e} - \ln\theta_{ABP}. \tag{20}$$

This result provides the identification of the macroscopic temperature jump $\overline{\vartheta}_c$ as perceived on the macroscale contact interface:

Identification of the Macroscopic Temperature Jump:

$$\overline{\vartheta}_c = \frac{\theta_c}{\theta_{ABP}} \exp\left(\langle \ln \theta \rangle_{\partial \mathscr{C}_o^e}\right) - \theta_c. \tag{21}$$

Hence, the characterization of $\overline{\vartheta}_c$ entails the solution of the original problem and a comparison problem (ABP), the latter being computationally inexpensive due to the one-dimensional nature of the associated solution fields (see Figure 2). While the motivation for this identification is based on that of sufficiently large sample heights, the identification itself is thermodynamically consistent for arbitrary sample dimensions in the sense that the transition from the original heterogeneous sample to the homogenized bar setting (micro-to-macro transition) preserves the total dissipation within the original sample. Moreover, via the expression for Φ^s_{HBP}, the knowledge of $\overline{\vartheta}_c$ provides the macroscopic thermal contact dissipation $\overline{\mathscr{D}}_c$ per unit contact area of the homogenized BVP:

Thermal Dissipation per Unit Nominal Macroscopic Contact Area:

$$\overline{\mathscr{D}}_c = -\overline{h} \ln\left(1 + \overline{\vartheta}_c / \theta_c\right) \geq 0. \tag{22}$$

Accordingly, the total thermal dissipation across the macroscopic contact interface that is important from an engineering point of view is $\overline{\Phi}_c := \int_{\partial \mathscr{R}^c} \overline{\mathscr{D}}_c \, da$. This dissipation is interpreted as the macroscale reflection of the bulk dissipation within the thin heterogeneous boundary layer.

Using the macroscopic temperature jump $\overline{\vartheta}_c$, one identifies

$$\overline{k}_c = -\overline{h} / \overline{\vartheta}_c \tag{23}$$

as the macroscopic thermal contact conductance parameter $\overline{k}_c > 0$.

3.6 Representative Contact Element

The test sample is now defined to qualify as an RCE if, under fixed BCs and a given topography on $\partial \mathscr{C}_o^r$, increasing the periodic sample height beyond a sufficiently large value leads to insignificant changes in the measured jump $\overline{\vartheta}_c$. In this case, the macroscopic temperature jump is the *effective* one ($\vartheta^*{}_c$) that is sought for the formulation of the macroscale problem. In view of previous discussions, $\theta \approx$ *constant* $= \theta_{RCE}$ on $\partial \mathscr{C}_o^e$ for such a sample and hence the identification takes the simpler expression

$$\vartheta^*{}_c = \frac{\theta_c}{\theta_{ABP}}(\theta_{RCE} - \theta_{ABP}). \tag{24}$$

Through Equation (23), $\overline{k}_c$ approaches the effective thermal conductance $k^*{}_c$ for increasing sample heights. For samples with random surfaces, the identification of $k^*{}_c$ would conceptually additionally entail the monitoring of ensemble averaged $\overline{k}_c$ for increasingly larger lateral sample dimensions so as to ensure the statistical representativity of the sample topography. See [15] for an example procedure in the context of contact micromechanics.

On the other hand, the classical macroscopic temperature jump definition $\overline{\vartheta}_c^{lim}$ for a test sample of any given size is essentially equivalent to (see [9, chapter 2])

$$\overline{\vartheta}_c^{lim} = \langle \theta \rangle_{\partial \mathscr{C}_o^e} - \theta_{ABP}. \tag{25}$$

If the temperature changes are small throughout the sample ($|\vartheta| \ll \theta_o$), which necessitates that the initial temperature of the medium is close to the contact surface temperature ($\theta_o \approx \theta_c$), then $\theta_c/\theta_{ABP} \approx 1$ and hence the present formulation recovers the classical definition with an RCE. However, for large temperature deviations, the classical definition is not thermodynamically consistent since it effectively reinterprets the thermal dissipation within the sample as that in a linearized theory. Explicitly, assuming $|\vartheta| \ll \theta_o \approx \theta_c$ from the outset in Section 3.5.1, one can *approximate* the dissipation via

$$\mathscr{D}_o = -\frac{\mathbf{q}_o \cdot \mathbf{g}_o}{\theta} \approx -\frac{\mathbf{q}_o \cdot \mathbf{g}_o}{\theta_o}$$

as in the linearized theory of thermoelasticity. A straightforward application of the derivation steps leading to the identification (21) would then directly yield $\overline{\vartheta}_c = \langle \theta \rangle_{\partial \mathscr{C}_o^e} - \theta_{ABP}$ *independent of the sample size*. Therefore, $\overline{\vartheta}_c^{lim}$ can be interpreted as a consistency check that should be satisfied by a nonlinear theory in the limit of small deviations from the reference temperature of the medium.

Remark 1. For a random contact topography, the identification process for the RCE must be appended by a sample enlargement procedure in the lateral dimension together with designing appropriate BCs.

Remark 2. The average temperature jump $\langle \theta - \theta_c \rangle_{\overline{\partial \mathscr{C}_o^r}}$ on the rough surface was employed in the definition of the thermal contact conductance in [16] and [12]. Via numerical investigations, it can be verified that this quantity is of the same order of magnitude with $\overline{\vartheta}_c$, i.e. a significant temperature jump indeed takes place across the two surfaces on the original heterogeneous macroscale problem. However, the thermodynamically consistent measure of this fast variation in the temperature field within a homogenization approximation is the macroscopic temperature jump $\overline{\vartheta}_c$.

4 Numerical Investigations

4.1 Modeling Aspects

In this section, representative numerical investigations within the finite element method framework will be presented that highlight major aspects of the thermal contact homogenization technique developed. For this purpose, the bulk constitutive response will be limited to isotropic finite thermoelasticity where the Helmholtz free energy function takes the form

$$\widetilde{\Psi}(\mathbf{F},\theta) = \frac{\theta}{\theta_o}\widetilde{\Psi}_o(\mathbf{F}) - \frac{\vartheta}{\theta_o}\widetilde{e}_o(\mathbf{F}) + \int_{\theta_o}^{\theta}\left(1-\frac{\theta}{\theta'}\right)\widetilde{c}(\mathbf{F},\theta')\,\mathrm{d}\theta', \tag{26}$$

with $\vartheta := \theta - \theta_o$ as the deviation from a reference temperature θ_o, $\widetilde{[\cdot]}_o(\mathbf{F})$ denotes the functional $\widetilde{[\cdot]}(\mathbf{F},\theta_o)$, $\widetilde{e}_o(\mathbf{F}) = 3\,\kappa\,\alpha_o\theta_o \ln J$ is the assumption of the modified entropic theory of elasticity [3] (κ: bulk modulus, α_o: thermal expansion coefficient) and c is the specific heat capacity (assumed to be a constant). The isothermal response Ψ_o is chosen as that of an Ogden material based on a volumetric-deviatoric decoupling via

$$\Psi_o = \Psi_o^{vol} + \Psi_o^{dev} \tag{27}$$

where, using $\mathbf{F}^{dev} = J^{-1/3}\mathbf{F}$ and Λ_i ($\prod_{i=1}^{3}\Lambda_i = 1$) for the principal stretches of $\mathbf{F}^{dev}$,

$$\widetilde{\Psi}_o^{vol}(J) = \frac{\kappa}{4}(J^2 - 2\ln J - 1), \quad \widetilde{\Psi}_o^{dev}(\mathbf{F}^{dev}) = \sum_{p=1}^{M}\frac{\mu_p}{\beta_p}\left(\sum_{i=1}^{3}\Lambda_i^{\beta_p} - 3\right) \tag{28}$$

subject to the linearization consistency $\sum_{p=1}^{M}\mu_p\beta_p = 2\mu$ with μ as the shear modulus. The reference temperature θ_o will conveniently be taken to match the temperature of the medium in the undeformed configuration $\mathscr{R}_o$. For the constitutive formulation of the isotropic heat flux response, the classical Fourier's law of heat conduction $\mathbf{q} = -k\mathbf{g}$, equivalently $\mathbf{q}_o = -Jk\,\mathbf{C}^{-1}\mathbf{g}_o$, is employed where $k = \widetilde{k}(\mathbf{F},\theta,\|\mathbf{g}\|)$ is assumed to be a constant. Simulation parameters employed in the investigations are summarized in Table 1.

Within the numerical investigations, it will be assumed that the topography of the heterogeneous boundary is an artificial periodic rough surface. In particular, the rough surface will be modeled as sinusoidal with a RMS roughness ρ_o in its undeformed configuration. A unit-cell of periodicity with equal lateral dimensions L_o for such a surface is depicted in Figure 3. Additionally, for simplicity, the total surfacial deformation will be parameterized along a single chosen path within subsequent numerical simulations via a parameter $\mathscr{H}$:

$$\overline{F}^{c}_{\alpha\beta} = \delta_{\alpha\beta} + \mathscr{H}. \tag{29}$$

The test space parametrization can be expressed by the variable string

Table 1 Simulation parameters employed in the numerical investigations are summarized. Values in parenthesis will be varied for several examples.

Bulk modulus	[N/mm^2]	κ	4.0		
Shear modulus	[N/mm^2]	μ	1.0		
Ogden material parameters		μ_1	0.660	β_1	1.8
$(M = 3 \Rightarrow \sum_{p=1}^{M} \mu_p \beta_p = 2\mu)$		μ_2	−0.231	β_2	−2.0
		μ_3	0.050	β_3	7.0
Volumetric thermal expansion coefficient	[K^{-1}]	α_o	0.001		
Thermal conductivity	[N/sK]	k	1.0		
Reference temperature	[K]	θ_o	293.15		
Contact surface temperature	[K]	θ_c	(293.15)		

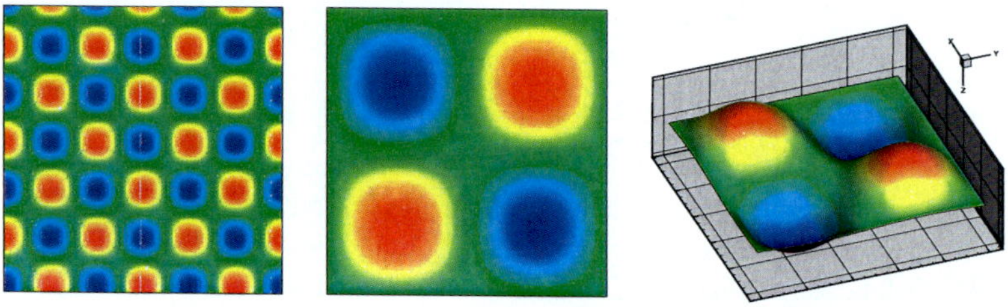

Fig. 3 Left: Height encoded representation of a sinusoidal surface. Middle: A unit-cell of the sinusoidal surface. Right: Three-dimensional view of the sinusoidal surface unit-cell.

$$\underline{\Lambda} = \{L_o^{-1} H_o, L_o^{-1} \rho_o, \theta_c, \mathscr{H}, \bar{p}, L_o \bar{h}\}. \qquad (30)$$

4.2 Dependence on Primary Macroscopic Variables

The primary macroscopic variables of the present homogenization framework are $\{\theta_c, \overline{\mathbf{FD}}_T, \bar{p}, \bar{h}\}$. Figure 4 summarizes the effect of varying the contact heat flux on the macroscopic response at a fixed macroscopic surfacial deformation and contact pressure. It is observed that the macroscopic contact conductance is nonlinearly dependent on the contact heat flux and is strongly influenced by the contact surface temperature. The unsymmetric behavior of the response with respect to a reversal of the flux direction is called *rectification*. This behavior has been observed in experiments as well, particularly for contact between dissimilar materials at large real contact areas, and can be argued analytically for simplified contact geometries (see [9, chapter 7.3]). Rectification is attributed to the distortion of the material near the contact interface due to thermal effects. In the present case, the material expands significantly when $\bar{h} < 0$ and contracts otherwise with respect to $\bar{h} = 0$ which results in different temperature gradients near the contact interface.

The variation of the macroscopic contact pressure at a fixed surfacial deformation and contact heat flux leads to stronger variations of the macroscopic contact

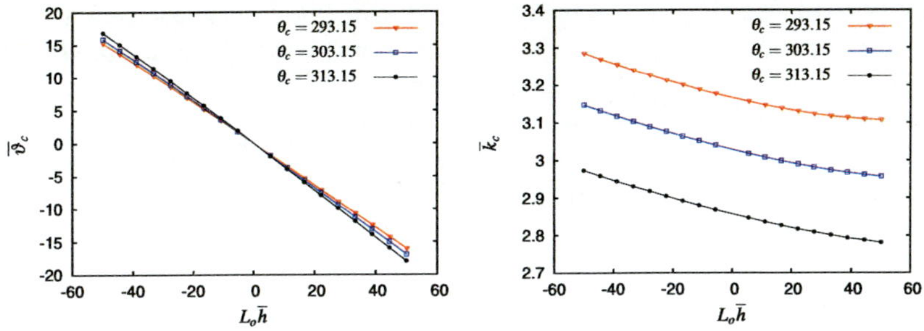

Fig. 4 The macroscopic response is monitored for variations of the macroscopic contact heat flux and contact surface temperature under the input $\underline{\Lambda} = \{1, 0.05, 0, 293.15, 0, 0.02\,\kappa, L_o\overline{h}\}$.

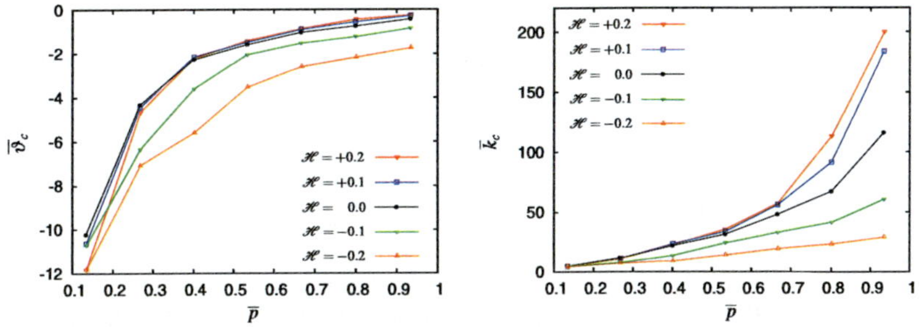

Fig. 5 The macroscopic response is monitored for variations of the macroscopic contact pressure and the surfacial deformation under the input $\underline{\Lambda} = \{1, 0.05, 0, 293.15, \mathscr{H}, \overline{p}, 50k\}$.

conductance as summarized in Figure 5. Larger pressures cause a higher degree of flattening of the asperities and hence lead to large real contact areas so that $\overline{k}_c$ appears higher. On the other hand, surfacial compressive deformation ($\mathscr{H} < 0$) effectively increases the asperity heights in the intermediate configuration, while tensile deformation ($\mathscr{H} > 0$) has the adverse effect. Consequently, negative stretches of the surface leads to a lower conductivity while positive stretches increase the conductivity. The present finite thermoelasticity example predicts an increase in $\overline{k}_c$ without bound as the pressure is increased, in the limit capturing the theoretical response of perfectly conforming surfaces. This observation is consistent with the enforcement of temperature continuity across microscale contact interfaces which were assumed perfectly conforming in Section 2.

5 Conclusion

In order to predict the macroscopic thermal contact conductance of rough contact interface topographies, a computational contact homogenization technique was developed within a finite deformation framework. Numerical investigations conducted revealed a strong dependence of the computed thermal contact conductance on macroscopic control variables. Although an experimental validation of the predictions is not sought at this stage of the developments, the implications of the developed framework are found to be consistent with practical experience. However, the particular form of the macroscopic contact conductance response is at most qualitatively meaningful within this framework where, in particular, a detailed consideration of possible inelasticity and anisotropy effects was not pursued. The consideration of highly localized large deformation plasticity at the asperities via appropriate models for the physically relevant length scales is an important next step towards a quantitative comparison of the computational results with laboratory experiments on metallic and non-metallic thermal interface materials. From a theoretical point of view, a complementary step is the ability to accurately represent actual contact interface topographies via artificially generated or computer scanned images. As remarked throughout the developments, modifications to the proposed computational contact homogenization framework are required to render the overall scheme appropriate for such applications and these are currently being pursued by the authors.

References

1. Ayers, G.H.: Cylindrical thermal contact conductance. Master's Thesis, Texas A&M University, College Station, Texas, USA (2003)
2. Bahrami, M., Yovanovich, M.M., Marotta, E.E.: Thermal joint resistance of polymer-metal rough interfaces. Journal of Electronic Packaging 128, 23–29 (2006)
3. Chadwick, P., Creasy, C.F.M.: Modified entropic elasticity of rubberlike materials. Journal of the Mechanics and Physics of Solids 32(5), 337–357 (1984)
4. Chung, D.D.L.: Factors that goven the performance of thermal interface materials. Journal of Electronic Materials 38(1), 175–192 (2009)
5. Gibbins, J.: Thermal contact resistance of polymer interfaces. PhD thesis, University of Waterloo, Waterloo, Ontario, Canada (2006)
6. Holzapfel, G.A.: Nonlinear Solid Mechanics: A Continuum Approach for Engineering. Wiley, Chichester (2001)
7. Jackson, R.L., Bhavnani, S.H., Ferguson, T.P.: A multiscale model of thermal contact resistance between rough surfaces. Journal of Heat Transfer 130, 81301 (2008)
8. Johnson, K.L.: Contact Mechanics. Cambridge University Press, Cambridge (1987)
9. Madhusudana, C.V.: Thermal Contact Conductance. Springer, New York (1996)
10. Prasher, R.: Thermal interface materials: Historical perspective, status and future directions. Proceedings of the IEEE 94(8), 1571–1586 (2006)
11. Prasher, R.S., Matayabas, J.C.: Thermal contact resistance of cured gel polymeric thermal interface material. IEEE Transactions on Components and Packaging Technologies 27(4), 702–709 (2004)
12. Sadowski, P., Stupkiewicz, S.: A model of thermal contact conductance at high real contact area fractions. Wear (2009), doi:10.1016/j.wear.2009.06.040

13. Salti, B., Laraqi, N.: 3-D numerical modeling of heat transfer between two sliding bodies: temperature and thermal contact resistance. International Journal of Heat and Mass Transfer 42, 2363–2374 (1999)
14. Song, S., Yovanovich, M.M., Nho, K.: Thermal gap conductance of conforming surfaces in contact. Journal of Heat Transfer 115, 533–540 (1993)
15. Temizer, I., Wriggers, P.: A multiscale contact homogenization technique for the modeling of third bodies in the contact interface. Computer Methods in Applied Mechanics and Engineering 198, 377–396 (2008)
16. Thompson, M.K.: A multi-scale iterative approach for finite element modeling of thermal contact resistance. PhD Thesis, Massachusetts Institute of Technology, Boston, Massachusetts, USA (2007)
17. Wriggers, P.: Computational Contact Mechanics, 2nd edn. Springer, Berlin (2006)
18. Wriggers, P., Reinelt, J.: Multi-scale approach for frictional contact of elastomers on rough rigid surfaces. Computer Methods in Applied Mechanics and Engineering 198, 1996–2008 (2009)
19. Zavarise, G., Borri-Brunetto, M., Paggi, M.: On the reliability of microscopical contact models. Wear 257, 229–245 (2004)
20. Zavarise, G., Wriggers, P., Stein, E., Schrefler, B.A.: Real contact mechanisms and finite element formulation – A coupled thermomechanical approach. International Journal for Numerical Methods in Engineering 35, 767–785 (1992)
21. Zhang, X., Chong, P., Fujiwara, S., Fujii, M.: A new method for numerical simulation of thermal contact resistance in cylindrical coordinates. International Journal of Heat and Mass Transfer 47, 1091–1098 (2004)
22. Temizer, İ., Wriggers, P.: Thermal contact conductance characterization via computational contact homogenization: A finite deformation theory framework. International Journal for Numerical Methods in Engineering (1), 24–58, doi:10.1002/nme.2822
23. Temizer, I.: Thermomechanical contact homogenization with random rough surfaces and microscopic contact resistance. Tribology International 44, 114–124, doi:10.1016/j.triboint.2010.09.011

Analysis of Granular Chute Flow Based on a Particle Model Including Uncertainties

F. Fleissner, T. Haag, M. Hanss and P. Eberhard

Abstract In alpine regions human settlements and infrastructure are at risk to be hit by landslides or other types of geological flows. This paper presents a new approach that can aid the design of protective constructions. An uncertainty analysis of the flow around a debris barrier is carried out using a chute flow laboratory model of the actual debris flow. A series of discrete element simulations thereby serves to assess barrier designs. In this study, the transformation method of fuzzy arithmetic is used to investigate the influence of epistemically uncertain model parameters. It turns out that parameter and modeling uncertainties can have a tremendous influence on the predicted efficiency of protective structures.

1 Introduction

Landslides are prevalent geological phenomena in mountain regions. The term landslide comprises different phenomena such as debris flows, earth flows and debris avalanches. Geological flows are commonly made up of more or less unconsolidated polydisperse rock material [6]. Often, the saturation of rock material with water is the determining factor to trigger such flow events. However, water does not necessarily need to be involved, as e.g. sturzstroms demonstrate [16]. Geological flow events are difficult to foresee. If at all, they can be predicted with a higher precision in space than in time. Along with this quasi unpredictability, the fact that landslides can be huge dangerous events, makes them a terrible threat for human beings.

F. Fleissner · P. Eberhard
Institute of Engineering and Computational Mechanics, University of Stuttgart, 70569 Stuttgart, Germany; e-mail: {fleissner,eberhard}@itm.uni-stuttgart.de

T. Haag · M. Hanss
Institute of Applied and Experimental Mechanics, University of Stuttgart, 70569 Stuttgart, Germany; e-mail: {haag,hanss}@iam.uni-stuttgart.de

G. Zavarise & P. Wriggers (Eds.): Trends in Computational Contact Mechanics, LNACM 58, pp. 121–134.
springerlink.com

Geotechnical engineers globally cooperate with engineering geologists to design dams and barriers that are intended to protect settlements, roads and other vital infrastructure from the effects of geological flow phenomena. Nowadays, this engineering work is more and more aided by computer simulation. Simulation approaches for discontinuous rock material, e.g. those applied in [1] or [3] are based on empirically derived laws that describe the dynamics of sliding material. Typically, a number of simplifying assumptions must be made to gain a simulation model with acceptable complexity. Common simplifications, such as the assumption of the rock material being homogeneous, along with the difficult choice of model parameters may impose a high degree of modeling uncertainty.

A common way to deal with uncertainties is to introduce safety factors for all crucial design variables. The computationally required cross section of a dam, e.g., is scaled with a factor greater than one to increase the likelihood that the dam does not break in case of an impact that is more powerful than anticipated in the modeling assumptions. However, this kind of consideration is only feasible if the result of parameter changes is obvious. A survey on how to estimate safety factors for structures that are exposed to debris flows is given in [19], based on a very coarse and verbal description of the system.

But is the respective design effective in any case, e.g. for all types of rock material? This question casts into doubt the assumption that a construction is generally safe if its design is based on just a certain set of parameters and assumptions. The objective of the design process might thus be reformulated: The best design of a protective structure should not only be safe for one set of model parameters, the nominal set, but it must also be safe for all values of the input parameters that seem to be possible. Moreover, the preferable design should not be sensitive to variations of all uncertain input parameters of the design process. In order to reach these properties within the computational process, a methodology is needed, that is capable of handling uncertain model parameters. In the following, different principles in modeling uncertainty are given and the one which is used in this work, namely the transformation method of fuzzy arithmetics, is explained. Then the method is applied to a chute flow described by using the discrete element method and the results are discussed.

2 Classification, Representation and Propagation of Uncertainty

2.1 Uncertainty Classification and Representation

In general, non-determinism in numerical models may arise from different sources, motivating some categorization of uncertainties. Although other classifications are possible, the following categorization of uncertainties [15] proves to be well-suited in this context: aleatory uncertainties that can be measured, such as variability or scatter caused by irregularities in fabrication, on the one side, and on the other side, epistemic uncertainties, which arise from an absence of information, rare data, vagueness in parameter definition, subjectivity in numerical implementation, or simplification and idealization processes employed in the modeling procedure. All these conditions are condensed to uncertain model parameters of a structurally invariant model. The structure of the model itself is potentially also uncertain but for practical reasons it is assumed that this structural uncertainty is captured by the uncertain model parameters. The uncertainties in the modeling procedure entail that the results that are obtained from simulations that only use one specific set of values as the most likely ones for the model parameters cannot be considered as representative of the whole spectrum of possible model configurations. Furthermore, this fake exactness offered by the numerical simulation of models with actually uncertain, but crisply quantified parameters can make the comparison between numerical simulations and experimental testing questionable. Such a comparison may be rated as unsatisfactory if the simulation results obtained with crisp, i.e., discrete and non-fuzzy values do not well match the experimental ones. It might be absolutely satisfactory, though, if the uncertainties inherent to the models would have been appropriately taken into account in the simulation procedure.

While aleatory uncertainties have successfully been taken into account by the use of probability theory [18, 20] and, in practice, by Monte Carlo simulation, the additional modeling of epistemic uncertainties still remains a challenging topic. As a practical approach to address this issue, a special interdisciplinary methodology to comprehensive modeling and analysis of systems is presented which allows for the inclusion of uncertainties - in particular of those of epistemic type - from the very beginning of the modeling procedure. This approach is based on fuzzy arithmetic, a special field of fuzzy set theory, which will be described in the following.

A special application of the theory of fuzzy sets, which is rather different from the well-established use of fuzzy set theory in fuzzy control, is the numerical implementation of uncertain model parameters as fuzzy numbers [17]. Fuzzy numbers are defined as convex fuzzy sets over the universal set $\mathbb{R}$ with their membership functions $\mu(x) \in [0,1]$, where $\mu(x) = 1$ is true only for one single value $x = \bar{x} \in \mathbb{R}$, the so-called center value or nominal value. For example, a fuzzy number $\widetilde{p}$ of triangular (linear) shape, expressed by the abbreviated notation [13]

$$\widetilde{p} = \mathrm{tfn}(\bar{x}, w_\mathrm{l}, w_\mathrm{r}) \,, \tag{1}$$

is defined by the membership function

$$\mu_{\widetilde{p}}(x) = \min\left\{ \max\left[0, 1-(\bar{x}-x)/w_\mathrm{l}\right], \max\left[0, 1-(x-\bar{x})/w_\mathrm{r}\right]\right\} \quad \forall x \in \mathbb{R} \,. \tag{2}$$

A triangular fuzzy number is shown in Figure 1. However, any other shape of membership function may be selected if appropriate to quantify the uncertainty of a specific model parameter. The calculation with fuzzy numbers is referred to as fuzzy arithmetic and proves to be a non-trivial problem, especially with regard to the evaluation of large mathematical models with fuzzy-valued operands.

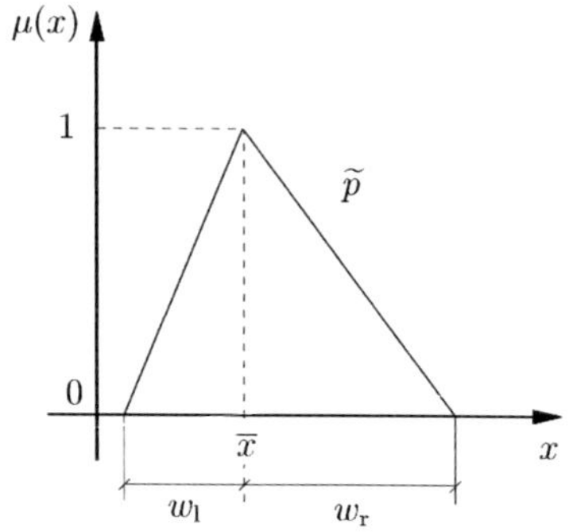

Fig. 1 Triangular fuzzy number $\widetilde{p}$.

2.2 Uncertainty Propagation Based on the Transformation Method

As a successful practical implementation of fuzzy arithmetic, which allows the evaluation of arbitrary systems with uncertain, fuzzy-valued model parameters, the transformation method [11] is used. Alternative methods to numerically handle uncertainties are, for example, presented in [14]. The transformation method is available in a general, a reduced and an extended form, with the most appropriate form to be selected depending on the type of model to be evaluated [11–13]. It can be applied to any existing software environment as it uses crisp-valued parameter samples and is thus explicitly well-suited for the Finite Element Method or, as in this paper, for the Discrete Element Method. The parameter samples are generated on the basis of a factorial-design sampling strategy. In comparison to a conventional sensitivity analysis, the results of the transformation method are based on a cuboid in the parameter space, rather than on a single point and the derivatives in this point. Thus the results can be considered more comprehensive and more reliable.

Assuming the uncertain system to be characterized by n fuzzy-valued model parameters $\widetilde{p}_i$, $i = 1, 2, \ldots, n$, the major steps of the method can briefly be described as follows.

In the first step, each fuzzy number $\widetilde{p}_i$ is discretized into a number of nested intervals $X_i^{(j)} = [a_i^{(j)}, b_i^{(j)}]$, assigned to the membership levels $\mu_j = j/m$, $j = 0, 1, \ldots, m$, that result from subdividing the possible range of membership equally spaced by $\Delta\mu = 1/m$, see Figure 2. In a second step, the input intervals $X_i^{(j)}$, $i = 1, 2, \ldots, n$, $j = 0, 1, \ldots, m$, are transformed to arrays $\widehat{X}_i^{(j)}$ that are computed from the upper and lower interval bounds after the application of a well-defined combinatorial scheme [11, 13]. Each of these arrays represents a specific sample of possible parameter combinations and serves as an input parameter set to the problem to be evaluated. As a result of the evaluation of the model for the input arrays $\widehat{X}_i^{(j)}$, output arrays $\widehat{Z}^{(j)}$ are obtained which are then retransformed to the output intervals $Z^{(j)} = [a^{(j)}, b^{(j)}]$

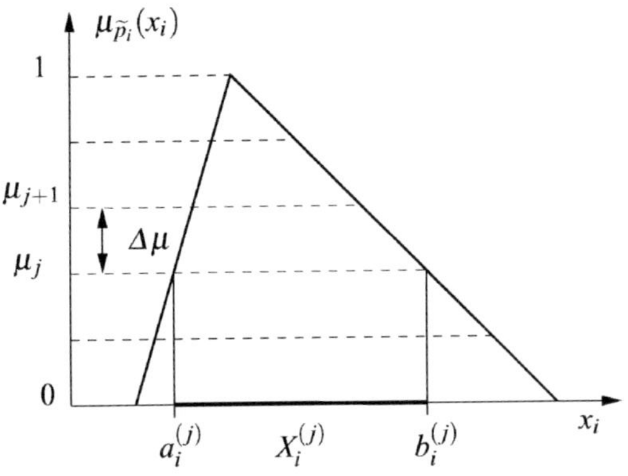

Fig. 2 Decomposition of a fuzzy number $\tilde{p}_i$ into intervals $X_i^{(j)}$, $j = 0, 1, \ldots, m$.

for each membership level μ_j and finally recomposed to the fuzzy-valued output $\tilde{q}$ of the system.

In addition to the simulation part of the transformation method described above, the analysis part of the method can be used to quantify the influence of each fuzzy-valued input parameter $\tilde{p}_i$ on the overall fuzziness of the model output $\tilde{q}$. For these purposes, the standardized mean gain factors κ_i and φ_i, and the normalized degrees of influence ρ_i and ω_i have been introduced [10, 11, 13], quantifying in an absolute and in a relative character, respectively, the effect of the uncertainty of the ith model parameter $\tilde{p}_i$ on the overall uncertainty of the model output $\tilde{q}$. In [11, 13], a standardization with respect to the nominal values is incorporated into the computation of the standardized mean gain factors κ_i and of the normalized degrees of influence ρ_i, whereas the approach proposed in [10] considers the influences of the overall input uncertainty on the overall output uncertainty.

Among other advantages of the transformation method, its characteristic property of reducing fuzzy arithmetic to multiple crisp-number operations entails that the transformation method can be used with an existing software environment for system simulation even if its internals are not known or accessible [13]. Expensive rewriting of the program code is not required. Instead, the steps of decomposition and transformation as well as of retransformation and recomposition can be coupled to an existing software environment by a separated pre- and postprocessing tool.

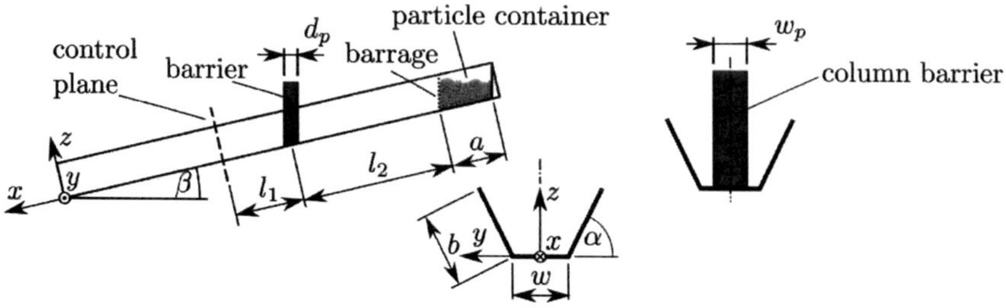

Fig. 3 Setup of the chute flow laboratory model.

Table 1 Geometric parameters of the chute flow laboratory model.

Parameter		Value	Parameter		Value
chute wall height	b [m]	0.15	distance barrage-barrier	l_2 [m]	0.5
chute base width	w [m]	0.05	barrier width	w_p [m]	0.03
chute wall slope	α [–]	60°	barrier length	d_p [m]	0.03
chute slope	β [–]	20°	number of spheres	n [–]	5000
distance barrier-control plane	l_1 [m]	0.17	sphere diameter	d [mm]	6
container length	a [m]	0.153			

3 Chute Flow Studies Using Particle Methods

3.1 Laboratory Model

To demonstrate the potentially high sensitivity of numerical debris flow models, a virtual laboratory representation is generated, that serves as basis for a numerical uncertainty analysis. As debris flows tend to follow natural or artificial grooves, the flow-bed of the slide is simplified towards a prismatic chute, see Figure 3. To allow for the simulations to be reproducible, the rock material is replaced by glass beads, thus considering only dry debris. Initially, the glass beads are stored in a reservoir at the upper end of the chute. At the beginning of the test the downstream facing barrage of the particle container is instantaneously removed. A resulting particle wave then starts to slide downstream. After a short period of free flow, the wave hits a barrier, see Figure 3. The barrier consists of a single column and represents a geotechnical engineering construction that is designed to reduce the impulse of the debris flow and to smooth its peak. Such constructions are typically placed above human settlements or infrastructure. To assess the efficiency of the barrier, the impulse of the flow wave is measured at a control plane behind the barrier. The geometric parameters of the chute flow laboratory setup are summarized in Table 1.

3.2 Numerical Simulations

Using a computational model of the chute-flow laboratory model, one is able to perform a sensitivity analysis. To describe the motion of the particles, the Discrete Element Method (DEM) [4] is chosen. This approach models the glass spheres as very stiff bodies with unconstrained dynamics. The spheres exchange impulse through surface contacts. Following the DEM approach, the particle dynamics is time-integrated on force-acceleration level [8]. This requires a suitable model for the particle contact forces. In a chute flow, particle contact forces are relatively small as they only result from inertial forces due to the absence of external compression. Therefore, a linear elastic contact model is well suited to compute the elastic contact force $\mathbf{f}_e^{ij}$ acting on a sphere i at the position $\mathbf{r}^i$ in case of a contact with another sphere j at position $\mathbf{r}^j$. This force depends on the depth of spheres' common surface interpenetration δ^{ij} and the sphere radii R^i and R^j as

$$\triangle\mathbf{r}^{ij} = \mathbf{r}^i - \mathbf{r}^j, \tag{3}$$

$$\mathbf{n}^{ij} = \frac{\triangle\mathbf{r}^{ij}}{||\triangle\mathbf{r}^{ij}||}, \tag{4}$$

$$\delta^{ij} = R^i + R^j - \mathbf{n}^{ij} \cdot \triangle\mathbf{r}^{ij}, \tag{5}$$

$$\mathbf{f}_e^{ij} = k_n \delta^{ij} \mathbf{n}^{ij}. \tag{6}$$

The parameter k_n is the stiffness of the normal contact. The coefficient of restitution for contacts between glass spheres is usually quite close to one. The resulting dissipation is modeled through a linear damper that is introduced in parallel to the linear elastic contact force element. The resulting dissipative force acting on sphere i is computed from the velocities of the spheres $\mathbf{v}^i$ and $\mathbf{v}^j$ as

$$\mathbf{f}_d^{ij} = d_n \mathbf{n}^{ij} \cdot (\mathbf{v}^j - \mathbf{v}^i)\mathbf{n}^{ij}. \tag{7}$$

Though this approach is widely used, the damping parameter d_n is usually determined by matching experiments and simulations. For alpine debris flows, such experiments are hardly possible. Contact damping is thus a highly epistemically uncertain parameter which, for this study, was estimated by educated guess.

Contacts between dry glass spheres cannot be modeled adequately without considering slipping and sticking friction forces $\mathbf{f}_f^{ij}$. Thereby, sticking friction requires a special treatment as it is not an applied force. The tangential elasticity of the contact region that results from surface roughness is modeled via a linear elastic tangential element [2, 5]. However, due to a lack of knowledge about the tribologic conditions in a real debris flow, the stiffness of the respective tangential element is another epistemically uncertain parameter.

The resulting overall force on a sphere i is computed as an accumulation of the applied forces that result from contacts with other spheres as

Table 2 Parameters of the sphere contact model.

Parameter		Value
material density (nominal)	ρ_{mass} [kg/m^3]	2500
slipping friction coefficient	μ [-]	0.6
sticking friction coefficient	μ_0 [-]	0.8
normal stiffness	k_n [10^5 N/m]	7.85
normal damping (nominal)	d_n [10^{-1} Ns/m]	1.4
tangential stiffness (nominal)	k_t [10^4 N/m]	1

$$\mathbf{f}^i = \sum_j \mathbf{f}^{ij} = \sum_j \mathbf{f}_e^{ij} + \mathbf{f}_d^{ij} + \mathbf{f}_f^{ij}. \tag{8}$$

All contact forces $\mathbf{f}^{ij}$ on sphere i are thereby also considered as counter forces acting on spheres j. To resolve contacts between the glass spheres and the boundary geometry of chute and barrier, the entire boundary geometry is considered as rigid, following the approach introduced in [9].

Another parameter that is regarded as epistemically uncertain is the material density of the debris. Only the density of the glass beads, used for this study, is measurable. Therefore, it is chosen as the nominal value for the stiffness uncertainty analysis. All important material and contact parameters of the simulation model are listed in Table 2.

Figures 4 and 5 depict the particle motion in the chute. The displayed snapshots exhibit how the barrier causes a stagnation of the overall particle motion above the barrier. At the barrier the spheres are deflected from their initial trajectory, which causes a loss of kinetic energy and impulse of the particle ensemble. The particle wave is thus effectively smoothed by the barrier. This becomes even more evident if the particle ensemble motion is compared to the motion of the same ensemble in case the barrier is omitted, see Figures 4(a) and 4(b).

3.3 Uncertainty and Sensitivity Analysis

The sensitivity of a parametric system is characterized by the magnitude of the variation of specific output parameters with respect to variations of specific input parameters. Three epistemically uncertain input parameters are investigated in a sensitivity analysis. The chosen parameters are the material density ρ_{mass}, the tangential stiffness k_t of the sticking friction model and the normal contact damping parameter d_n. The density is chosen as an example for a parameter that is uncertain as it is very difficult to estimate for a large amount of real debris. The latter two parameters are examples for the type of epistemically uncertain parameters that are part of a simplified physical model. This type of parameters is usually determined by curve fitting of experimental data, a method that is inherently subjective and thus

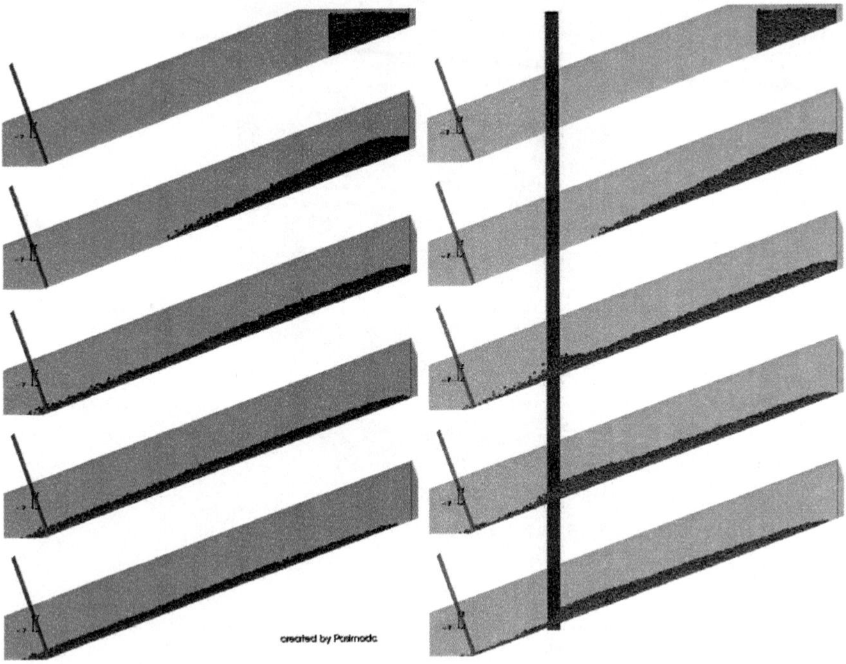

(a) Free flow (b) Column barrier

Fig. 4 Snapshots from chute flow particle simulations.

Table 3 Input parameters for the uncertainty analysis.

Parameter $\widetilde{p}_i = \mathrm{tfn}(\bar{x}_i, w_{l,i}, w_{r,i})$		$\bar{x}_i$	$w_{l,i}$	$w_{r,i}$
material density (nominal)	ρ_{mass} [kg/m^3]	2500	250	250
normal damping (nominal)	d_n [10^{-1} Ns/m]	1.4	0.7	0.7
tangential stiffness (nominal)	k_t [10^4 N/m]	1	0.5	0.5

uncertain. Even a curve fitting requires a number of experiments and simulations. As the resulting overhead is often unacceptable, parameters of the respective type are determined by educated guess. In Table 3, the parameters of the triangular fuzzy input parameters $\widetilde{p}_i$ that are used for the uncertain simulations are given.

A uniform grid is used to generate sets of parameters from the three dimensional parameter space of ρ_{mass}, k_t and d_n. This sampling process results in an array $\widehat{X}_i^{(j)}$ with 189 sets of input parameters. All simulations are carried out using the particle simulation program Pasimodo [7], which is developed at the Institute of Engineering and Computational Mechanics of the University of Stuttgart. One simulation run of the chute flow using Pasimodo on a 3.2 GHz Pentium IV takes approximately three hours. Thus the overall computation time is about 24 days. As the computations for

Fig. 5 Debris deflection at the column barrier

different parameter sets are independent, the overall simulation process is accelerated by distributing the simulations to several processors on a computer cluster. The results of all simulations are gathered to serve as input for a post processing that reassembles the actual uncertainty analysis.

4 Results

From a practical point of view, it is interesting to consider the impulse that affects a structure that is hit by a debris flow. In this work, the debris flow is approximated by the previously described chute flow and the affected structure is represented by a virtual control plane that is located below the debris flow barrier and oriented perpendicular to the flow. The quantity that is used to assess the potential damage of the structure is the accumulated impulse that acts on the control plane. This quantity is used in favor of the maximum impact force or the maximum impulse which are both highly scattered quantities as they result from individual particle impacts. Figure 6 shows the accumulated impulse on the control plane for two nominal simulations, one without any barrier and the other one with a column barrier. The accumulated impulse is reduced by about 90% through the barrier which shows its effectiveness.

In Figure 7, the accumulated impulse is shown for a debris flow barrier. The dashed line depicts the result of a nominal simulation, i.e., the result that emerges if no uncertainty is considered. The other lines reflect the ranges of possible outputs for different membership levels or, equivalently, for different degrees of possibility. It is evident, in fact, that it makes sense to consider uncertainty, as the worst-case

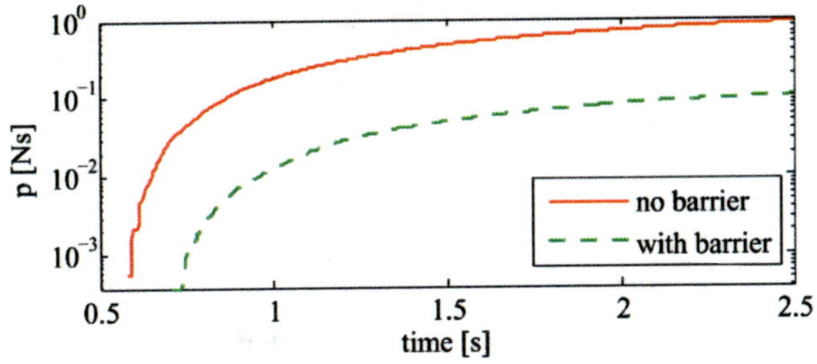

Fig. 6 Accumulated impulse on the control plane for the nominal input parameter set.

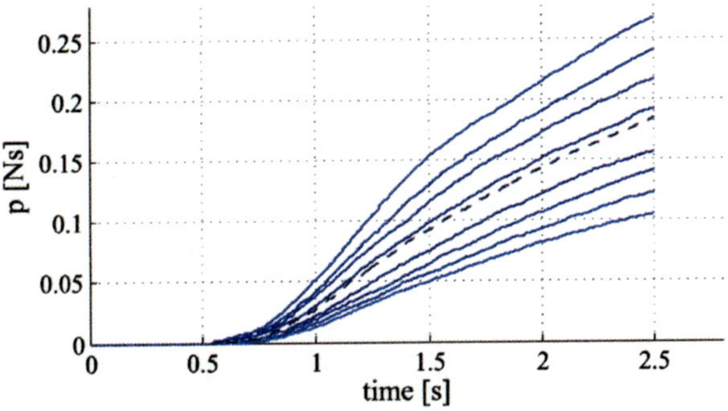

Fig. 7 Uncertain accumulated impulse over time for the column barrier.

output deviates about ±50% from the nominal configuration. For assessing the risk of an exposed structure, this variance definitely has to be taken into account.

In order to reduce the output uncertainty and to achieve a better understanding of the system under consideration, it is important to know which uncertain input parameter causes which amount of output uncertainty. This question is answered by the so-called measures of influence which are a by-product of the transformation method. Figure 8 shows the values of ρ_i for the accumulated impulse over time. The influence measure ρ_i reflects the relative effect of an unitary percentaged variation of the input $\widetilde{p}_i$ on the output. Recalling the definition of the impulse as the product of velocity and mass, it appears plausible that the uncertainty of the impulse is predominantly governed by the uncertainty of the material density, but it is also clear that the parameters of the contact model do have a non-negligible effect on the dynamic behavior of the particle set. Figure 8 quantifies these competing influences and reveals that the uncertainty of the density makes up about 80% of the output uncertainty while the two parameters of the contact model make up about 10% each. The short-term fluctuations that can be observed for points in time which are smaller than about 0.8 s are attributed to the normalization with very small values and cannot be considered as reliable.

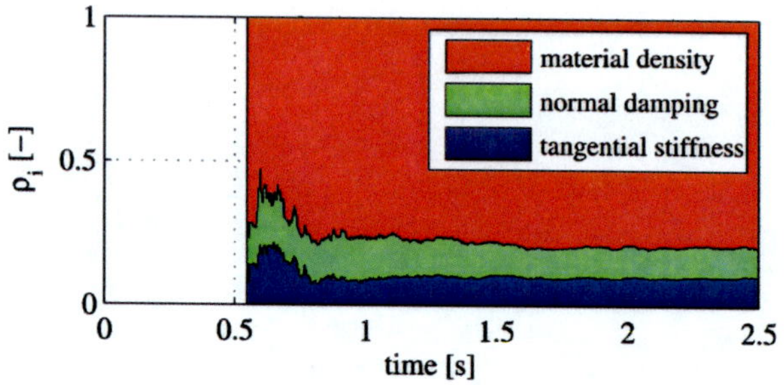

Fig. 8 Normalized relative measure of influence ρ_i for the accumulated impulse on the control plane behind the column barrier.

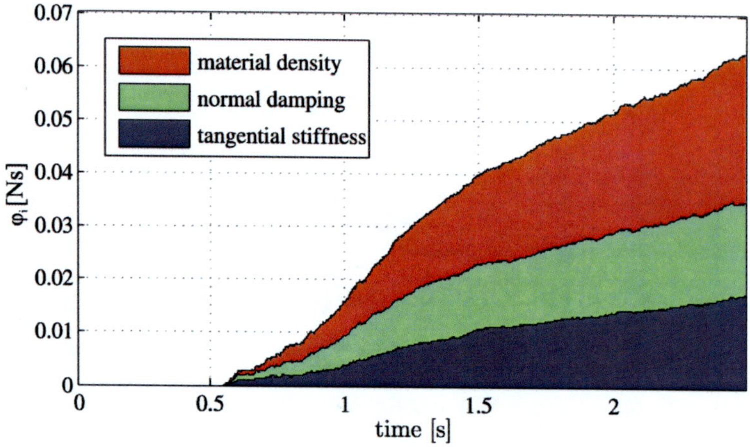

Fig. 9 Absolute measure of influence φ_i for the accumulated impulse.

The use of a unitary percentaged variation of the input parameters is indicated when a rather theoretical analysis of a system is performed. In the previous paragraph, this is done for the accumulated impulse, and the acquired influence measures are compared to an estimation based on the underlying mathematical equations.

In engineering applications, however, some parameters allow for a much larger relative worst-case deviation than others. As a consequence, it is indicated to consider the direct contribution φ_i of the input uncertainty to the uncertainty of the output on an absolute scale. In Figure 9, these absolute measures of influence φ_i are shown for the accumulated impulse. As the estimation of the uncertainties of the inputs is done to the authors' best knowledge, it becomes clear that the influence of the uncertainty of the material density loses its dominant role and in reality only makes up about a third of the uncertainty of the output and thus is not more important than the two uncertain contact parameters tangential stiffness and normal damping.

Based on the computation of worst-case bounds, the uncertainty analysis based on the transformation method offers the possibility to assess the reliability of the

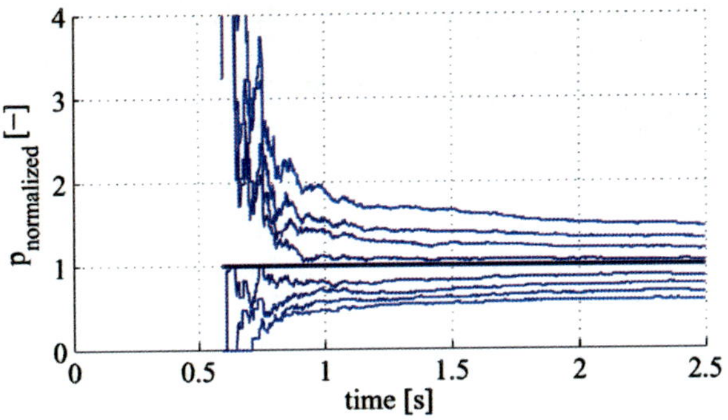

Fig. 10 Relative deviation of the accumulated impulse for a column barrier.

computational results. For this reason, Figure 7 is normalized with respect to their nominal value and shown in Figure 10. It can be seen that the uncertainty amounts to roughly 50% in the stationary phase. Wheather this amount of uncertainty can be tolerated or not depends on the decision that has to be based on the simulation results.

5 Conclusions

For the particle-flow problem considered in this work, it is strongly recommended to consider uncertainties in the simulation process, as precise values for the contact parameters as well as for the mass properties of the system are unknown for the most part. Moreover, the uncertainties inherent to the particle-flow model have an enormous effect on the output of interest which emphasizes their importance. With respect to the issue of risk analysis and safety assessment, the inclusion of uncertainties in analysis and design of protective structures using a fuzzy arithmetical approach represents a comprehensive and well-defined strategy. Moreover, advanced and better-founded safety factors can be derived which are based on a numerical analysis of the problem, rather than on educated guesses only.

References

1. Bourrier, F., Dorren, L., Nicot, F., Berger, F., Darve, F.: Toward objective rockfall trajectory simulation using a stochastic impact model. Geomorphology 110, 68–79 (2009)
2. Brendel, L., Dippel, S.: Lasting contacts in molecular dynamics simulations. In: Herrmann, H.J., Hovi, J.-P., Luding, S. (eds.) Physics of Dry Granular Materials. NATO ASI Series E, pp. 313–318. Kluwer Academic Publishers, Dordrecht (1998)

3. Chao-Lung, T., Jyr-Ching, H., Ming-Lang, L., Lacques, A., Chia-Yu, L., Yu-Chang, C., Hao-Tsu, C.: The Tsaoling landslide triggered by the Chi-Chi earthquake, Taiwan: Insights from a discrete element simulation. Engineering Geology 106, 1–19 (2009)
4. Cundall, P.A.: A computer model for simulating progressive, large-scale movements in blocky rock systems. In: Proceedings of the Symposium of the International Society of Rock Mechanics, Nancy (1971)
5. Cundall, P.A., Strack, O.D.L.: A discrete numerical model for granular assemblies. Géotechnique 29, 47–56 (1979)
6. Easterbrook, D.J.: Surface Processes and Landforms, 2nd edn. Prentice-Hall, Upper Saddle River (1999)
7. Fleissner, F.: Parallel Object Oriented Simulation with Lagrangian Particle Methods. In: Schriften aus dem Institut für Technische und Numerische Mechanik der Universität Stuttgart, Shaker Verlag, Aachen (2010)
8. Fleissner, F., Eberhard, P.: Examples for modeling, simulation and visualization with the discrete element method in mechanical engineering. In: Talaba, D., Amditis, A. (eds.) Product Engineering: Tools and Methods Based on Virtual Reality, pp. 419–426. Springer, Heidelberg (2008)
9. Fleissner, F., Gaugele, T., Eberhard, P.: Applications of the discrete element method in mechanical engineering. Multibody System Dynamics 18(1), 81–94 (2007)
10. Gauger, U., Turrin, S., Hanss, M., Gaul, L.: A new uncertainty analysis for the transformation method. Fuzzy Sets and Systems 159, 1273–1291 (2007)
11. Hanss, M.: The transformation method for the simulation and analysis of systems with uncertain parameters. Fuzzy Sets and Systems 130(3), 277–289 (2002)
12. Hanss, M.: The extended transformation method for the simulation and analysis of fuzzy-parameterized models. International Journal of Uncertainty, Fuzziness and Knowledge-Based Systems 11(6), 711–727 (2003)
13. Hanss, M.: Applied Fuzzy Arithmetic – An Introduction with Engineering Applications. Springer, Berlin (2005)
14. Hanss, M., Herrmann, J., Haag, T.: Vibration analysis of fluid-filled piping systems with epistemic uncertainties. In: Proceedings of the IUTAM Symposium on the Vibration Analysis of Structures with Uncertainties. St. Petersburg, Russia (2009)
15. Hofer, E.: When to separate uncertainty and when not to separate. Reliability Engineering and System Safety 54(2-3), 113–118 (1996)
16. Hsü, K.J.: Catastrophic debris streams (sturzstroms) generated by rockfalls. Geological Society of America Bulletin 86, 129–140 (1975)
17. Kaufmann, A., Gupta, M.M.: Introduction to Fuzzy Arithmetic. Van Nostrand Reinhold, New York (1991)
18. Loeven, G.J.A., Bijl, H.: Probabilistic collocation used in a two-step approach for efficient uncertainty quantification in computational fluid dynamics. CMES – Computer Modeling in Engineering & Sciences 36(3), 193–212 (2008)
19. Proske, D., Kaitna, R., Suda, J., Hübl, J.: Abschätzung einer Anprallkraft für murenexponierte Massivbauwerke. Bautechnik 85(12), 803–811 (2008)
20. Jefferson Stroud, W., Krishnamurthy, T., Smith, S.A.: Probabilistic and possibilistic analyses of the strength of bonded joint. CMES – Computer Modeling in Engineering & Sciences 3(6), 755–772 (2002)

Soft Soil Contact Modeling Technique for Multi-Body System Simulation

Rainer Krenn and Andreas Gibbesch

Abstract In the context of planetary exploration with mobile robots a soil contact model (SCM) for prediction and assessment of locomotion performance in soft uneven terrain has been developed. The SCM approach provides a link between the classical, semi-empirical terramechanics theory of Bekker and the capabilities of multi-body system (MBS) simulation technique for general, full 3D simulations of soil contact dynamics problems. Beyond the computation of contact forces and torques SCM keeps track of the plastic soil deformation during simulation. For this purpose it comprises features such as generation of ruts and displacement of soil material that allow computing typical terramechanical contact phenomena like bulldozing, multi-pass effects and drawbar-pull–slippage relations. Unlike volumetric, Finite Element/Discrete Element Method-like approaches SCM applies exclusively surface oriented algorithms with relatively small complexity constants. Moreover, most of the algorithms are of linear complexity. Therefore, the computational efficiency is quite high and adequate for MBS simulation requirements.

1 Introduction and Motivation

Robotic mobility became an increasingly important issue for current and future planetary exploration missions. Unmanned, remotely controlled locomotion and navigation in rough, unstructured planetary terrain with wheeled, tracked and legged rovers or with hybrid mobility system concepts are major challenges in this field of research (see Figure 1). In particular, the capabilities for prediction and assessment of the locomotion performance are crucial skills for successful rover missions. They support both, the definition of an optimal locomotion subsystem design during the

Rainer Krenn · Andreas Gibbesch
Institute of Robotics and Mechatronics, Deutsches Zentrum für Luft- und Raumfahrt (DLR), Münchner Straße 20, 82234 Oberpfaffenhofen-Wessling, Germany;
e-mail: {rainer.krenn, andreas.gibbesch}@dlr.de

G. Zavarise & P. Wriggers (Eds.): Trends in Computational Contact Mechanics, LNACM 58, pp. 135–155.
springerlink.com

Fig. 1 Mobile robot examples with different locomotion subsystems. Top left: Wheeled system (ESA, ExoMars breadboard B1, [5]). Top right: Tracked system (ESA, Nanokhod breadboard A, [8]). Bottom left: Legged system (DLR, Crawler, [3]). Bottom right: Hybrid mobility system (NASA, Athlete, [9]).

mission preparation phase as well as adequate accomplishment of navigation tasks during the operational mission phase. The knowledge is mainly taken from three types of sources: Precursor missions, experimental testing and numerical simulations. The latter method offers significant advantages in terms of economic and time resources due to several reasons like reduced number of required prototypes, easy system parameter variation and parallel simulation opportunities. However, the reliability simulation results and finally the mapping of computed data to the real physical system depends strongly on the quality of the applied models inside the simulation framework.

2 MBS Based Soil Contact Modeling Technique

The preferred simulation technique for locomotion subsystem related investigations is multi-body system (MBS) simulation. This technique can be characterized as a very good compromise between system complexity reduction, simulation accuracy

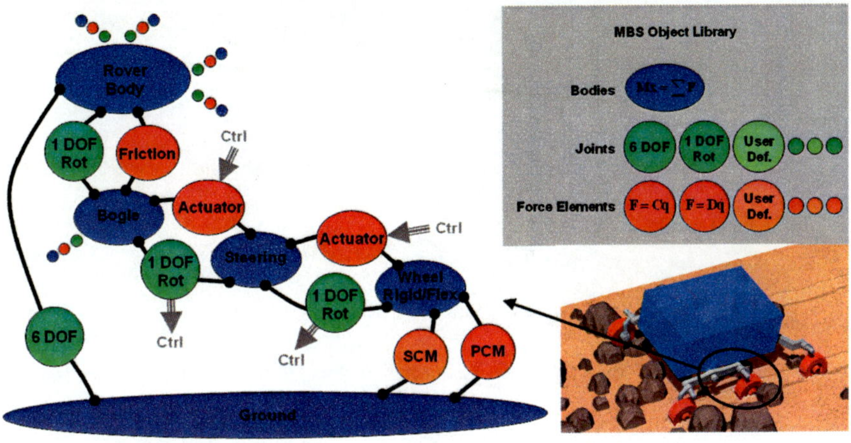

Fig. 2 Object based structure of an MBS model of wheeled rover chassis.

and computational performance. Moreover, MBS simulation offers typically an easy way to integrate sub-models from different simulation domains like mechanics and electronics in one top-level model. In Figure 2 a simplified structure of a MBS model is pictured, which is assembled from specific objects of a MBS-library. It shows the locomotion subsystem structure of a rover exemplarily for one single wheel. The major object types of the structure are body objects (blue) and joint objects (green), which together define the kinematics tree of the system, and force objects (red), which are used to apply forces and torques to the bodies depending on their specific implementation (e.g. friction, actuators, contact forces).

For the specific problem of mobility simulation on sandy, planetary terrain (see applications in Figure 3) a new modeling and simulation technique called Soil Contact Model (SCM) has been developed. It makes the classical, semi-empirical terramechanics theory of Bekker [1] and Wong [10] compatible with the specific requirements of multi-body system simulation regarding modularity, fidelity and computational efficiency. The efficiency of SCM is mainly based on the empirical-phenomenological modeling approach that provides an adequate level of abstraction for the implementation of complex dynamics problems like soil contact within multi-body systems.

From the object based modeling point of view SCM can be characterized as a specific force object for computing full 3D contact forces and torques between arbitrarily shaped contact objects and a terrain of plastically deformable soil. In the sample MBS of Figure 2 the SCM object is highlighted in orange. Other contact dynamics objects like the Polygonal Contact Model (PCM) [4], which is a modeling technique for rigid body contact dynamics (e.g. wheel-rock interaction), may optionally work in parallel with SCM without any need for model modification.

From the implementation point of view SCM can be generally described by the function

$$[\mathbf{F},\mathbf{T}] = \mathrm{SCM}(\mathbf{x},\mathbf{A},\mathbf{v},\boldsymbol{\omega},\mathrm{Parameters}) \qquad (1)$$

Fig. 3 Rover locomotion in sandy terrain. Left: Mars Exploration Rover (NASA). Right: Nanokhod (ESA).

with the following interface arguments:

- Input Arguments: Relative position **x**, relative attitude **A**, relative linear velocity **v** and relative angular velocity $\boldsymbol{\omega}$ between contact object and soil.
- Output Arguments: Contact forces **F** and contact torques **T** between contact object and soil.
- The model specific parameters contain the geometric descriptions surfaces of contact object and soil and the dynamics parameters of the soil.

This interface definition is a kind of standard interface for force objects inside an object based MBS. Accordingly, SCM is typically called by the MBS solver at each time integration step of the MBS simulation run (see Figure 4). Nevertheless, SCM can be utilized outside the MBS context as well, e.g. as an algorithm inside a model-based predictor of a controller or for offline computations of specific contact configurations.

3 Details of SCM Implementation

This paper provides a closer insight of the architecture and implementation details of SCM. A general overview of the major functional components and the corresponding data flow is presented in Figure 4. It groups the implemented algorithms into two functional blocks:

1. The computation of contact forces and torques between contact object and soil as function of the contact parameters and the relative kinematics between the contact object and the soil (green block).
2. The computation of the plastic soil deformation during contact (yellow block). This feature plays an essential role for the correct representation of terramechnical phenomena like bulldozing or multi-pass effects and their dynamics impact during simulation.

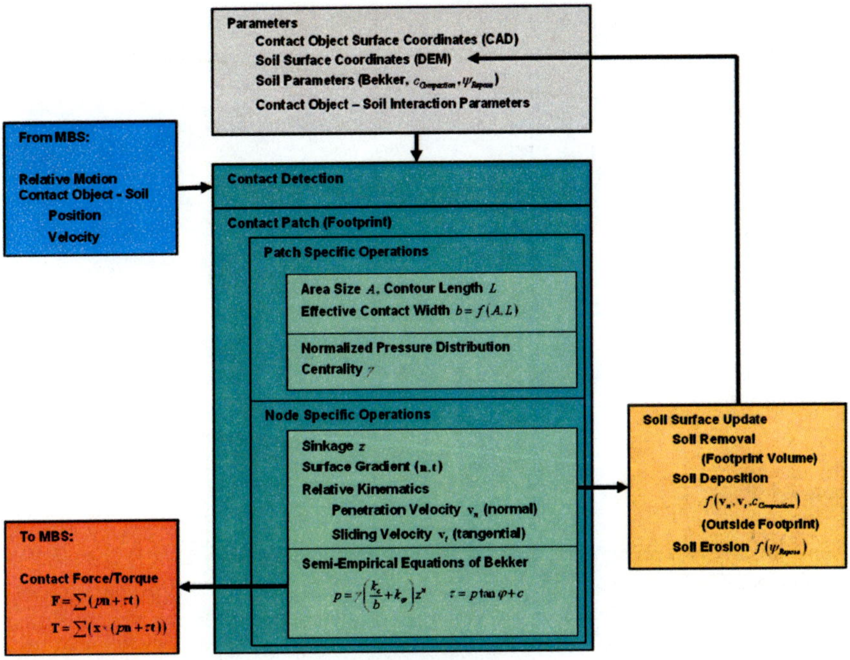

Fig. 4 SCM architecture.

3.1 Contact Force Computation

The central task of SCM is the computation of contact forces and torques between the contact body surface and the soil surface. For generality reasons SCM is designed to work with arbitrarily shaped contact surfaces and to handle single point contact problems in the same way as multiple point contact problems. Thus, the application of SCM is by no means limited to specific body topologies, respectively to specific mobility systems.

3.1.1 Function Parameters

The parameters of the SCM-module can be arranged into two basic groups: The description of the contact surfaces and the dynamics parameters for the interaction of contact body and soil.

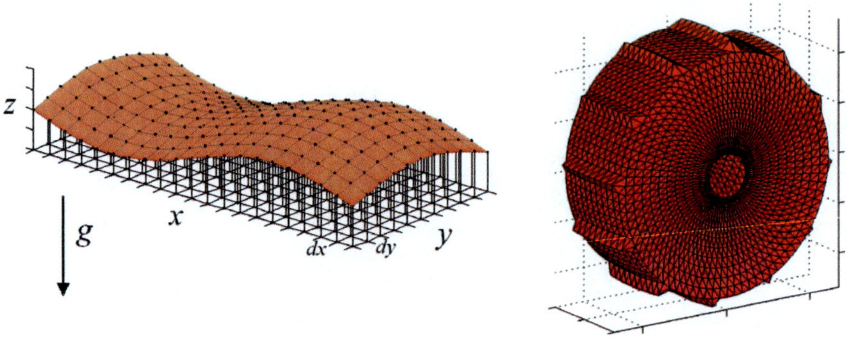

Fig. 5 Geometry definition of contact surfaces. Left: Digital elevation map of soil surface. Right: Polygonal mesh of wheel.

Soil Topology

The soil surface topology is defined by a Digital Elevation Map (DEM) as shown in (Figure 5 left). This type of 2.5-D surface description fits very well to the physical realities of sandy terrain, where overhangs do not occur and where the maximum surface inclination is limited by the soil specific maximum angle of repose ψ. The DEM is based on a uniformly spaced $(dx = dy = ds)$ horizontal mesh grid of dimension (nx, ny) and constant mesh size $dA = ds^2$. Each node of the grid is representing a discrete part of the soil surface dO on top of dA by its elevation. The individual size of dO depends on the local surface inclination. The direction of the elevation coordinate z is in parallel with the gravity vector g. The geometrical granularity of the mesh grid should be fine enough in order to represent the soil topology with a sufficient accuracy, even after plastic deformation, and to describe a contact patch with sufficient resolution.

Surface Geometry of Contact Body

The surface geometry of the contact body is described by an ordinary, three-dimensional polygonal mesh (Figure 5 right), which is well known from CAD and computer graphics applications. It is defined by the surface vertices and triangular faces that define the surface topology between adjoining vertices. For compatibility reasons with the soil description the maximum face edge length should not exceed the mesh grid width ds. This is relevant for safe contact detection computation. Due to the general method used for surface description SCM can work with complex body geometries like profiled wheels exactly in the same manner as with simple, smooth surface shapes.

Table 1 Soil parameters of Bekker theory.

Parameter	Variable	Unit
Exponent of sinkage	n	[–]
Cohesive modulus	k_c	[N/m^{n+1}]
Frictional modulus	k_ϕ	[N/m^{n+2}]
Cohesion	c	[Pa]
Angle of internal friction	ϕ	[–]

Table 2 Additional soil parameters for SCM.

Parameter	Variable	Unit
Maximum angle of repose	ψ	[–]
Surface friction coefficient	μ	[–]

Contact Dynamics Parameters

In addition to the elevation data each node of the soil mesh grid is associated with a number of contact dynamics parameters. Most of the parameters are required to apply the semi-empirical terramechanics theory of Bekker [1] within SCM. The remaining ones consider the maximum slope angle of piled soil and the Coulomb friction between the contact body and the soil surfaces. The complete list of Bekker parameters is presented in Table 1. In Table 2 two additional soil parameters used within SCM are listed. If the soil properties are constant over the entire mesh grid the parameters can be globally defined by scalars. In case of location dependent parameters, they are provided as function $f(x,y)$, respectively as matrix of dimension (nx,ny).

3.1.2 Force Computation Algorithm

The SCM contact pressure computation is based on Bekker's well-known pressure-sinkage relationship

$$p = \left(\frac{k_c}{b} + k_\phi\right) z^n, \tag{2}$$

which is verified by experimental results from bevameter tests. A modern, electric driven version of a bevameter is shown in Figure 6 left. In Figure 6 top right the scheme of the original bevameter test setup with hydraulic actuated probe is presented. The variables p, b and z, which are used in (2), are derived from the hydraulic bevameter setup as follows:

- The variable p is the mean pressure in the contact zone of probe and soil. It is proportional to the hydraulic pressure p_{Cyl} inside the cylinder, which is actually measured.

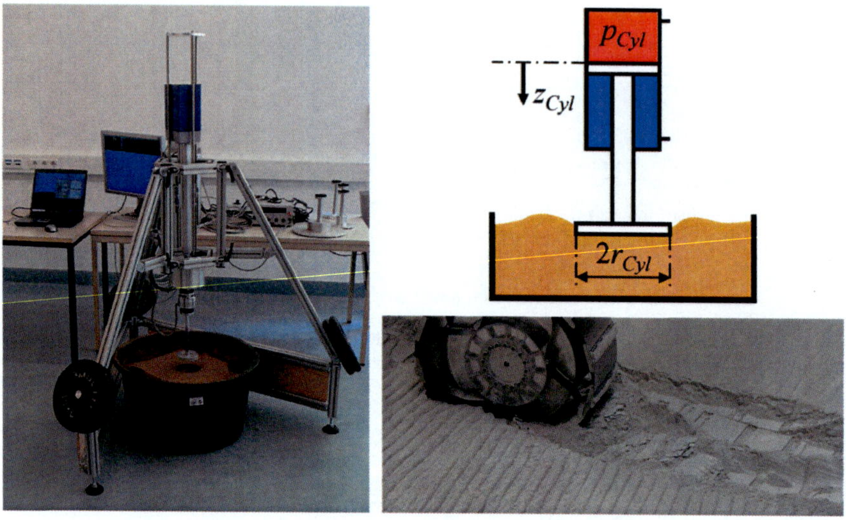

Fig. 6 Bevameter-experiment versus general soil contact case.

- The variable z is the sinkage of the probe inside the soil. It is measured via the position z_{Cyl} of the bevameter piston.
- The variable b is the contact width of the probe. In case of circular bevameter probes it is equal to the radius r_{Cyl} of the probe.

The aim of the SCM algorithm is the extension of the pressure sinkage relationship (2), which was originally developed for the one-dimensional contact case with a mechanical system of one degree of freedom, for general, three-dimensional contact cases between soil and arbitrarily shaped contact bodies that have six degrees of freedom each (Figure 6 bottom right). In other words: SCM has to be able to compute the local, discrete contact pressure p_i at any node of the soil mesh grid that is involved in contact with the surface of a contact body.

Contact Detection and Local Sinkage

The first section of the contact force computation is the contact detection algorithm. It starts with a transformation of the contact body vertices (Figure 5 right) according to the contact body pose, which is typically computed by the MBS solver during a running simulation. A potential relative pose of contact body and soil is exemplarily shown in Figure 7a. It is taken as initial configuration for the following contact detection steps. The top view on this configuration (Figure 7a bottom) shows the initial, vertical projection of the body vertices onto the soil grid. In the following step (Figure 7b) the transformed vertices are mapped into the soil grid space such that they are arranged in vertical columns at soil grid node locations. In computer graphics this process is referred to as Spatial Binning and Z-Buffering. Supposed that the

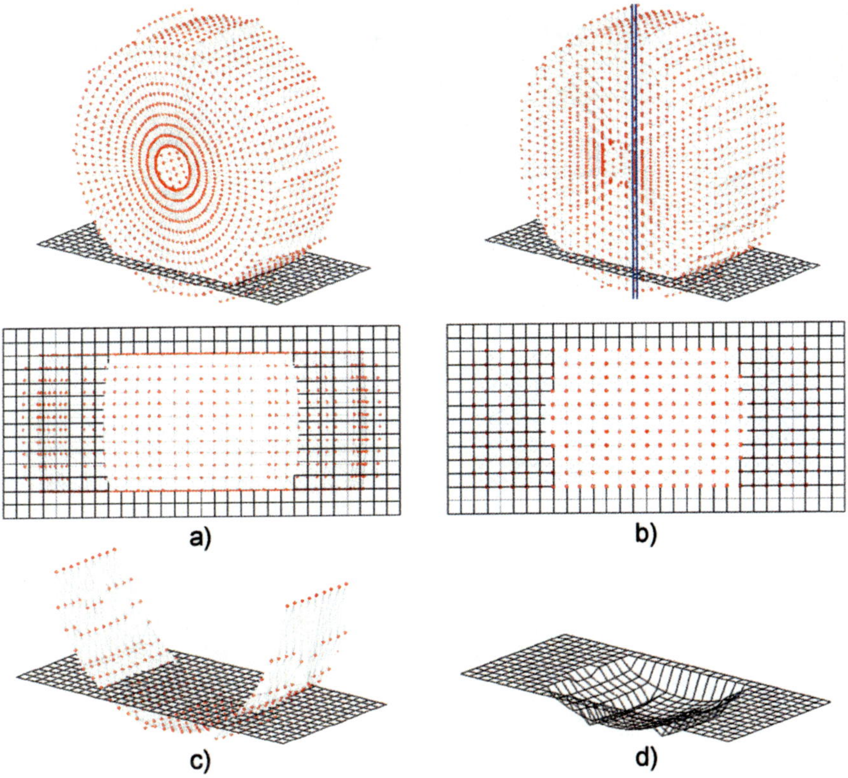

Fig. 7 Contact detection by vertex mapping

mesh grid resolution is adequately chosen, the reduction of the geometrical preciseness caused by the vertex mapping is negligible. A subset of the mapped vertices, exactly spoken the minima of each column (Figure 7c) are potential candidates for contact with the soil. The actual contact detection is simply comparing the vertical coordinates of the column minima with the elevation at the corresponding soil grid nodes.

Due to the point oriented contact detection method with mapped vertices the computational load of this algorithm is basically very low. Nevertheless, for further speed-up the algorithm is embedded in a hierarchical bounding volume (BV) tree contact detection algorithm. An overview and details of computational, BV based contact detection methods can be found in [11]. In SCM the selected bounding volumes are axis aligned bounding boxes (Figure 8). The general advantage of the BV based contact detection is the pre-selection of potential contact vertices that are actually processed, respectively transformed, mapped and compared with the soil elevation. However, the transformation of a bounding box and the collision check of boxes take significantly more computation time than the corresponding operations applied to points. Therefore, the number of hierarchical BV tree layers

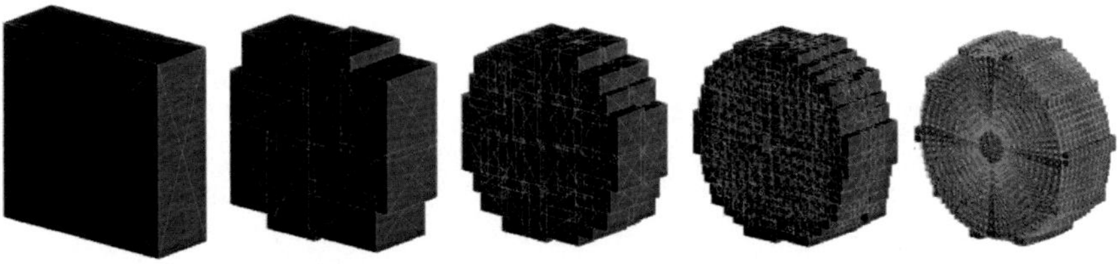

Fig. 8 Sample levels of bounding box tree.

and implicitly the number of vertices per leaf box of the BV tree have to be chosen carefully. The optimal choice depends on both, the number of body and soil vertices and the typical number of contact points during simulation.

The output of the contact detection procedure is a discrete contact patch, respectively the intersection volume (footprint) of the contact object in the soil (Figure 7d). This result includes implicitly the local sinkage z_i at each single contact node of the patch, which is the first required component for the general formulation of the pressure sinkage relationship (2). Additionally, the individual contact velocity vector $\mathbf{v}_i$ and the individual contact surface normal and tangential vectors, $\mathbf{n}_i$ and $\mathbf{t}_i$ can be computed directly from the relative kinematics of the contact surfaces.

Effective Contact Width

The contact width b of the pressure-sinkage relationship (2) is primary defined for circular contact patches (Figure 6 left and Figure 9 left top) as the radius of the circle. In a further definition b is the smaller side of rectangular patch with $b \ll a$ (Figure 9 left bottom). In order to define the effective contact width b_{eff} for a generally shaped contact patch as presented in Figure 9 right, the common characteristics of the two initial definitions has to be identified and applied to the general case.

For the two simple cases b can be expressed as a relation of patch area size A and contour length L along the patch border as introduced in (3) and (4):

$$\text{Circular Patch:} \qquad A = b^2\pi; \quad L = 2b\pi; \quad b = 2\frac{A}{L}. \tag{3}$$

$$\text{Rectangular Patch:} \quad A = ab; \quad L = 2(a+b); \quad b \approx 2\frac{A}{L}; \quad \text{with } b \ll a. \tag{4}$$

The identified relation of A and L is than applied in SCM as the effective contact width

$$b_{\text{eff}} = 2\frac{A}{L}, \tag{5}$$

which is the second required component for the general formulation of relationship (2).

The advantage of this solution is the fact that the area size A and contour length L are always numerically computable during simulation. They can be expressed as

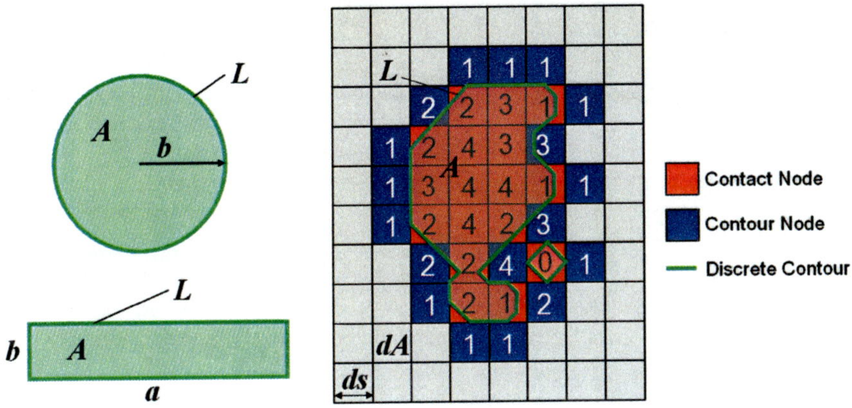

Fig. 9 Contact patches characterization. Left: Circular and rectangular patches. Right: Discrete contact patch with node classification.

Table 3 Weighting factors for area size and contour length.

Node classification (number of contact node neighbors)	$c_{A,\text{contact}}$	$c_{A,\text{contour}}$	$c_{L,\text{contact}}$	$c_{L,\text{contour}}$
1	0.5	0	0.5	0.5
2	0.625	0.125	$\sqrt{0.5}$	$\sqrt{0.5}$
3	0.875	0.375	$\sqrt{0.5}$	$\sqrt{0.5}$
4	0.1	0.5	0	0.5

functions of the number contact nodes n_{contact}, the number of contour nodes n_{contour} and individual weighting factors c_A and c_L for each node. The variables $c_{A,\text{contact}}$ and $c_{A,\text{contour}}$ are area size weighting factors of the nodes to be multiplied by the node area size dA. The variables $c_{L,\text{contact}}$ and $c_{L,\text{contour}}$ are the contour length weighting factors to be multiplied by the length ds. The assignment of the weighting factors depends on the individual node type (contact node or contour node) and on the individual node classification (number of neighbor nodes in contact). In Table 3 the weighting factors used in SCM are documented. An example for the node classification is given in Figure 9 right.

The mathematical formulation of the functions A and L can be found in equations (6) and (7). A graphical example how to compute A and L in the general case is given in Figure 9 right.

$$A = \sum_{i=1}^{n_{\text{contact}}} c_{A,\text{contact},i} dA + \sum_{j=1}^{n_{\text{contour}}} c_{A,\text{contour},j} dA; \quad c_A = [0 \ldots 1]. \tag{6}$$

$$L = \sum_{i=1}^{n_{\text{contact}}} c_{L,\text{contact},i} ds + \sum_{j=1}^{n_{\text{contour}}} c_{L,\text{contour},j} ds; \quad c_L = \left[0 \ldots \frac{1}{\sqrt{2}}\right]. \tag{7}$$

Pressure Distribution

In order to be able to correctly compute the local contact pressure p_i the pressure distribution function inside the contact patch must be considered. The basic assumption for computing the pressure distribution in the context of SCM is that the pressure drops from central regions of the contact patch towards the border in a parabolic shape. This assumption is confirmed by experimental testing and measurements of the pressure bulb underneath tires in terramechanical applications [2]. The parabolic characteristics of the pressure distribution for generally shaped contact patches is generated by node specific weighting factors γ_i to be multiplied with the corresponding node specific, sinkage dependent contact pressure as formulated in (8).

$$p_i = \gamma_i \left(\frac{k_c}{b_{\text{eff}}} + k_\phi \right) z_i^n. \tag{8}$$

For consistency reasons the mean value of γ has to be 1. The introduction of γ is the final SCM specific adaptation of relationship (2).

In the simplest case the weighting factor γ_i is equal to the centrality. It describes the contact node location within the entire contact patch according to (9) by a function that takes the distances $\mathbf{r}_j - \mathbf{r}_i$ to all other contact nodes into account.

$$\gamma_i = \frac{n_{\text{contact}} w_i}{w_{\text{total}}}; \quad \text{with } w_i = \sum_{j=1, j\neq i}^{n_{\text{contact}}} \frac{1}{(\mathbf{r}_j - \mathbf{r}_i)^T (\mathbf{r}_j - \mathbf{r}_i)}; \quad w_{\text{total}} = \sum_{i=1}^{n_{\text{contact}}} w_i. \tag{9}$$

The corresponding function for a circular contact patch is shown in Figure 10 left. For more precise solutions the balance of soil cohesion c and internal friction angle ϕ has to be considered and the intermediate result w_i of equation (9) has to be extended accordingly by

$$w_i = \sum_{j=1, j\neq i}^{n_{\text{contact}}} \frac{1}{f(|\mathbf{r}_j - \mathbf{r}_i|, c, \phi)}. \tag{10}$$

In Figure 10 middle and right the pressure distribution functions under consideration of different soil conditions are presented. In case of dominant internal friction (e.g. dry sand) the pressure distribution is flat with a maximum value γ_{max} of approximately 1. In case of dominant cohesion (e.g. saturated clay) the function rises continuously from border to center and obtains its maximum γ_{max} close to 2. These numerical results match quite well with the theory as published in [7].

An example for the SCM computed pressure distribution of a non-circular, discontinuous footprint area is shown in Figure 11, which should represent the general case (e.g. wheel profile, track profile).

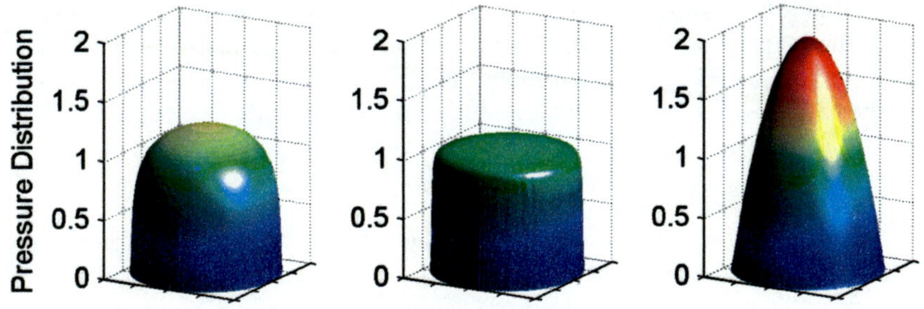

Fig. 10 Pressure distribution in circular contact patch. Left: Centrality only. Middle: Dominant internal friction. Right: Dominant cohesion.

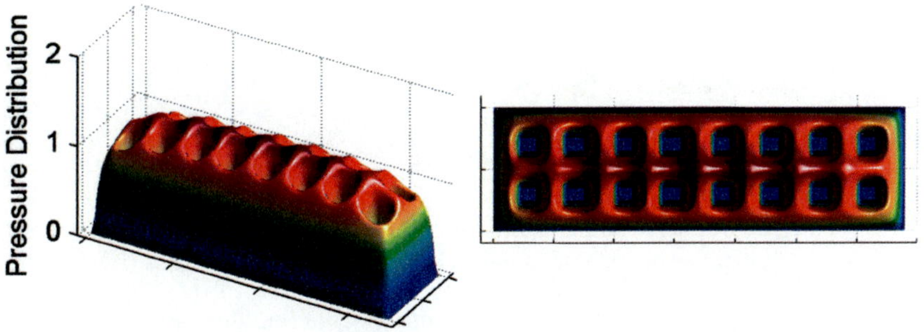

Fig. 11 Pressure distribution in a generally shaped contact patch.

Contact Forces

The relationship (8) for the local contact pressure p_i leads to the formulation of the discrete force vector $\mathbf{F}_i$ for a single contact node, expressed in the soil reference frame as introduced in paragraph soil topology:

$$\mathbf{F}_i = \left(\begin{pmatrix} c \\ c \\ 0 \end{pmatrix} + p_i \begin{pmatrix} \tan\phi & 0 & 0 \\ 0 & \tan\phi & 0 \\ 0 & 0 & 1 \end{pmatrix} \right) dA\mathbf{n}_i + \mu p_i dA\mathbf{t}_i. \tag{11}$$

The force consists of two basic components: The first one is caused by normal soil penetration and acts in the normal direction $\mathbf{n}_i$ (Figure 12) of the contact patch. The z-component is the vertical sinkage resistance force. The horizontal components include the Mohr–Coulomb failure criterion for soil, which is a function of cohesion c and the internal friction angle of the soil ϕ. It defines the maximum applicable shear stress the soil can resist by

$$\tau_{\text{Mohr-Coulomb}} = c + p\tan\phi. \tag{12}$$

The second force component is caused by lateral sliding along the contact patch. It acts in the tangential direction $\mathbf{t}_i$ (Figure 12) during contact and reflects the shear stress due Coulomb friction between the contact body surface and the soil surface. The shear stress is a function of the friction coefficient μ with

$$\tau_{\text{Coulomb}} = p\mu. \tag{13}$$

The total contact forces and their corresponding torques are finally returned to the MBS (Figure 4, orange block) and applied to the contact body at the body reference frame. They are computed by discrete integration (summation) over the entire contact patch area as given in (14).

$$\mathbf{F}_{\text{total}} = \sum_{i=1}^{n_{\text{contact}}} \mathbf{F}_i; \quad \mathbf{T}_{\text{total}} = \sum_{i=1}^{n_{\text{contact}}} (\mathbf{r}_i \times \mathbf{F}_i). \tag{14}$$

3.2 Plastic Soil Deformation

The computation of plastic soil deformation is an essential pre-requisite for correct simulation of typical terramechanical phenomena like

- Increased rolling resistance caused by humps in front (bulldozing),
- Reduced rolling resistance in pre-deformed ruts (multi-pass),
- Lateral guidance inside ruts,
- Drawbar pull as function of slippage.

Typically, soil volume deformation problems as well known from civil engineering are solved using Finite Element Method or Discrete Element Method techniques. However, due to the extreme computational load of these methods they are not applicable fast MBS simulations. In SCM a computationally very efficient technique for manipulating the soil DEM is implemented. It is purely based on surface oriented operations and is fast enough to be applied at each sample time of the MBS solver.

3.2.1 Soil Displacement

For soil displacement purposes it is supposed that at each contact node a discrete amount of soil

$$dV_i = z_i dA \tag{15}$$

has to be replaced that depends on the local footprint depth. The applied displacement law is a function of the local contact velocity as explained in Figure 12. However, only normal components of the contact velocity are related to soil penetration and therefore only this component is relevant for plastic soil deformation and soil displacement, respectively.

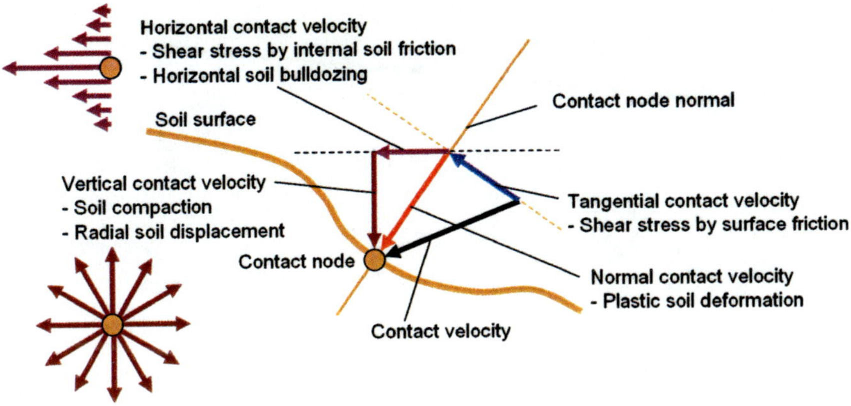

Fig. 12 Contact velocity components and corresponding soil displacement effects at contact node.

The normal velocity can be further divided in two parts, while each of them has specific effects in term of soil displacement: The horizontal component causes bulldozing effects and the corresponding local soil flow field is therefore modeled as a horizontal parallel field according to Figure 12 top left. The vertical component of normal velocity causes the vertical displacement of the soil. Under the assumption of incompressible or semi-compressible soils we consequently obtain a horizontal radial soil flow field as pictured in Figure 12 bottom left. The ratio of the field strengths is supposed to be equal to the ratio of the corresponding penetration velocity components.

3.2.2 Soil Deposition

The deposition of the displaced soil, respectively the manipulation of the corresponding DEM regions is implemented in two steps: In a first temporary step the soil is banked up at the border of the footprint as shown in Figure 13 left. Hereby, each border node (marked with blue dots) obtains a portion of soil from each contact node (marked with red dots), depending on the local intensities of the emitting, radial and parallel soil flow fields at the contact patch border.

The DEM manipulation is finished after applying an iterative erosion algorithm (second step of soil deposition). It smooths the soil in the vicinity of the contact patch under the constraint that the maximum DEM inclination is smaller than the maximum soil specific angle of repose ψ. It can be formulated by

$$\frac{|z_j - z_i|}{ds} \leq \tan \psi. \tag{16}$$

For this purpose the algorithm computes the erosion potentials at each investigated node (index i) with respect to its neighbor nodes (index j) by

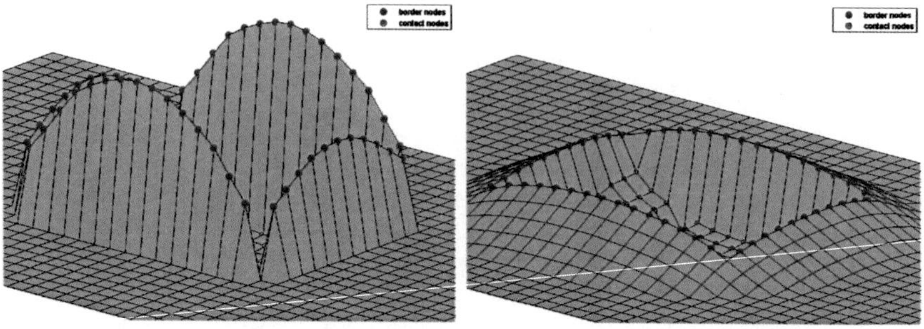

Fig. 13 Soil deposition around the contact patch.

$$dz_j = z_i - z_j; \quad j = 1 \ldots n_{\text{neighbor}}. \tag{17}$$

If the maximum local erosion potential exceeds the given limit, expressed by

$$\max(dz) > ds \tan \psi, \tag{18}$$

the DEM is updated according to the following rules for the investigated node and its neighbors:

$$z_i = z_i - \frac{\max(dz)}{2}, \tag{19}$$

$$z_j = z_j + \frac{\max(dz)}{2} \frac{dz_j}{\sum_{k=1}^{n_{\text{neighbor}}} dz_k}; \quad j = 1 \ldots n_{\text{neighbor}}. \tag{20}$$

The result of the soil erosion is shown in Figure 13 right. The updated DEM is finally stored by SCM as a function (see Figure 4) and applied at the next call as initial soil surface description parameter for contact force computations according to Section 3.1.2.

4 Verification of SCM

In the model verification process the correctness regarding the fidelity of the implemented theory is proven. In the kernel of the SCM the pressure-sinkage relationship (2) as proposed by Bekker is implemented, which is originally derived from results of bevameter experiments (see Figure 6). Accordingly, for verification purposes it is recommended and appropriate to correlate the ideal results of relationship (2) with results of SCM simulations that feature bevameter tests with circular probes. The SCM modeling technique is supposed to be verified if the reproduction of the original pressure sinkage relationship and the identification of the applied soil parameters are possible with sufficient accuracy.

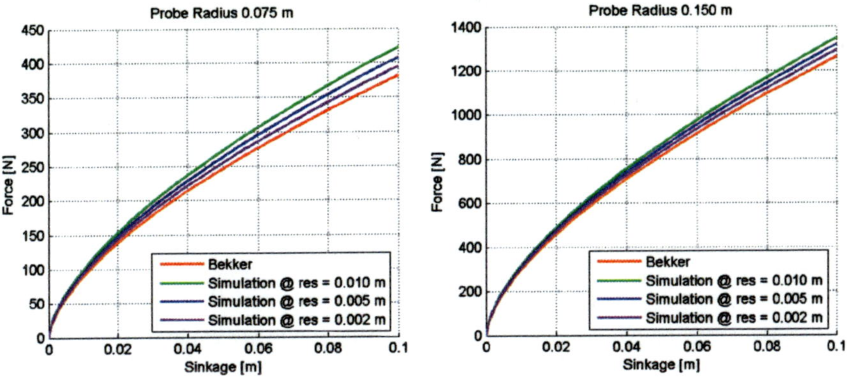

Fig. 14 Ideal and simulated force-sinkage relationship $F(z)$. Applied soil parameters: Martian soil simulant DLR-A [6].

In Figure 14 the simulation results of bevameter experiments for different probe radii and soil grid resolutions are presented and compared with the ideal force-sinkage relationship of Bekker. The discrepancies between the simulation results and the ideal reference are caused by the discrete computation of the contact patch. It affects the maximum achievable accuracy of the patch area size A and contour length L. Both variables influence the accuracy of the effective contact width b_{eff} to that the relationship (8) is very sensitive. Therefore, the best results can be obtained if the number of grid nodes, which define the outer contour of the contact patch, is relatively small compared with the number of grid nodes inside the contact patch (large footprint at high grid resolution, see Figure 14 bottom right).

In the second verification step the soil parameters of (2), k_c, k_ϕ and n are identified based on the simulation results as shown in Figure 14. In order to identify them individually two simulated force-sinkage functions $F_1(z)$ and $F_2(z)$, obtained at identical soil grid resolution but with different probe radii r_1 and r_2, have to be considered within a minimum search algorithm. The corresponding objective function is

$$f_{\min} = \mathbf{g}(\mathbf{x})^T \mathbf{g}(\mathbf{x}) \tag{21}$$

with

$$\mathbf{g}(\mathbf{x}) = \left(\left(\frac{1}{r_1^2 \pi} \right) \mathbf{F}_1 - \left(\frac{x(1)}{b_{1,\text{eff}}} + x(2) \right) \mathbf{z}^{x(3)} \right) + \left(\left(\frac{1}{r_2^2 \pi} \right) \mathbf{F}_2 - \left(\frac{x(1)}{b_{2,\text{eff}}} + x(2) \right) \mathbf{z}^{x(3)} \right) \tag{22}$$

and

$$\mathbf{x} = \begin{pmatrix} k_c & k_\phi & n \end{pmatrix}. \tag{23}$$

Table 4 Identified soil parameters.

Soil parameter	k_c	k_ϕ	n
Applied soil parameters (Simulant DLR-A, [6])	2370	60300	0.63
Identified parameters, ds=0.010 m	2398	60159	0.63
Identified parameters, ds=0.005 m	2377	60260	0.63
Identified parameters, ds=0.002 m	2371	60293	0.63

The results of the identification are collected in Table 4. Obviously, the identified soil parameters match quite well with the actual applied ones and therefore it is valid to state that the implementation comprises the essentials of Bekker's theory for application in MBS models, which is actually the intension of SCM.

5 Profiling of SCM Code

In order to evaluate the compatibility with MBS requirements and the computational efficiency the profiling of the SCM code is a useful method. It helps to identify time-critical sections of the code and potential resources for model optimization. A representative profile of the SCM code running on a standard desktop PC is presented in Figure 15. It is recorded for a single wheel-soil interaction application with 16860 vertices per wheel and 40401 vertices per soil DEM (wheel diameter is 0.25 m, wheel width is 0.1 m, see Figure 1 top left and Figure 6 bottom right). However, in terms of computation time the topology of the bodies and the numbers of vertices are only relevant until the contact detection is finished. Then, for the rest of the SCM code only the number of detected soil contact nodes inside the contact patch affects the computational load. Accordingly, the time profile is plotted against the number of contact nodes. The profile in Figure 15 presents the accumulated computation time after each major step of the SCM code, which are listed in the legend from top to bottom.

Obviously, not all of the functions are explicitly visible since some of the operations take just negligible computation time and their corresponding functions hide the previous ones. Nevertheless, four costly operations can be identified that are listed in Table 5. The most expensive operations are algorithms of complexity N^2, which cover the pressure distribution (see Section 3.1.2) and the soil deposition (see Section 3.2.2). Thus, in terms of code optimization these parts that occupy about 90% of the computation time will be subject of future improvements or model simplifications if the implemented level of detail is not required.

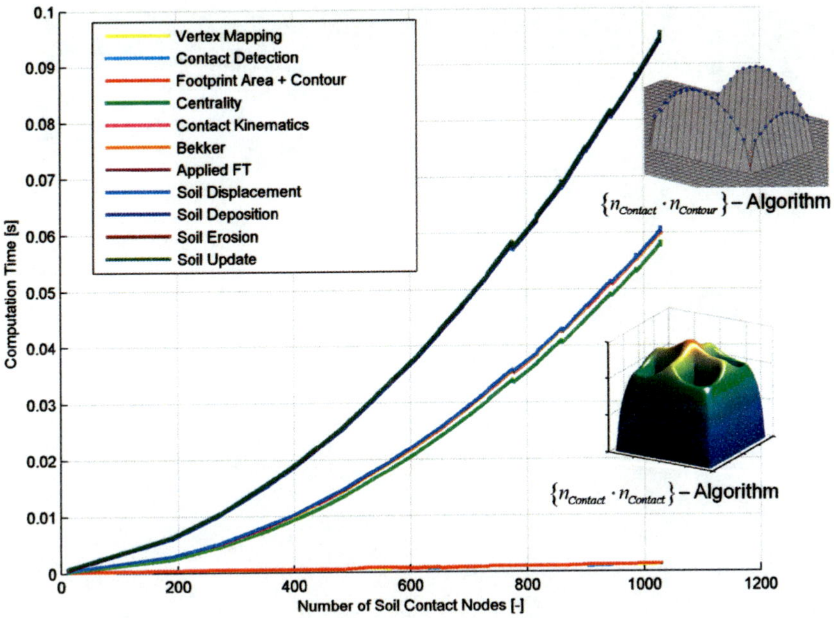

Fig. 15 Profiling of SCM code.

Table 5 Complexity of algorithms.

Operation	Complexity	Parameters
Vertex mapping (see Fig. 7)	N	Number of contact nodes, Number of BV-tree layers
Pressure distribution (centrality, see Fig. 11)	N^2	Number of contact nodes
Contact kinematics	N	Number of contact nodes
Soil deposition (see Fig. 13)	N^2	Number of contact nodes, Number of contour nodes

6 Conclusion

The SCM modeling technique extends the MBS simulation capabilities in terms of soft soil contact dynamics by a novel empirical-phenomenological approach. Due to its general, module like implementation as a MBS library object it is applicable for many different mobility systems that are in interaction with loose soil ground conditions. This is typically the case for planetary rovers as well. A visual conclusion is presented in the following clips (Figures 16 to 18) that consist of movie frames taken from animations of rover locomotion simulations. The first ones (Figures 16 and 17) show benchmark tests regarding uphill and cross-hill locomotion capabilities of a ExoMars-class rover [5].

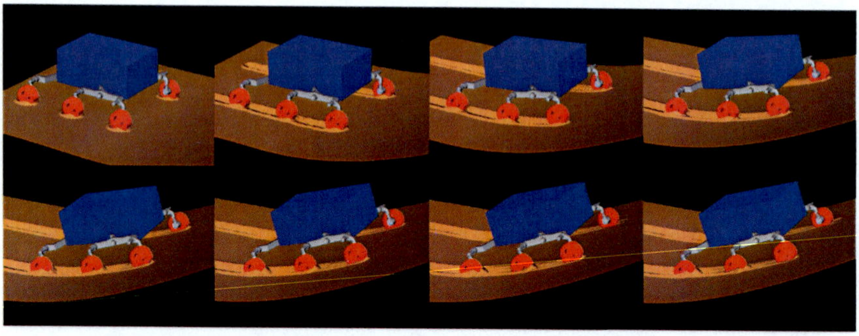

Fig. 16 Movie sequence: Simulation of uphill locomotion of rover.

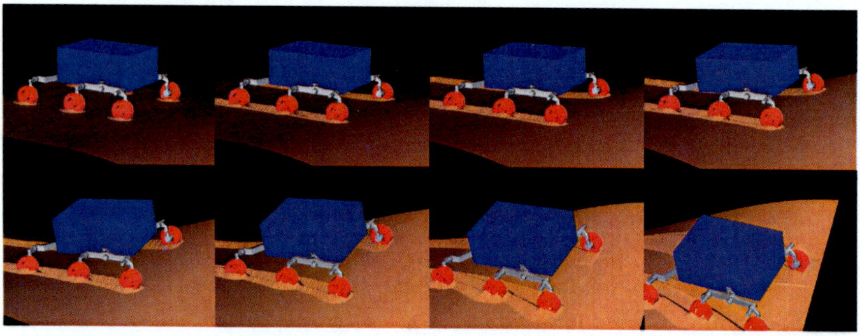

Fig. 17 Movie sequence: Simulation of cross-hill locomotion of rover.

Fig. 18 Movie sequence: Simulation result of rover locomotion in rocky planetary terrain.

In Figure 18 the parallel operation of different contact models according to Figure 2 is demonstrated. It is a sample result for simulation of rover mobility in a representative planetary terrain with rocky obstacles inside a sandy matrix, which is currently the most ambitious implementation and the initial status for further developments regarding MBS related soil contact modeling.

References

1. Bekker, M.G.: Introduction to Terrain-Vehicle Systems. The University of Michigan Press, Ann Arbor (1969)
2. Bolling, I.: Bodenverdichtung und Triebkraftverhalten bei Reifen – Neue Meß- und Rechenmethoden. PhD Thesis, TU München (1987)
3. Görner, M., Wimbäck, T., Baumann, A., Fuchs, M., Bahls, T., Grebenstein, M., Borst, C., Butterfass, J., Hirzinger, G.: The DLR-Crawler: A testbed for actively compliant hexapod walking based on the fingers of DLR-Hand II. In: Proceedings of International Conference on Intelligent Robots and Systems, Nice, France (2008)
4. Hippmann, G.: Modellierung von Kontakten komplex geformter Körper in der Mehrkörpersimulation. PhD Thesis, TU Wien (2004)
5. Michaud, S., Höpflinger, M., Thuer, T., Lee, C., Krebs, A., Despont, B., Gibbesch, A., Richter, L.: Lessons learned from ExoMars locomotion system test campaign. In: Proceedings of 10th Workshop on Advanced Space Technologies for Robotics and Automation, ESTEC The Netherlands (2008)
6. Scharringhausen, M.: Martian, terrestrial and lunar soil parameters. Technical Report, DLR Internal Report (2008)
7. Söhne, W.: Agricultural engineering and terramechnics. Journal of Terramechanics 9(4) (1969)
8. Steinmetz, B., Arbter, K., Brunner, B., Landzettel, K.: Autonomous vision-based navigation of the Nanokhod rover. In: Proceedings of 6th International Symposium on Artificial Intelligence, Robotics and Automation in Space, Montreal, Canada (2001)
9. Wilcox, B.: ATHLETE, An option for mobile lunar habitats. In: Proceedings of International Conference on Robotics and Automation, Pasadena, CA, USA (2008)
10. Wong, J.Y.: Theory of Ground Vehicles, 4th edn. Wiley, New York (2008)
11. Zachmann, G.: Virtual reality in assembly simulation – Collision detection, simulation algorithms and interaction techniques. PhD Thesis, TU Darmstadt (2000)

A Semi-Explicit Modified Mass Method for Dynamic Frictionless Contact Problems

David Doyen, Alexandre Ern and Serge Piperno

Abstract We propose a semi-explicit numerical method for solving dynamic frictionless contact problems. This method combines a modified mass matrix, a central difference time-integration scheme, and an exact enforcement of the contact condition. At each time step, the displacements of the nodes in the interior of the domain are computed in an explicit way, while the displacements of the nodes at the contact boundary are computed by solving a nonlinear problem. Such a method presents the same stability condition as the central difference scheme without contact and achieves a tight energy behavior. We perform numerical simulations, in 1D and 2D, to illustrate the behavior of the proposed method. In particular, we compare our method to a method combining a central difference scheme and an implicit contact condition.

1 Introduction

Solving dynamic contact problems with an explicit time-integration scheme is not straightforward. Firstly, it is not possible to enforce an exact contact condition in an explicit way; the problem would be ill-posed [3]. An option is then to enforce implicitly the contact condition as in [3] or [12, 13]. This approach provides a semi-explicit method. Another option is to replace the exact contact condition with an approximate contact condition (allowing an inter-penetration). The use of a penalty contact condition gives a fully explicit method, but worsens the stability condition

David Doyen
EDF R&D, 1 avenue du Général de Gaulle, 92141 Clamart Cedex, France;
e-mail: david.doyen@gmail.com

Alexandre Ern · Serge Piperno
CERMICS, Université Paris-Est, Ecole des Ponts, 77455 Marne la Vallée Cedex 2, France;
e-mail: ern@cermics.enpc.fr, e-mail: sergio.piperno@enpc.fr

G. Zavarise & P. Wriggers (Eds.): Trends in Computational Contact Mechanics, LNACM 58, pp. 157–168.
springerlink.com

on the time step [1]. The use of a contact condition in velocity as in [1] provides a semi-explicit method.

In this paper, we propose a semi-explicit method for solving dynamic frictionless contact problems. This method combines a modified mass matrix, a central difference time-integration scheme, and an exact enforcement of the contact condition. In the modified mass matrix, the entries associated with the (normal) displacements at the contact boundary are set to zero. This idea of modifying the mass matrix for contact problems has been introduced in [10]. With the modified mass matrix, we obtain a well-posed space semi-discrete problem. This semi-discrete problem can then be discretized with various time-integration schemes. In [6, 10], implicit schemes have been employed. If we choose a central difference scheme, we obtain a semi-explicit method. At each time step, the displacements of the nodes in the interior of the domain are computed in an explicit way, while the displacements of the nodes at the contact boundary are computed by solving a nonlinear problem. Such a method presents the same stability condition as the central difference scheme without contact and achieves a tight energy behavior.

The paper is organized as follows. In Section 2, we formulate the space semi-discrete problem with the modified mass matrix. We briefly recall the favorable properties of this formulation. In Section 3, we discretize in time using central differences leading to the semi-explicit modified mass method; we also describe a semi-explicit scheme with implicit contact [3, 12, 13] to be used for comparison. We examine the energy balance and the stability of these two schemes. In the last section, 1D and 2D numerical simulations illustrate the behavior of the two schemes.

2 Space Semi-Discretization by the Modified Mass Method

2.1 Governing Equations

We consider the infinitesimal deformations of a body occupying a reference domain $\Omega \subset \mathbb{R}^d$ ($d \in \{1,2,3\}$) during a time interval $[0,T]$. The tensor of elasticity is denoted by $\mathscr{A}$ and the mass density is denoted by ρ. An external load f is applied to the body. Let $u : (0,T) \times \Omega \to \mathbb{R}^d$, $\varepsilon(u) : (0,T) \times \Omega \to \mathbb{R}^{d,d}$, and $\sigma(u) : (0,T) \times \Omega \to \mathbb{R}^{d,d}$ be the displacement field, the linearized strain tensor, and the stress tensor, respectively. Denoting time-derivatives by dots, the momentum conservation equation is

$$\rho \ddot{u} - \operatorname{div} \sigma = f, \quad \sigma = \mathscr{A} : \varepsilon, \quad \varepsilon = \frac{1}{2}(\nabla u + {}^{\mathrm{T}}\nabla u) \quad \text{in } \Omega \times (0,T). \tag{1}$$

The boundary $\partial\Omega$ is partitioned into three disjoint open subsets Γ^D, Γ^N, and Γ^c. Dirichlet and Neumann conditions are prescribed on Γ^D and Γ^N, respectively,

$$u = u_D \quad \text{on } \Gamma^D \times (0,T), \quad \sigma \cdot n = f_N \quad \text{on } \Gamma^N \times (0,T), \tag{2}$$

where n denotes the outward unit normal to Ω. We set $u_n := u_{|\partial\Omega} \cdot n$ and $\sigma_n := n \cdot \sigma_{|\partial\Omega} \cdot n$, the normal displacement and the normal stress on $\partial\Omega$, respectively. On Γ^c, a unilateral contact condition, also called Signorini condition, is imposed,

$$u_n \le \gamma_0,\ \sigma_n(u) \le 0,\ \sigma_n(u)(u_n - \gamma_0) = 0 \quad \text{on } \Gamma^c \times (0,T), \tag{3}$$

where γ_0 is the initial gap. At the initial time, the displacement and velocity fields are prescribed,

$$u(0) = u^0, \quad \dot{u}(0) = v^0 \quad \text{in } \Omega. \tag{4}$$

2.2 Space Semi-Discrete Formulation

Firstly, we discretize the problem in space with a finite element method. We assume that the mesh is compatible with the above partition of the boundary. The number of degrees of freedom is denoted by N_d. Let K, M, and $F(t)$ be the stiffness matrix, the mass matrix, and the column vector of the external forces, respectively. Let N_c be the number of nodes lying on the contact boundary. Set $N_d^* := N_d - N_c$. For the sake of simplicity, suppose that the degrees of freedom associated with normal displacements at the contact boundary are numbered from $N_d^* + 1$ to N_d. The modified mass matrix is defined as

$$M^* = \begin{pmatrix} M_{**} & 0 \\ 0 & 0 \end{pmatrix}.$$

Many choices are possible to build the block M_{**}. In [6, 10], the authors devise various methods to preserve some features of the standard mass matrix (the total mass, the center of gravity, and the moments of inertia). We can also simply keep the corresponding block in the standard mass matrix.

We define the linear normal trace operator on Γ^c, $g : v \longmapsto -v_{|\Gamma} \cdot n$ and the associated matrix G. Note that the size of G is $N_c \times N_d$. We denote by $\{G_i\}_{1 \le i \le N_c}$ the rows of the matrix G. Thus, $G_i u$ yields the value of the normal displacement at the ith node of the contact boundary. Let G_0 be the column vector associated with the initial gap γ_0 and denote by $\{G_{0i}\}_{1 \le i \le N_c}$ the entries of G_0.

We can now formulate the space semi-discrete problem

$$M^* \ddot{u}(t) + Ku(t) = F(t) + {}^{\mathrm{T}}Gr(t), \tag{5}$$

$$Gu(t) \ge G_0, \quad r(t) \ge 0, \quad {}^{\mathrm{T}}r(t)(Gu(t) - G_0) = 0. \tag{6}$$

Here, $r(t)$ stands for the contact pressure equal which is the opposite of the normal stress (with outward normal) at $x = 0$. If we set $u(t) = \begin{pmatrix} lu_*(t) \\ u_c(t) \end{pmatrix}$, $K = \begin{pmatrix} llK_{**} & K_{*c} \\ K_{c*} & K_{cc} \end{pmatrix}$, $F(t) = \begin{pmatrix} lF_*(t) \\ F_c(t) \end{pmatrix}$, and $G = (0\ G_c)$, then equations (5) and (6) can be recast as

$$M_{**}\ddot{u}_*(t) + K_{**}u_*(t) + K_{*c}u_c(t) = F_*(t), \tag{7}$$

$$K_{c*}u_*(t) + K_{cc}u_c(t) = F_c(t) + {}^{\mathrm{T}}G_c r(t), \tag{8}$$

$$G_c u_c(t) \geq G_0, \quad r(t) \geq 0, \quad {}^{\mathrm{T}}r(t)(G_c u_c(t) - G_0) = 0. \tag{9}$$

For a given t and a given $u_*(t)$, there exists one and only one $u_c(t)$ satisfying (8) and (9). Denote by $Q : [0,T] \times \mathbb{R}^{N_d^*} \to \mathbb{R}^{N_c}$ the non-linear map such that $u_c(t) = Q(t, u_*(t))$. Then, problem (7)–(9) amounts to seeking a displacement $u : [0,T] \to \mathbb{R}^{N_d}$ such that, for all $t \in [0,T]$,

$$M_{**}\ddot{u}_*(t) + K_{**}u_*(t) + K_{*c}Q(t, u_*(t)) = F_*(t), \tag{10}$$

$$u_c(t) = Q(t, u_*(t)), \tag{11}$$

with the initial conditions $u(0) = u^0$ and $\dot{u}(0) = v^0$ discretizing u^0 and v^0. The operator $Q(t,\cdot)$ is Lipschitz continuous at each time t, so that equation (10) is a Lipschitz system of ordinary differential equations. Therefore, it has a unique solution u_*, twice differentiable in time. Owing to (11), u_c is differentiable in time almost everywhere. Furthermore, for all $t_0 \in [0,T]$, the following energy balance holds:

$$E^*(u(t_0)) - E^*(u(0)) = \int_0^{t_0} F(t) \cdot \dot{u}(t)dt, \tag{12}$$

where

$$E^*(v) := \frac{1}{2}{}^{\mathrm{T}}\dot{v}M^*\dot{v} + \frac{1}{2}{}^{\mathrm{T}}vKv.$$

The detailed mathematical analysis of the space semi-discrete modified mass formulation can be found in [10]. A result of convergence of the space semi-discrete solutions to a continuous solution is proven for viscoelastic materials in [4].

Remark 1. With a standard mass term, proving the existence of a semi-discrete solution to the dynamic contact problem is quite delicate. It is necessary to add an impact law and to work with BV and measures spaces [2, 15]. The modification of the mass term greatly simplifies the analysis.

3 Semi-Explicit Time Schemes

For simplicity, the interval $[0,T]$ is divided into equal subintervals of length Δt. We set $t^n = n\Delta t$ and denote by u^n, $\dot{u}^n$, and $\ddot{u}^n$ the approximations of $u(t^n)$, $\dot{u}(t^n)$, and $\ddot{u}(t^n)$, respectively. We also set $F^n := F(t^n)$.

3.1 The Semi-Explicit Scheme with Modified Mass

The modified mass matrix is built from the lumped mass matrix, so that M_{**} is diagonal. We propose the following semi-explicit discretization with modified mass:

Discretization 3.1 (Central differences-modified mass) *Seek* $u^{n+1} \in \mathbb{R}^{N_d}$ *such that*

$$M_{**}\left(\frac{u_*^{n+1} - 2u_*^n + u_*^{n-1}}{\Delta t^2}\right) + K_{**}u_*^n + K_{*c}Q(t^n, u_*^n) = F_*^n, \tag{13}$$

$$u_c^{n+1} = Q(t^{n+1}, u_*^{n+1}). \tag{14}$$

In practice, the equations are solved in the following way:

1. Seek $u_*^{n+1} \in \mathbb{R}^{N_d^*}$ such that

$$M_{**}\left(\frac{u_*^{n+1} - 2u_*^n + u_*^{n-1}}{\Delta t^2}\right) + K_{**}u_*^n + K_{*c}u_c^n = F_*^n. \tag{15}$$

2. Seek $u_c^{n+1} \in \mathbb{R}^{N_c}$ and $r^{n+1} \in \mathbb{R}^{N_c}$ such that

$$K_{c*}u_*^{n+1} + K_{cc}u_c^{n+1} = F_c^{n+1} + {}^{\mathrm{T}}G_c r^{n+1}, \tag{16}$$

$$G_c u_c^{n+1} \geq G_0, \quad r^{n+1} \geq 0, \quad {}^{\mathrm{T}}r^{n+1}(G_c u_c^{n+1} - G_0) = 0. \tag{17}$$

The first step is explicit. Since the mass matrix M_{**} is diagonal, it requires only a matrix-vector product. The second step is a constrained problem, similar to a static contact problem, it concerns only the variable u_c^{n+1}, and thus only a few degrees of freedom. This problem can be numerically solved with efficient methods, such as the primal-dual active set method [9].

We notice that Discretization 3.1 amounts to

$$M^*\left(\frac{u^{n+1} - 2u^n + u^{n-1}}{\Delta t^2}\right) + Ku^n = F^n + {}^{\mathrm{T}}Gr^n,$$

$$Gu^n \geq G_0, \quad r^n \geq 0, \quad {}^{\mathrm{T}}r^n(Gu^n - G_0) = 0.$$

3.2 The Semi-Explicit Scheme with Implicit Contact

A detailed review of time-integration schemes, including Discretization 3.1, for the finite element dynamic Signorini problem can be found in [5]. For brevity, we only consider herein a single scheme for comparison combining a central difference scheme with implicit contact [3, 12, 13]. The reason is that this scheme also achieves an exact enforcement of the contact condition, is parameter-free, and allows for a similar CFL condition to that of Discretization 3.1.

Discretization 3.2 (Central differences-implicit contact) *Seek* $u^{n+1} \in \mathbb{R}^{N_d}$ *and* $r^{n+1} \in \mathbb{R}^{N_c}$ *such that*

$$M\left(\frac{u^{n+1} - 2u^n + u^{n-1}}{\Delta t^2}\right) + Ku^n = F^n + {}^{\mathrm{T}}Gr^{n+1}, \tag{18}$$

$$Gu^{n+1} \geq G_0, \quad r^{n+1} \geq 0, \quad {}^{\mathrm{T}}r^{n+1}(Gu^{n+1} - G_0) = 0. \tag{19}$$

In practice, with a lumped mass matrix, this scheme is solved in the following way (observe that the first step is again explicit and requires a single matrix-vector product):

1. Seek $u^{n+1} \in \mathbb{R}^{N_d}$ such that

$$M\left(\frac{u^{n+1} - 2u^n + u^{n-1}}{\Delta t^2}\right) + Ku^n = F^n. \tag{20}$$

2. If at the ith node of the contact boundary, $G_i u^{n+1} < G_{0i}$, then u^{n+1} is modified so that $G_i u^{n+1} = G_{0i}$.

3.3 Energy Balance

For the central difference scheme, the discrete velocity is

$$\dot{u}^n := \frac{u^{n+1} - u^{n-1}}{2\Delta t}$$

and the discrete acceleration

$$\ddot{u}^n := \frac{u^{n+1} - 2u^n + u^{n-1}}{\Delta t^2}.$$

We define the energy of the system at time t^n as

$$E^n := \frac{1}{2}{}^{\mathrm{T}}\dot{u}^n M \dot{u}^n + \frac{1}{2}{}^{\mathrm{T}}u^n K u^n. \tag{21}$$

In linear elastodynamics, the central difference scheme does not preserve the energy. Nevertheless, the scheme preserves the following quadratic form, referred to as a shifted energy:

$$\mathscr{E}^n := E^n - \frac{\Delta t^2}{8}{}^{\mathrm{T}}\ddot{u}^n M \ddot{u}^n. \tag{22}$$

We also define the modified energy E_*^n and the modified shifted energy $\mathscr{E}_*^n$, where the standard mass matrix is replaced by the modified mass matrix. Proceeding as in [15], we obtain the following modified shifted energy balance for Discretization 3.1:

$$\mathscr{E}_*^{n+1} - \mathscr{E}_*^n = {}^{\mathrm{T}}\left(\frac{r^{n+1}+r^n}{2}\right)(Gu^{n+1} - Gu^n) + {}^{\mathrm{T}}\left(\frac{F^{n+1}+F^n}{2}\right)(u^{n+1} - u^n), \quad (23)$$

and the following shifted energy balance for Discretization 3.2:

$$\mathscr{E}^{n+1} - \mathscr{E}^n = {}^{\mathrm{T}}\left(\frac{r^{n+2}+r^{n+1}}{2}\right)(Gu^{n+1} - Gu^n) + {}^{\mathrm{T}}\left(\frac{F^{n+1}+F^n}{2}\right)(u^{n+1} - u^n). \quad (24)$$

For Discretization 3.1, when a node comes into contact ($G_i u^n > G_{0i}$, $r^n = 0$, $G_i u^{n+1} = G_{0i}$, $r^{n+1} > 0$), the work of the contact pressure is negative; when a node is released ($G_i u^n = G_{0i}$, $r^n > 0$, $G_i u^{n+1} > G_{0i}$, $r^{n+1} = 0$), the work of the contact pressure is positive. For Discretization 3.2, the work of the contact pressure is always non-positive. Our numerical tests (see Figures 3 and 7 below) indicate that Discretization 3.1 yields a much more satisfying energy behavior than Discretization 3.2 especially in cases with multiple impacts.

3.4 Stability

In linear elastodynamics, the central difference scheme presents a sufficient stability condition (CFL condition) of the form

$$\Delta t \leq 2\sqrt{\frac{\lambda_M}{\Lambda_K}}, \quad (25)$$

where Λ_K is the largest eigenvalue of K and λ_M is the lowest eigenvalue of M [8]. In particular, in 1D, with linear finite elements and a lumped mass matrix, the CFL condition is $c_0 \Delta t \leq \Delta x$, where Δx is typically the smallest mesh element diameter and $c_0 := \sqrt{E/\rho}$ is the wave speed in 1D. More generally the CFL condition can be formulated in the form $c_d \Delta t \leq O(\Delta x)$, where

$$c_d := \sqrt{\frac{E(1-\nu)}{\rho(1+\nu)(1-2\nu)}}$$

is the speed of dilatational waves. Here, E is the Young modulus and ν the Poisson ratio.

We observe numerically that the stability condition for Discretization 3.1 takes the same form as in the linear case. A rigorous proof of this fact is still open. For Discretization 3.2, this fact is proven in [14], exploiting that the work of the contact pressure is always non-positive.

Remark 2. As mentioned in the Introduction, the penalty contact condition worsens the CFL condition. It introduces a constraint on the time step of the form $\Delta t \leq O(\sqrt{\rho\varepsilon}\Delta x_c)$, where Δx_c is the mesh size near the contact boundary and ε the penalty parameter [1].

3.5 Gibbs Phenomenon

When approximating a solution presenting a shock or a sharp wave front with finite elements in space and a time-stepping scheme, spurious oscillations are observed. This is the well-known Gibbs phenomenon which is due to the poor approximation of eigenmodes associated with the high frequencies. In the elastodynamics Signorini problem, we must deal with shock waves after the impacts. If we plot the stress distribution computed with Discretization 3.1, we observe this phenomenon (Figure 4). These oscillations can be damped using explicit dissipative schemes, such as the Tchamwa–Wielgosz scheme or the Chung–Hulbert scheme (see [11] and references therein). The damping obtained with a Chung–Hulbert scheme is presented in Figure 4.

4 Numerical Simulations

4.1 1D Simulations

We present two 1D problems. Both problems can be formulated in the same setting. We consider an elastic bar dropped against a rigid ground (Figure 1). The bar is dropped, undeformed, from a height h_0, with an initial velocity $-v_0$, under a gravity acceleration $g_0 \geq 0$. The length of the bar is denoted by L, the Young modulus by E and the density by ρ. In the first problem, $v_0 > 0$ and $g_0 = 0$, so that there is just one impact. In the second problem, $v_0 = 0$ and $g_0 > 0$, so that the bar can make several bounces. The exact solutions to these problems can be found in [5]. The parameters used in the numerical simulations are $E = 900$, $\rho = 1$, $L = 10$, $h_0 = 5$. In the first benchmark $v_0 = 10$, $g_0 = 0$; in the second benchmark $v_0 = 0$, $g_0 = 10$. The bar is discretized with a uniform mesh size Δx and linear finite elements are used. We define the Courant number $\nu_c := c_0(\Delta t/\Delta x)$. Recall that the CFL condition is in this case $\nu_c \leq 1$.

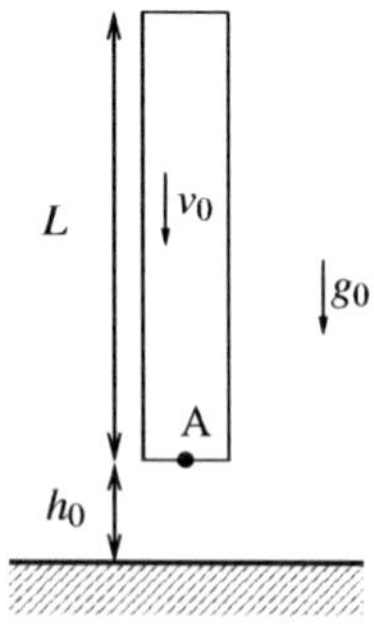

Fig. 1 Impact of an elastic bar. Initial configuration.

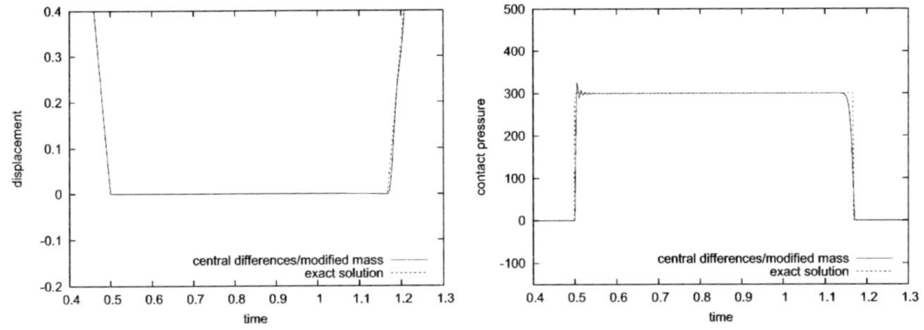

Fig. 2 Impact of an elastic bar. Displacement at point A (left) and contact pressure (right). Discretization 3.1. $\Delta x = 0.1$, $\Delta t = 0.0025$, $v_c = 0.75$.

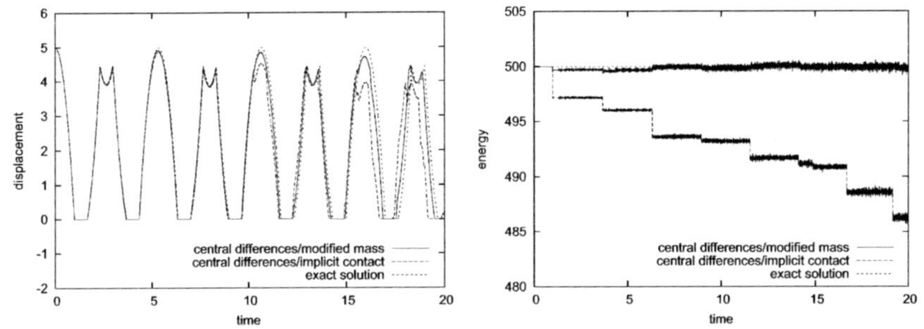

Fig. 3 Bounces of an elastic bar. Displacement at point A (left) and energy (right). Discretizations 3.1 and 3.2. $\Delta x = 0.1$, $\Delta t = 0.0025$, $v_c = 0.75$.

Figure 2 shows the displacement at point A and the contact pressure for the impact benchmark obtained with Discretization 3.1. In Figure 3, we compare Discretizations 3.1 and 3.2 (displacement at point A and energy) for the multiple bounce benchmark. What we call energy is the quantity $E_*^n - {}^{\mathrm{T}}Fu^n$ for Discretization 3.1 and the quantity $E^n - {}^{\mathrm{T}}Fu^n$ for Discretization 3.2, where F is the (constant) vector of external forces. For Discretization 3.1, the amplitude of energy oscillations is 1.3; this amplitude decreases to 0.5 by halving Δx and Δt so that the decrease appears to be at least linear with the time step at fixed Courant number. Finally, the Gibbs phenomenon and its treatment by a dissipative scheme is illustrated in Figure 4.

4.2 2D Simulations

Now we present a 2D numerical simulation: the bounces of an elastic disk against a rigid ground (Figure 5). The disk is dropped, undeformed, with a zero initial velocity, under a gravity acceleration g_0, its center being at a height h_0. The disk has a radius R. The material is supposed to be linear elastic (plane strain) with a Young modulus E, a Poisson ratio ν, and a mass density ρ. The contact bound-

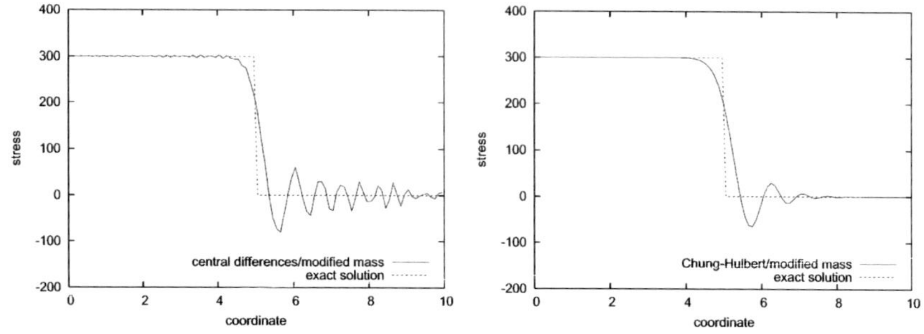

Fig. 4 Impact of an elastic bar. Stress in the bar at time t=1.0. $\Delta x = 0.1$, $\Delta t = 0.0025$, $\nu_c = 0.75$. Discretization 3.1 (left) and Chung–Hulbert scheme with maximal dissipation (right).

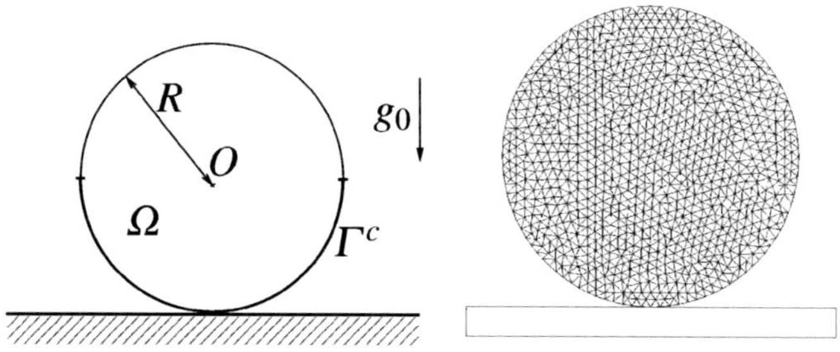

Fig. 5 Bounces of an elastic disk. Reference configuration (left), mesh (right).

ary is the lower half part of the disk boundary. We formulate the contact condition using the normal vector to the ground. The parameters used in the numerical simulations are $E = 4000$, $\nu = 0.2$, $\rho = 100$, $g_0 = 5$, $R = 1$, $h_0 = 0.1$. The present choice of parameters is relatively severe for energy behavior because of relatively high impact velocity and low Young modulus. The disk is meshed with triangles (100 edges on the boundary, 1804 triangles, 953 vertices, see Figure 5) and we use linear finite elements. The number of nodes lying on the contact boundary is 51. The simulations are performed with FREEFEM++ [7]. We define the Courant number $\nu_c := c_d(\Delta t/\Delta x)$, where Δx is the size of the edges at the boundary. Numerically, the observed CFL condition is $\nu_c \leq 0.65$.

In Figure 6 we present the deformation and von Mises stress at different times obtained with Discretization 3.1. The von Mises stress is given by

$$\sigma_{VM}^2 := \sum_{i,j=1}^{3} \left| \sigma_{ij} - \delta_{ij} \frac{1}{3} \mathrm{tr}(\sigma) \right|^2 .$$

In Figure 7, we compare Discretizations 3.1 and 3.2 (displacement of the disk center and energy using the same definitions as in the 1D case). In Figure 8, we consider a coarser discretization (50 edges on the boundary) for Discretization 3.1. The energy behavior is quite satisfactory, even with the coarse discretization. The amplitude of

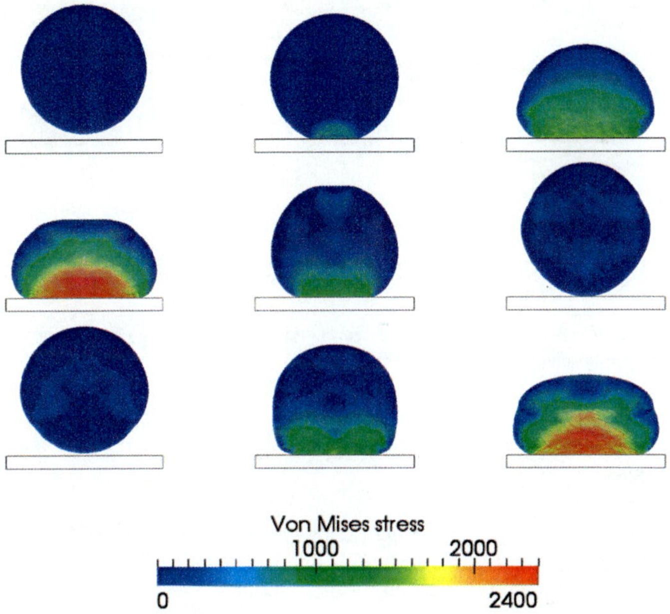

Fig. 6 Bounces of a disk. Discretization 3.1. $\Delta x = 0.0628$, $\Delta t = 0.005$, $\nu_c = 0.53$. Deformation and von Mises stress at times $t = 0$ s, $t = 0.25$ s, $t = 0.5$ s, $t = 0.75$ s, $t = 1.0$ s, $t = 1.25$ s, $t = 1.5$ s, $t = 1.75$ s, $t = 2.0$ s.

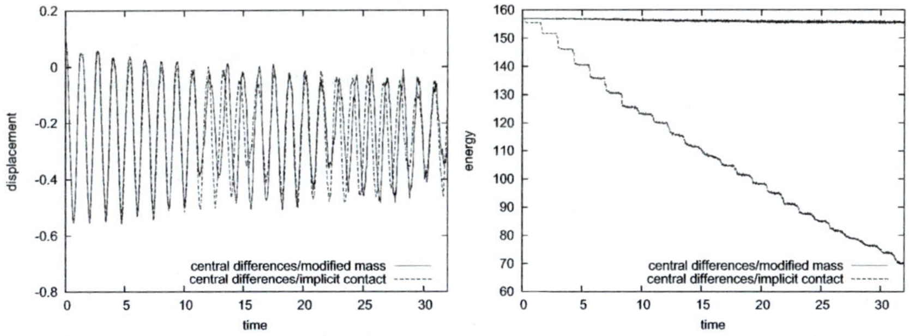

Fig. 7 Bounces of a disk. Displacement at the center of the disk (left) and energy (right). Discretizations 3.1 and 3.2. $\Delta x = 0.0628$, $\Delta t = 0.005$, $\nu_c = 0.53$.

the energy oscillations is, respectively, 1.0 and 0.3 with the fine and coarse discretizations; thus, the amplitude of energy oscillations appears to decrease at least linearly with the time step at fixed Courant number. The initial discrepancy between the two energies is due to the approximation of the disk by a polygon. We observe a shortening of the oscillation period because of the loss of mass in the modified mass matrix (2.75 and 1.37% mass loss, respectively, with the fine and coarse discretizations). This can be improved with a mass redistribution as in [6, 10].

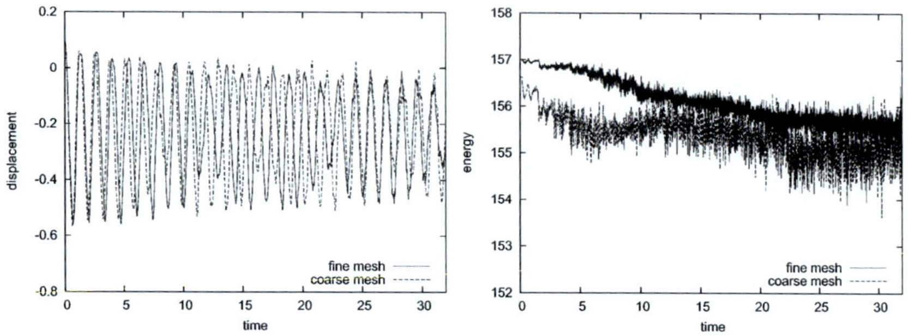

Fig. 8 Bounces of a disk. Displacement at the center of the disk (left) and energy (right). Discretization 3.1. Fine discretization: $\Delta x = 0.0628$, $\Delta t = 0.005$, $v_c = 0.53$. Coarse discretization: $\Delta x = 0.126$, $\Delta t = 0.01$, $v_c = 0.53$.

References

1. Belytschko, T., Neal, M.: Contact-impact by the pinball algorithm with penalty and Lagrangian methods. Int. J. Numer. Methods Engrg., 547–572 (1991)
2. Brogliato, B.: Nonsmooth Impact Mechanics. Springer, London (1996)
3. Carpenter, N., Taylor, R., Katona, M.: Lagrange constraints for transcient finite element surface contact. Int. J. Numer. Meth. Engng. 32, 103–128 (1991)
4. Doyen, D., Ern, A.: Convergence of a space semi-discrete modified mass method for the dynamic Signorini problem. Commun. Math. Sci. 7(4), 1063–1072 (2009)
5. Doyen, D., Ern, A., Piperno, S.: Time-integration schemes for the finite element dynamic Signorini problem. SIAM J. Sci. Comput. 33(1), 223–249 (2011)
6. Hager, C., Hüeber, S., Wohlmuth, B.I.: A stable energy-conserving approach for frictional contact problems based on quadrature formulas. Int. J. Numer. Methods Engrg. 73(2), 205–225 (2008)
7. Hecht, F., Pironneau, O.: Freefem++, `http://www.freefem.org`
8. Hughes, T.J.R.: The Finite Element Method. Prentice Hall, Englewood Cliffs (1987)
9. Ito, K., Kunisch, K.: Semi-smooth Newton methods for variational inequalities of the first kind. M2AN Math. Model. Numer. Anal. 37(1), 41–62 (2003)
10. Khenous, H.B., Laborde, P., Renard, Y.: Mass redistribution method for finite element contact problems in elastodynamics. Eur. J. Mech. A Solids 27(5), 918–932 (2008)
11. Nsiampa, N., Ponthot, J.-P., Noels, L.: Comparative study of numerical explicit schemes for impact problems. Int. J. Impact Engrg. 35(12), 1688–1694 (2008)
12. Paoli, L., Schatzman, M.: A numerical scheme for impact problems. I. The one-dimensional case. SIAM J. Numer. Anal. 40(2), 702–733 (2002)
13. Paoli, L., Schatzman, M.: A numerical scheme for impact problems. II. SIAM J. Numer. Anal. 40(2), 734–768 (2002)
14. Schatzman, M., Bercovier, M.: Numerical approximation of a wave equation with unilateral constraints. Math. Comp. 53(187), 55–79 (1989)
15. Stewart, D.E.: Rigid-body dynamics with friction and impact. SIAM Rev. 42(1), 3–39 (2000)

An Explicit Asynchronous Contact Algorithm for Elastic-Rigid Body Interaction

Raymond A. Ryckman and Adrian J. Lew

Abstract The use of multiple-time-step integrators can provide substantial computational savings over traditional one-time-step methods for the simulation of solid dynamics, while maintaining desirable properties, such as energy conservation. Contact phenomena generally require the adoption of either an implicit algorithm or the use of unacceptably small time steps to prevent large amount of numerical dissipation from being introduced. This paper introduces a new explicit dynamic contact algorithm that, by taking advantage of the asynchronous time-stepping of Asynchronous Variational Integrators (AVI), delivers an outstanding energy performance at a much more acceptable computational cost. We demonstrate the performance of the numerical method with several three-dimensional examples.

1 Introduction

Numerical methods for impact and contact between deformable and rigid bodies during transient dynamics is an area of computational mechanics that has not yet reached the maturity and robustness of others. While very complex simulations are nowadays performed, issues of robustness, accuracy, parallel scalability and computational efficiency still plague problems in which contact plays an important role. In this paper we address one particular issue, concerned with the construction of explicit contact algorithms that showcase good energy conservation properties through contact events at moderately large time steps.

Explicit in time approaches to contact between deformable bodies has been dominated by penalty methods. As described in many standard textbooks [10, 16], these methods consist of replacing the walls of the bodies by stiff potentials (often quadratic) that allow for some interpenetration between bodies. By making the poten-

Raymond A. Ryckman · Adrian J. Lew
Department of Mechanical Engineering, Stanford University, 496 Lomita Mall, Stanford, CA 94305, U.S.A.; e-mail: {rryckman, lewa}@stanford.edu

G. Zavarise & P. Wriggers (Eds.): Trends in Computational Contact Mechanics, LNACM 58, pp. 169–191.
springerlink.com

tials stiffer the representation of the walls is improved and the contact conditions are more accurately imposed. The main drawback of such an approach is that in increasing the stiffness of the penalty potentials to improve accuracy, the time step of the simulation needs to be severely reduced for stability reasons. So it is often challenging to find an appropriate value of the penalty parameter that strikes the necessary balance between accuracy and efficiency, and that is useful throughout an entire simulation involving material nonlinearities.

Implicit approaches to dynamic problems with contact often involve a more accurate imposition of the contact conditions, through for example Lagrange multipliers or augmented Lagrangian methods, although penalty approaches have also been proposed [1]. As always, for some problems it is convenient to solve the large systems of nonlinear equations characteristic of implicit schemes instead of stepping explicitly in time.

Recently variational integrators for contact problems have been formulated [4,5]. The resulting integrators are implicit, and exhibit outstanding energy conservation properties for very long integration times. The main drawback is that the impact time of each contact event need to be solved for. Arguing that for problems involving deformable bodies solving for each contact event could be a computationally overwhelming task, Cirak and West in [3] proposed relaxing this condition, as penalty methods do. In doing so, they lost the precise variational nature of the algorithm, but obtained an *explicit* contact algorithm with very good conservation properties as well. The algorithm can be summarized as follows: a momentum-preserving variational integrator is adopted to integrate the system forward for some given initial condition. If at a given time step interpenetration is detected, a projection is defined to move the configuration of the system slightly to a penetration-free one, and a new momentum for the system is computed to respect the appropriate energy and momentum balances. This new position and momentum for the system can be interpreted as new initial conditions to be integrated forward with the variational integrator until the following contact event. A similar design concept for purely rigid body interactions can be found in [2]. Finally, a nice aspect of the work in [3] is that they proposed an intrinsic decomposition of the generalized momentum of the system during a contact event that enables the extension of their algorithm to include a coefficient of restitution.

A drawback of the algorithm in [3] is that at each contact event an energy drift is observed. Consequently, the outstanding energy behavior of the variational integrator between contact events is deteriorated by energy loss or gain at each contact event (if the system is supposed to conserve energy). To obtain a satisfactory energy behavior very small time steps need to be adopted. The algorithm proposed here addresses this problem. The basic idea is to take advantage of the possibility of advancing each element with its own time step afforded by the Asynchronous Variational Integrators (AVI) [11–13] to take small time steps only in those elements that violate the contact constraints. As shown here, the resulting algorithm has a much better energy behavior at a small additional computational cost, because the time step is reduced for very few elements in the mesh during the contact event only. Since AVI are momentum-preserving and variational, they display the nice behavior we expect to have between two contact events. The basic idea of using AVIs to adopt smaller time steps on those elements that may be in contact to improve the energy conservation properties of the algorithm was first presented in [8] and published in [14].

A similar concept has been proposed independently in [7]. In this case, the authors constructed a multi-layer penalty potential with increasing stiffness as the depth of penetration grows, and use AVI to avoid having to reduce their time step in the entire mesh. If the time steps are chosen small enough, then the resulting algorithms are perfectly variational. If the time steps are not small enough, similar energy drift as that observed in the algorithm in [3] would be observed.

This rest of the paper is organized as follows: in Section 2 we review the decomposition of the momentum in [3] for frictionless contact problems in finite dimensional system. We do so because deformable bodies, once spatially discretized with finite elements, lead to finite dimensional mechanical systems. It is the evolution in time of this finite dimensional system that we are interested in computing. In Section 3 we then discuss the basic idea behind the integrators proposed here further, leading to a review of AVI in Section 4. The asynchronous contact algorithm is introduced in Section 5. In Section 6 we examine the numerical performance of the algorithm, and numerical examples are shown in Section 7.

2 Frictionless Contact Mechanics for Finite Dimensional Mechanical Systems

We briefly recall here elementary aspects of contact mechanics for finite dimensional mechanical systems. A more detailed description can be found elsewhere, e.g., [3] and references therein. We consider a finite dimensional mechanical system whose motion can be described with Cartesian coordinates $\mathbf{q} \in \mathbb{R}^d$. The equations of motion for this system written in these coordinates take the form

$$\mathbf{M}\ddot{\mathbf{q}} = \mathbf{F}, \tag{1}$$

at all times, where $\mathbf{F}$ denotes all generalized forces acting on the particles, and $\mathbf{M}$ denotes the symmetric and positive definite mass matrix of the system in these coordinates. Although not necessary, we have assumed that $\mathbf{M}$ in these coordinates does not depend on $\mathbf{q}$. This is the case for most mass matrices obtained as a result of a finite element discretization. For this set of coordinates the conjugate momenta are defined as

$$\mathbf{p} = \mathbf{M}\dot{\mathbf{q}}. \tag{2}$$

The generalized forces often satisfy $\mathbf{F}(\mathbf{q}) = -\nabla V(\mathbf{q})$, for some potential energy V. In this case, the finite dimensional mechanical system above is Hamiltonian, and among other things, the mechanical energy of the system

$$H(\mathbf{q},\mathbf{p}) = \frac{1}{2}\mathbf{p}^{\mathsf{T}} \cdot \mathbf{M} \cdot \mathbf{p} + V(\mathbf{q}) \tag{3}$$

is conserved throughout the trajectory.

A type of contact constraint is defined as a restriction of the mechanical system to have its coordinates satisfy $\mathbf{q} \in \overline{Q}$, where Q is an open subset of $\mathbb{R}^d$. The closure of Q, $\overline{Q}$, is termed the *admissible region*. We will assume that Q has a piecewise smooth boundary, with a uniquely defined outer normal $\mathbf{n}$ almost everywhere. We will also assume that $\overline{Q}$ can be specified implicitly through a function $g : \mathbb{R}^N \to \mathbb{R}$ as

$$\overline{Q} = \left\{\mathbf{q} \in \mathbb{R}^d \mid g(\mathbf{q}) \leq 0\right\}. \tag{4}$$

For example, for a system of N particles in two-dimensions with Cartesian coordinates $(x_1, y_1, x_2, y_2, \ldots, x_N, y_N)$ that need to move in the $\{y > 0\}$ region of the plane, the function g may take the form

$$g(x_1, y_1, x_2, y_2, \ldots, x_N, y_N) = -\sum_{i=1}^{N} \min(y_i, 0) \tag{5}$$

The function g needs to satisfy some smoothness requirements near the boundary of Q. In particular, we would like the one-sided gradient ∇g computed from either inside or outside Q at any point in the boundary to be well-defined and different than zero almost everywhere along ∂Q. This gradient may be defined and have different values on the two sides of the boundary, but either one will satisfy if it is parallel to the normal to the boundary at all such points.

Under the contact constraint, the evolution of the mechanical system may undergo impact events. An impact event occurs when at a time t_c the system satisfies $g(\mathbf{q}(t_c)) = 0$. Notice that it is possible to have a continuous time interval of contact events, during which the system moves along the boundary of the admissible region. For simplicity, the discussion below focuses on isolated contact events, in which the momentum of the system is continuous in an interval of time around t_c, except perhaps at t_c. In this context, we define

$$\mathbf{p}^{\pm} = \lim_{t \to t_c^{\pm}} \mathbf{p} \tag{6}$$

as the momentum of the system just before $(-)$ and just after $(+)$ the contact event. Since at each contact event the momentum of the system may be discontinuous, we may have $\mathbf{p}^- \neq \mathbf{p}^+$. The relation between the two is obtained from the contact conditions

$$p_n^+ = -e p_n^- \tag{7}$$

$$\mathbf{p}_t^+ = \mathbf{p}_t^-, \tag{8}$$

where the normal and tangential component of the momentum of the system are defined as

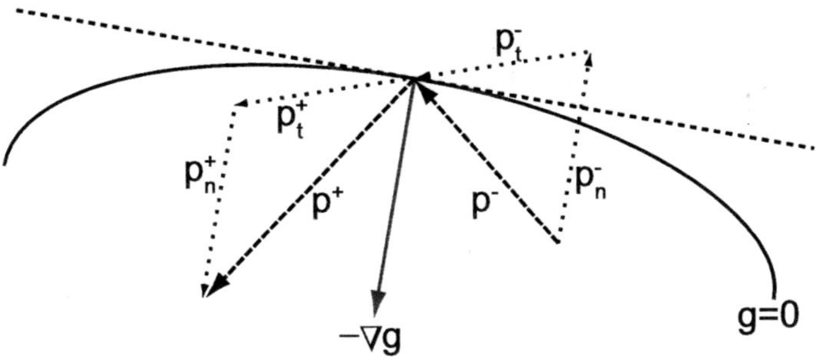

Fig. 1 When the system reaches the boundary of the admissible region Q, indicated here with the curve $g = 0$, the normal component of the momentum of the system in the scalar product defined by the inverse of the mass matrix is instantaneously changed. The normal to the boundary at the point of contact is parallel to ∇g. Here we show the case in which the collision is perfectly elastic, so energy is conserved ($e = 1$).

$$p_n^{\pm} = \mathbf{p}^{\pm} \cdot \mathbf{M}^{-1} \cdot \mathbf{n}(\mathbf{q}) \qquad \text{and} \qquad \mathbf{p}_t^{\pm} = \mathbf{p}^{\pm} - p_n^{\pm}\mathbf{n}(\mathbf{q}), \tag{9}$$

where the unit normal $\mathbf{n}(\mathbf{q})$ to ∂Q at $\mathbf{q}$ can be computed from g as

$$\mathbf{n}(\mathbf{q}) = \frac{\nabla g(\mathbf{q})}{\left[\nabla g(\mathbf{q}) \cdot \mathbf{M}^{-1} \cdot \nabla g(\mathbf{q})\right]^{1/2}}, \tag{10}$$

and $e \in [0,1]$ makes it possible to consider a coefficient of restitution. For elastic collisions, $e = 1$. These equations are graphically depicted in Figure 1. This normal and tangential decomposition of the momentum of the system was introduced in [3].

3 Symplectic Algorithms and Energy Conservation

Symplectic or variational integrators have a number of desirable properties that make them attractive for the simulation of problems in solid dynamics. Chief among these properties is exact linear and angular momentum conservation and excellent long-term energy conservation. Of course, this can only occur if the mechanical system has the same properties.

A simple way to regard the dynamics of the type of finite dimensional mechanical systems above is as a sequence of discrete contact events at time instants $t_1, t_2, \ldots, t_k$, ... The system evolves between times t_{k-1} and t_k according to (1). At time t_k, the contact conditions (6) and (7) compute a new set of initial conditions for the evolution between t_k and t_{k+1} (same coordinates $\mathbf{q}$, but the momentum changes from $\mathbf{p}^-$ to $\mathbf{p}^+$), and the evolution of the system then continues according to (1). This is sketched in Figure 2.

A natural way to think about the design of time integration algorithms when contact is involved is to select any standard time-integration scheme to evolve the

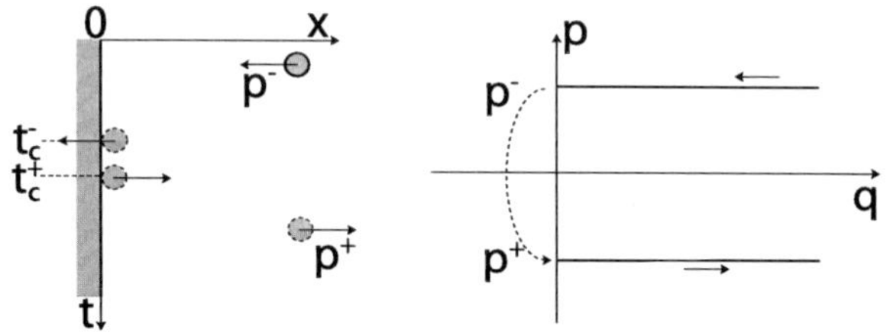

Fig. 2 Sketch of a collision event for a particle moving in one-dimension colliding against a wall at $x = 0$, as shown on the right in the position-time plane. On the left, phase space diagram of the same event. When the particle reaches $x = 0$, its momentum jumps from p^- to p^+.

system between times t_{k-1} and t_k, for any k, and then use the contact conditions (6) and (7) to reinitialize the simulation for the next integration interval, in this case (t_k, t_{k+1}). Consequently, if we select a symplectic time-integration scheme to perform the time integration between contact events, we should expect to obtain a time-integrator with outstanding energy and momentum conservation properties.

This ideal scenario has some drawbacks. First, it requires the precise detection of each contact event. Second, while due to the symplectic nature of the time-integration algorithm the energy between contact events is nearly exactly preserved, a noticeable energy change is observed after each contact event. Such energy drift can be analyzed and understood based on the computation of the shadow Hamiltonian of the underlying symplectic integrator, a task we shall not undertake here. However, it also emanates from the analysis that the size of the energy drift after each collision scales with the time step of the time integrator.

This is one key motivation behind the use of an asynchronous time integrator for a deformable body spatially discretized with finite elements. Since generally only a few elements are involved in a contact event at any time, it is possible to obtain a much smaller energy drift on each contact event by carefully reducing the size of the time steps on those elements only.

4 Asynchronous Variational Integrators (AVI)

Asynchronous variational integrators were introduced in [11, 12] as multisymplectic time integrators for deformable bodies. A parallel version of these integrators has been introduced in [9]. We briefly highlight some features below, and refer the reader to the earlier references for a thorough description.

AVI can be regarded as an extension of Newmark's second-order, explicit method (also known as central differences or velocity Verlet). This last algorithm is given by the following update equations that return $(\mathbf{q}^{i+1}, \mathbf{p}^{i+1})$ if $(\mathbf{q}^i, \mathbf{p}^i)$:

$$\mathbf{p}^{i+1/2} = \mathbf{p}^i + \frac{\Delta t}{2}\mathbf{F}(\mathbf{q}^i) \tag{11a}$$

$$\mathbf{q}^{i+1} = \mathbf{q}^i + \Delta t \mathbf{M}^{-1}\mathbf{p}^{i+1/2} \tag{11b}$$

$$\mathbf{p}^{i+1} = \mathbf{p}^{i+1/2} + \frac{\Delta t}{2}\mathbf{F}(\mathbf{q}^{i+1}) \tag{11c}$$

where $\Delta t = t_{i+1} - t_i$ is the time step, and for simplicity the vector of generalized forces $F(\mathbf{q})$ was assumed to depend on the coordinates of the particle only. Of course, other types of forces, such as dissipative forces, can be included in a standard way. With conservative forces only, this is a variational integrator that conserves linear and angular momentum, and nearly exactly preserves the value of the energy for very long times.

AVI extends this algorithm to multiple time steps by relying on an additive decomposition of the generalized forces, namely,

$$\mathbf{F}(\mathbf{q}) = \mathbf{F}_1(\mathbf{q}) + \cdots + \mathbf{F}_N(\mathbf{q}) \tag{12}$$

For example, for a deformable body spatially discretized with finite elements, $\mathbf{F}_i$ are the internal forces computed with element i. Each one of these forces is then assigned a possibly different time step (subjected to stability considerations), and the order in which each one of them needs to be computed is determined based on their time steps. Additionally, the mass matrix $\mathbf{M}$ is assumed to be diagonal, or in the case of finite elements, it is lumped. Equations (11) are then applied to advance the positions and velocities from the time step of one force to the next time step among all forces. More precisely, each node a that belongs to an element K whose internal forces are computed at time t_K are updated at this time according to

$$\mathbf{q}_a^{i_a} = \mathbf{q}_a^{i_a-1} + \Delta t_a^{i_a}\mathbf{M}_a^{-1}\mathbf{p}_a^{i_a-1/2} \tag{13a}$$

$$\mathbf{p}_a^{i_a} = \mathbf{p}_a^{i_a-1/2} + \frac{\Delta t_K}{2}\mathbf{F}_K^a \tag{13b}$$

$$\mathbf{p}_a^{i_a+1/2} = \mathbf{p}_a^{i_a-1/2} + \Delta t_K\mathbf{F}_K^a, \tag{13c}$$

where $\mathbf{M}_a$ is the diagonal entry in the mass matrix for node a, $\mathbf{q}_a^{i_a}$ denotes the coordinates of node a at its i_a-th update, $\Delta t_a^{i_a}$ is time interval between the i_a-th and $(i_a - 1)$-th updates of node a, and $\mathbf{p}_a^{i_a-1/2}$ is the momentum of the node in between two updates. Together with $\mathbf{q}_a^{i_a-1}$, $\mathbf{p}_a^{i_a-1/2}$ is assumed to be known at the beginning of the update. Additionally, $\mathbf{p}_a^{i_a}$ denotes the momentum of the node at the i_a-th update of node a. The internal forces $\mathbf{F}_K$ are computed with the updated positions at time t_K of all nodes in element K, and $\mathbf{F}_K^a$ denotes the internal forces acting on the momenta of node a. Time step Δt_K is computed as the time interval between t_K and the last time the internal forces for the same element were computed. A more detailed description of the algorithm can be found in [6].

In the context of deformable bodies, this reduces to advancing each element with its own time step, computing the internal forces on that element at each time step,

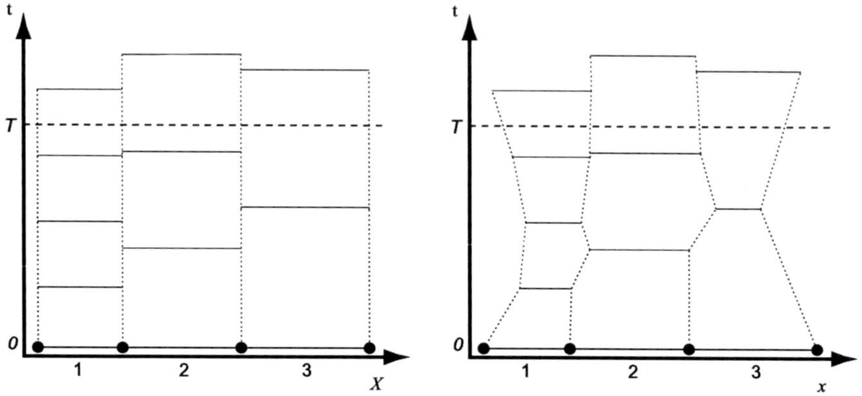

Fig. 3 Example of a one-dimensional, three-element mesh discretizing a deformable body that is advanced in time with AVI. The evolution in time of the reference configuration is shown on the left, while that of the deformed configuration is on the right. The horizontal lines represent each elemental update, while the dashed lines represent the trajectories of the nodes.

and imparting an impulse on each one of its nodes computed from these internal forces. The evolution of the elements in time for a one-dimensional bar is sketched in Figure 3. Each horizontal line above an element is an internal force computation. After each one of these computations, the slope of the trajectories of the nodes in spacetime changes, since the impulse exerted by the element has changed their velocities. Causality is respected by making sure that all operations are performed in the right order. We generally do it through a *priority queue*, although other alternatives are possible.

Therefore, in addition to the outstanding momentum and energy conservation properties, AVI has the desirable feature that the time step of each element can be selected independently. This is in contrast with synchronous explicit Newmark's methods where the smallest element in the mesh generally determines the time step for all elements in the simulation. The stability of asynchronous methods is a complex issue that is not fully resolved yet. For example, combinations of arbitrarily small time steps can generate instabilities, as discussed in [6]. In practice, however, for meshes that do not have very sharp spatial changes on element sizes or material constants, selecting each individual time step to be a fraction of the local CFL condition renders stable integrators. More precisely, the time step for an element with diameter h is set to be $\Delta t = fh/c$, where c is the maximum wave speed of the material and $f \in (0,1)$ is called the *time factor* and sets the time step to be a fraction of the CFL limit.

5 Explicit Asynchronous Contact Algorithm

In the following we introduce an asynchronous contact algorithm based on the ideas discussed in Sections 3 and 4. For simplicity, we restrict the discussion to the collision between a deformable and a rigid body. Self-contact and contact between deformable bodies requires a more complex contact detection algorithm, but the key ideas introduced here remain the same.

Our contact algorithm consists of essentially two main ideas:

1. The consideration of level sets of g near the level set $g(\mathbf{q}) = 0$ to have a collection of possible boundaries of the admissible region near the exact one. Instead of having a fixed admissible region Q, we consider the possibility of the system moving into coordinates outside of Q during a fraction of the time step, adopting as the boundary of the admissible region whatever level set of g the system fall on at that time. This idea lowers the order of convergence of the algorithm (and could make some contact events to be utterly missed in some cases). However, it is a computationally efficient approach because it circumvents the need to find the exact time at which a contact event happens, in the context of an explicit algorithm. A discussion about the convenience of a similar approach is provided in [3]. In contrast to the approach in this last reference, we do not project the configuration of the system back on ∂Q, although in some occasions it could be convenient. For example, when working with thin shells.
2. The creation of a zone near the boundary of the admissible region in which the time step of elements with a node inside it is reduced, termed the *time step refinement zone* (TSRZ). This zone serves to minimize the significance of the violations of the contact constraint introduced by the first idea. More importantly, however, this is what makes the energy behavior of the algorithm substantially better. We are able to do this because with AVI we can choose the time step for each element in the mesh.

We discuss the details of these ideas below.

5.1 Construction of the Admissible Region

Given a deformable body spatially discretized with finite elements, and a rigid body $\mathscr{R}$, we define the admissible region $\overline{Q}$ with

$$g(\mathbf{q}) = -\sum_{a \in \text{nodes}} \min(d(\mathbf{q}_a;\mathscr{R}),0), \tag{14}$$

where the sum is over all nodes in the mesh, $\mathbf{q}_a$ denotes the Cartesian coordinates of node a, and $d : \mathbb{R}^3 \to \mathbb{R}$ is the signed distance function of a point in $\mathbb{R}^3$ to the rigid body $\mathscr{R}$. The value of $d(\mathbf{q}_a;\mathscr{R})$ is negative if $\mathbf{q}_a \in \mathscr{R}$, zero if $\mathbf{q}_a \in \partial\mathscr{R}$, and positive otherwise. Its magnitude is equal to the distance of $\mathbf{q}_a$ to the closest point that belongs to $\overline{\mathscr{R}}$. In this way, g is equal to zero on $\overline{Q}$, and it is positive otherwise.

One key advantage of g defined in this way is that ∇g at a point in ∂Q can be computed by adding derivatives of the signed distance function evaluated at the nodes that instantaneously are inside the rigid body. This is important for the asynchronous algorithm, since in this way ∇g can be computed at each elemental update with only the nodal positions of the element's own nodes.

5.2 Time Step Refinement Zone

The second central element of our contact algorithm is using the possibility of adopting independent time steps on each element of AVI to improve the accuracy of our numerical results without vastly increasing the computational cost. This property not only prevents elements with poor aspect ratios or local refinements from affecting the time step of the entire mesh, but it also allows for time step refinement in regions where additional accuracy is required. One such case is for contact, where insufficiently small time steps can induce energy drifts at each contact event, as well as deep penetrations of the system into the inadmissible region. This is even more true for high velocity contact, where it becomes necessary to prevent elements from inverting as well as the associated onset of instabilities. Locally refining the time step in the region of the contact interface is a nice application of AVI.

The method proposed here takes full advantage of this property by creating a Time Step Refinement Zone (TSRZ). The TSRZ is defined as

$$\mathrm{TSRZ} = \left\{ \mathbf{x} \in \mathbb{R}^3 \mid d(\mathbf{x};\mathscr{R}) < C_{\mathrm{TSRZ}} \right\} \tag{15}$$

for some $C_{\mathrm{TSRZ}} \geq 0$. The calculation of (14) already involves computing $d(\mathbf{q}_a;\mathscr{R})$ for any node a, so the computation of the TSRZ does not involve any additional calculation, at least in the context of deformable and rigid body contact.

The way the TSRZ works is that whenever an element has any of its nodes within the TSRZ, the time factor f on the element is reduced to a fraction of its value, decreasing in this way the time step size. The time step size of an element is changed after the elemental update that follows the entrance of the first node of the element in the TSRZ, and remains that way as long as the above condition is satisfied. The full time step of the element is restored after the elemental update that follows the last node of the element leaving the TSRZ. A sketch of these ideas is shown in Figure 4.

The size of the TSRZ defined by C_{TSRZ} and the time factor within it should be sized so as to prevent nodes from entirely bypassing the zone. Given foreknowledge of the expected maximum velocity attainable by a node, the TSRZs are designed to prevent that. A few iterations might be needed to tune this value for a given problem, so as to avoid defining a very large TSRZ.

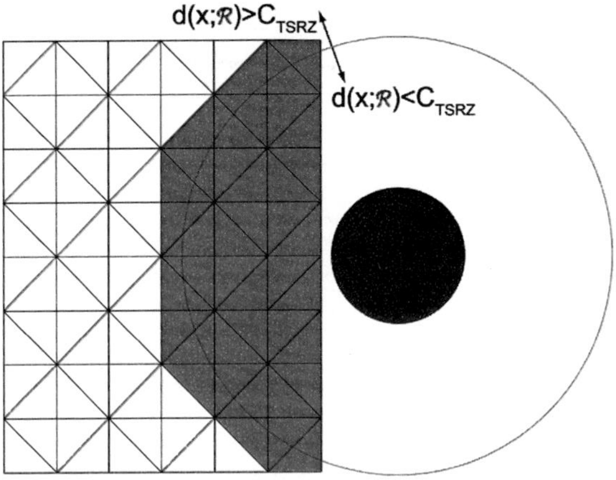

Fig. 4 An example of the Time Step Refinement Zone for the case of a circular rigid body approaching a deformable one. Elements in gray have had their time step reduced, since they have at least one node within the TSRZ.

5.3 Contact Detection and Momentum Reflection

As mentioned earlier, we adopt a strategy in which we do not detect the exact time at which our system reaches the boundary of the admissible region. Instead, we allow the system to attain values outside the admissible region, and define a new admissible region boundary wherever the system falls. Consider for simplicity the case in which all elements have the same time step so that AVI reduces to (11), for which this idea is illustrated in Figure 5. The system satisfies $\mathbf{q}^{i-1} \in Q$, but $\mathbf{q}^i \notin Q$. At this point we compute the value of $\mathbf{p}^{i+}$ based on the value of $\mathbf{p}^{i-}$ following (7), (8), (9), and (10). More generally, the contact conditions are used to compute a reflected value for the momentum any time that

$$g(\mathbf{q}^i) \geq 0 \qquad \text{and} \qquad (p^{i-})_n = \mathbf{p}^{i-} \cdot \mathbf{M}^{-1} \cdot \mathbf{n}(\mathbf{q}^i) > 0. \tag{16}$$

In this way, if $\mathbf{q}^i \notin Q$ but the momentum $\mathbf{p}^i$ is making the system travel towards Q the contact conditions are not applied.

This algorithm is nearly identical to the one proposed in [3]. The only differences are that in the aforementioned reference the authors do not need the second condition in (16), since they map $(\mathbf{q}^{i-}, \mathbf{p}^{i-})$ to $(\mathbf{q}^{i+}, \mathbf{p}^{i+})$ by setting $\mathbf{q}^{i+}$ to be a projection of $\mathbf{q}^{i-}$ on ∂Q. They then compute $\mathbf{p}^{i+}$ using the normal $\mathbf{n}(\mathbf{q}^{i+})$ in (7), (8) and (9). For a smooth enough g, the normals $\mathbf{n}(\mathbf{q}^{i-})$ we adopt and $\mathbf{n}(\mathbf{q}^{i+})$ adopted in [3] become progressively similar as the time step is reduced.

For small enough time steps, the violations of the contact constraint are akin to what is encountered in adopting a penalty method, with the difference that it takes the penalty approach many time steps to eventually revert the direction of the normal component of the momentum, while here it is done in a single time step.

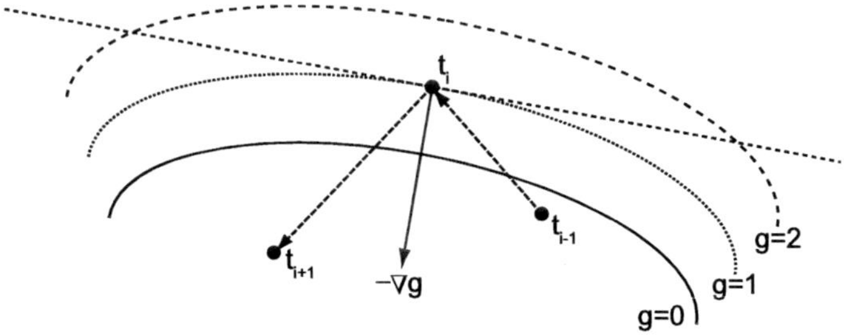

Fig. 5 Sketch of one of the ideas behind the proposed contact algorithm. At time t_{i-1} the system is within the admissible region, but in the following time step the system has violated the contact constraint. Instead of projecting back to the $g = 0$ surface, or backtracking to find the precise time at which contact happens, we adopt as the instantaneous contact boundary the level set of g that goes through the position of the system at time t_i. Contact conditions are applied with respect to this new boundary and the simulation is continued with a new momentum for the system at time t_i. The system may or may not be within the admissible region at time t_{i+1}.

Additionally, in this case we do not need to define what the penalty parameter(s) should be. However, a nice feature of a penalty approach is that, if a small enough time step is chosen, the resulting mechanical system is perfectly Hamiltonian ($e = 1$) and hence by adopting a symplectic integrator like (11) the energy through each contact event is nearly perfectly conserved (as shown in [7]). As we shall see in the numerical examples, the approach we adopt here does introduce a drift of the energy at each contact event, the size of which can be controlled by reducing the time step size (a similar behavior would be observed if a penalty approach was adopted and the time step was not small enough).

To formulate the asynchronous contact algorithm we adopt the simplest extension of these ideas. To this end, we define the elemental version of (14) to verify whether nodes in an element are within or without the admissible region. For element K, set

$$g_K(\mathbf{q}_K) = -\sum_{a \in \text{nodes in } K} \min(d(\mathbf{q}_a;\mathscr{R}),0), \tag{17}$$

where $\mathbf{q}_K$ is are the coordinates of all nodes in element K. Following the notation in Section 4, after the elemental update of element K according to (13a) and (13b), we check whether

$$g_K(\mathbf{q}_K^j) \geq 0 \qquad \text{and} \qquad (p^{j-})_n = \mathbf{p}_K^{j-} \cdot \mathbf{M}_K^{-1} \cdot \mathbf{n}_K(\mathbf{q}_K^j) > 0, \tag{18}$$

with

$$\mathbf{n}_K(\mathbf{q}_K) = \frac{\nabla g_K(\mathbf{q}_K)}{\left[\nabla g_K(\mathbf{q}_K) \cdot \mathbf{M}_K^{-1} \cdot \nabla g_K(\mathbf{q}_K)\right]^{1/2}}. \tag{19}$$

Here $\mathbf{q}_K^j$ denotes the values of all coordinates of the nodes in element K at its j-th update. For node a in K this corresponds to its i_a-th update, so the updated coordinates of a in $\mathbf{q}_K^j$ are $\mathbf{q}_a^{i_a}$ computed from (13a). Similarly, $\mathbf{p}_K^{j-}$ denotes the momenta of all nodes in the element after this update, which contains $\mathbf{p}_a^{i_a}$ computed with (13b) for each node in the element. Additionally, $\mathbf{M}_K$ is the diagonal mass matrix for the nodes in element K, as extracted from $\mathbf{M}$, so it contains mass contributions from neighboring elements as well.

If conditions (18) are met, then the momenta of the nodes in the element are changed according to

$$\mathbf{p}_K^{j+} = \mathbf{p}_K^{j-} - (1+e)(p^{j-})_n \mathbf{n}_K(\mathbf{q}_K^j). \tag{20}$$

Equation (20) is a consequence of contact conditions (7) and (8). The momentum $\mathbf{p}_a^{i_a+}$ of each node a after the contact conditions have been applied are part of $\mathbf{p}_K^{j+}$. These are the momenta that participate in (13c) to compute $\mathbf{p}^{i_a+1/2}$ for each node.

This algorithm can be interpreted as a special case of the synchronous one described at the beginning of this section. The asynchronous algorithm follows after assuming that all nodes that do not belong to the element being updated are outside the rigid body, even though some of them may be inside because we are allowing for some temporary penetration. Because of this, when the time step for each element is the same, the asynchronous contact algorithm does not reduce to the synchronous contact algorithm described at the beginning of the section.

6 Performance

In the following, we demonstrate the performance of the proposed algorithm and highlight its main properties through numerical examples. All simulations were performed by choosing a nonlinear elastic model for the continuum. More precisely, a neo-Hookean material model extended to the compressible range. Under these conditions, the mechanical energy (3) should be constant as the body deforms. We look at how well this is reproduced by the algorithm below.

6.1 AVI's Computational Efficiency

It is possible with AVI to adopt a different time step for virtually every element in the mesh. For graded meshes in which a distribution of element sizes is present, significant computational savings can be obtained, see [12]. One example of a mildly graded mesh is shown in Figure 6, in which the time step size of each element is shown with a contour plot. A more striking example can be found in [9].

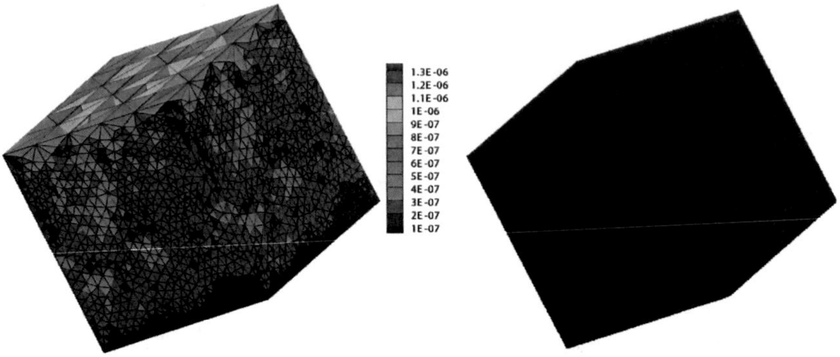

Fig. 6 A cube meshed with a graded unstructured mesh, with element sizes differing by a factor of ten between the bottom and the top of the cube. The contours illustrate the time step of each element when asynchronous time stepping is adopted. In contrast, the smallest of these time steps would have to be used for every element in the mesh in the synchronous case.

6.2 Energy Conservation

We showcase next the main benefit of the proposed asynchronous contact algorithm: the possibility of resolving contact events by adopting smaller time steps in those elements that may have contact interactions only. This benefit is directly manifested in a substantial reduction of the drift in the energy of the system before and after a contact event.

We consider a nonlinear elastic cylinder impacting a rigid flat wall parallel to one of its circular caps. The mesh of the cylinder is shown in Figure 7, in which the CFL limit for some elements on its outer surface as well as some elements in its interior are shown, see Section 4. Time steps vary by a factor of 3 throughout the mesh.

We simulate this problem in two ways. In the first case we adopt the same time step for each element, equal to a fraction f of the minimum value of the CFL limit throughout the mesh. We denote it the synchronous case. The simulated energy evolution through the collision and bounce back from the wall are shown in Figure 8(a) for $f = 0.1$ and $f = 0.01$. Making the collision perfect elastic ($e = 1$), the energy of the system should be conserved throughout the simulation. This is clearly the case before and after the collision, a characteristic of our symplectic integrator. The remarkable feature in this example is, however, the fact that as f becomes smaller the energy drift (or loss in this case) shrinks with it.

Introducing the asynchronicity does not substantially change the energy behavior, as shown in Figure 8(b). In this second case all elements outside the TSRZ are simulated with $f = 0.1$. The TSRZ was chosen here so that C_{TSRZ} is smaller than one element size away from the wall. In one of the simulations the elements in the TSRZ have also been simulated with $f = 0.1$, so that effectively there is no TSRZ. The time steps in this case are only mildly different that those in Figure 8(a) with $f = 0.1$, but asynchronous. The energy drift through the collision is, as mentioned

Fig. 7 A coarse (3D) mesh for a cylinder. The contour plot shows the CFL limit for each element. On top, we show elements on the lateral surface of the cylinder, while a slice through the mesh showing elements in its interior is displayed at the bottom.

earlier, essentially the same, with the asynchronous case displaying slightly larger oscillations of the energy after the collision is over. The difference however, is found in the computational time, with the synchronous case taking 93% more time than the asynchronous one.

More remarkable results are found when we change the time factor in the TSRZ to $f = 0.01$, as shown in Figure 8(b). The energy drift in this case is much smaller than in the last case, and roughly the same to that obtained in the synchronous case by adopting $f = 0.01$ for all elements in the mesh (Figure 8(a)). The asynchronous algorithm with $f = 0.01$ in the TSRZ only is approximately 20 times faster than the synchronous algorithms with the same time factor throughout the mesh, rendering the same energy drift after the collision. Changing the time factor from $f = 0.1$ to $f = 0.01$ increased the computational cost of the asynchronous algorithm by approximately 19%, but rendered a much smaller energy drift.

It is somewhat remarkable that the energy drifts in Figures 8(a) and 8(b) seem to depend exclusively on the time steps of the elements that enter in contact with the wall. We investigated this some more. We first examined the effect of changing the time factor in those elements outside the TSRZ, while leaving the time factor in element in the TSRZ fixed. The results shown in Figure 9(a) demonstrate that the energy drift after the collision is over essentially does not depend on the time step outside the TSRZ. The size of the oscillations in energy do increase markedly with it. In contrast, by fixing the time factor outside the TSRZ and changing the one inside it, we find that the energy drift decays linearly with the time step inside the TSRZ, and that the magnitude of the energy oscillations after impact remains unchanged, see Figure 9(b).

A more comprehensive perspective of the effect of changing the time factors within and without the TSRZ is shown in Figure 10. Therein we plotted the mean value and standard deviation of the energy drift after the collision is over. Remarkably, the figure shows that the energy drift seems to depend essentially on the time step inside the TSRZ only, while the magnitude of the oscillations after impact

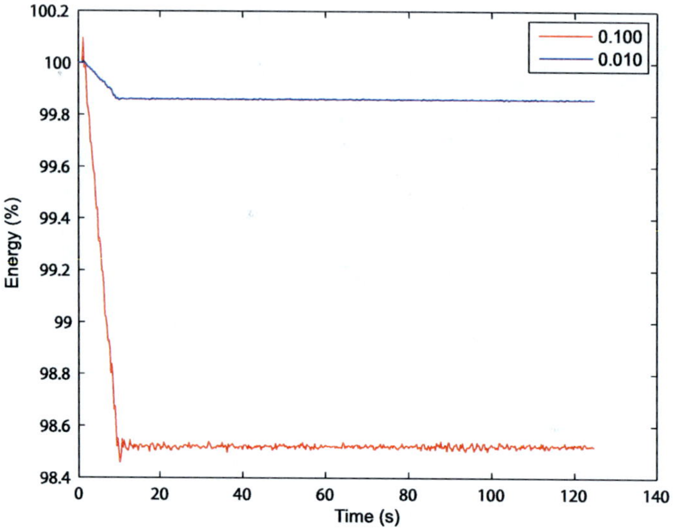

(a) The same time step was selected for all elements in the mesh, equal to a fraction f of the minimum of the CFL limit among all elements in the mesh. Simulations for $f = 0.1$ and $f = 0.01$ are shown, as indicated in the legend.

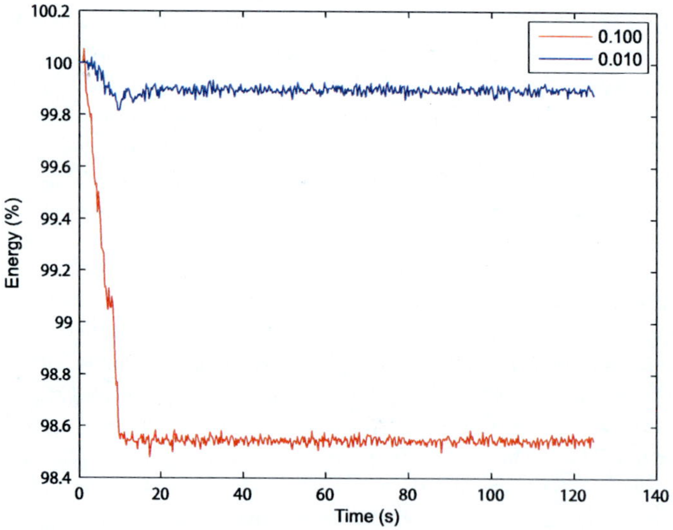

(b) Asynchronous time stepping with a time factor $f = 0.1$ for all elements outside the TSRZ. Elements in the TSRZ have $f = 0.1$ or $f = 0.01$, as indicated in the legend.

Fig. 8 Evolution of the mechanical energy of an elastic cylinder impacting into a rigid wall. The simulations used linear tetrahedral elements with standard mass lumping. The maximum sound speed in the undeformed cylinder is approximately 1, and the initial velocity of the cylinder with respect to the wall is 0.1.

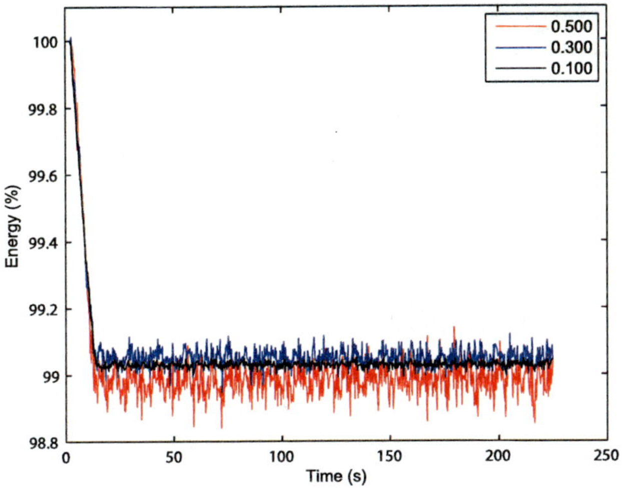

(a) Changing the time step outside the TSRZ, as indicated by the time factor in the legend.

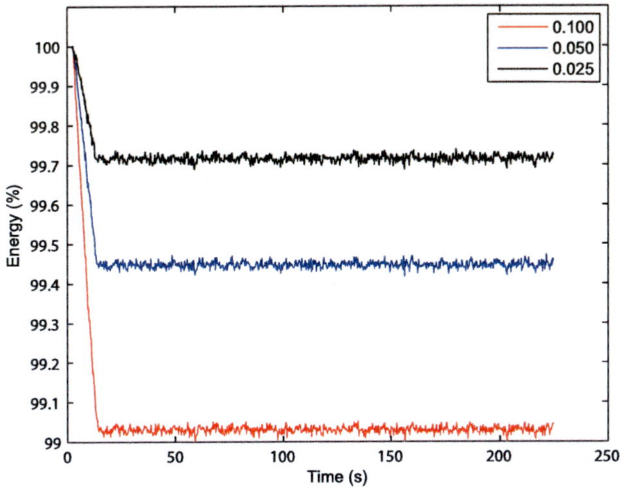

(b) Changing the time step in the TSRZ, as indicated by the time factor in the legend.

Fig. 9 Effect of changing the time steps on the energy evolution in the simulation of the elastic cylinder impacting against a flat rigid wall of Figure 8.

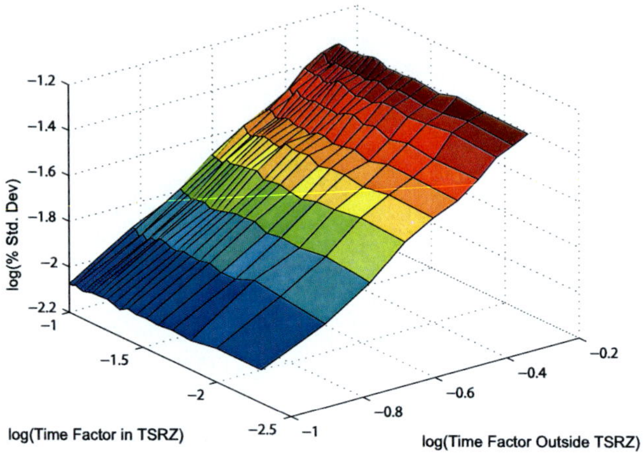

(a) Standard deviation

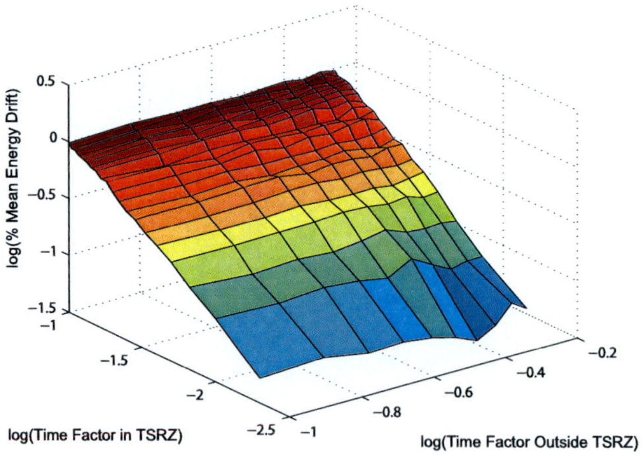

(b) Mean value

Fig. 10 Dependence on the time factors within and without the TSRZ of the mean value and standard deviation of the energy drift after the collision is over, for the elastic cylinder impacting a rigid flat wall. Both the mean energy drift and the standard deviation are normalized with the initial total energy The simulations were performed with quadratic tetrahedral elements, and the ratio formed by the velocity of the cylinder with respect to the flat wall over the bulk sound speed is approximately 0.16.

(measured by the standard deviation) seem to depend only on the time factor in the elements outside the TSRZ, with an approximate first-order dependence in both cases.

Several of the advantages mentioned during the algorithm description have been illustrated here: the energy level is well-conserved both before and after the impact; and the time step in the TSRZ reduces the energy drift through the collision. This last benefit is gained by changing the time step for only a relatively small number of elements, resulting in substantially better energy behavior without a large increase in the computational time.

7 Numerical Examples

Here we consider two illustrative examples of slightly more complicated contact constraints. The first one consists of a sphere elastically bouncing inside a rigid cube, showcasing multiple consecutive impacts. The second case shows the impact and bounce back of a rigid sphere against a nonlinear elastic block. In this case the dynamics of the sphere is solved for together with that of the block.

7.1 Multiple Impacts

The case of repeated impacts of an elastic sphere against multiple walls serves to illustrate the consistent energy behavior of the algorithm. In this example, the sphere starts by traveling along a line forming angles of 45 degrees with two walls of the cube, and orthogonal to their common edge. The sphere initially impacts these two walls simultaneously. It then bounces back and impacts the walls adjacent to the opposite edge of the cube. Due to the non-symmetric spatial discretization of the sphere, at later times the sphere impacts the two walls adjacent to each edge at different times. Eventually the sphere leaves the edge entirely, impacting only a single wall at each impact, see Figure 11 for some snapshots. Somewhat surprisingly, Figure 12 shows that the energy drift resulting from each impact for this example happens to always be negative, a feature that we did not observe in all cases we tested. Each impact against walls can be detected in this plot as a sharp drop in the energy value. Single wall interactions have roughly half the energy effect of the initial simultaneous impacts against two walls.

7.2 Impact of a Rigid Body on an Elastic Block

We consider next a rigid sphere with finite mass impacting an elastic block. In this case we solve for both the dynamics of the block and the sphere in an asynchronous

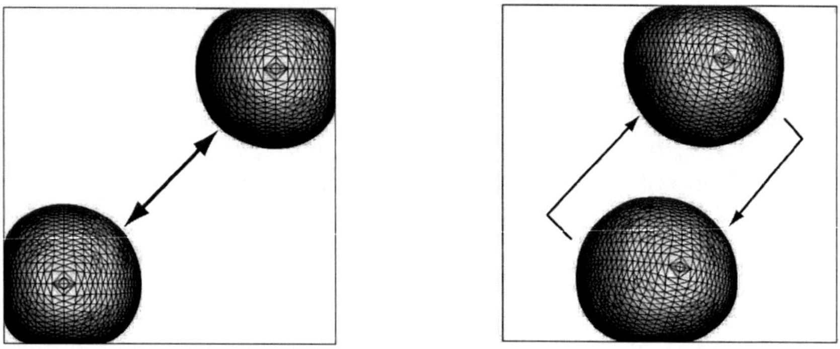

(a) The initial trajectory of the sphere, impacting and rebounding against opposing corners.

(b) The final trajectory of the sphere, due to the spatial discretization of the sphere.

Fig. 11 Impact of an elastic sphere against the walls of a rigid cube in a simulation that includes 16 separate impacts against walls.

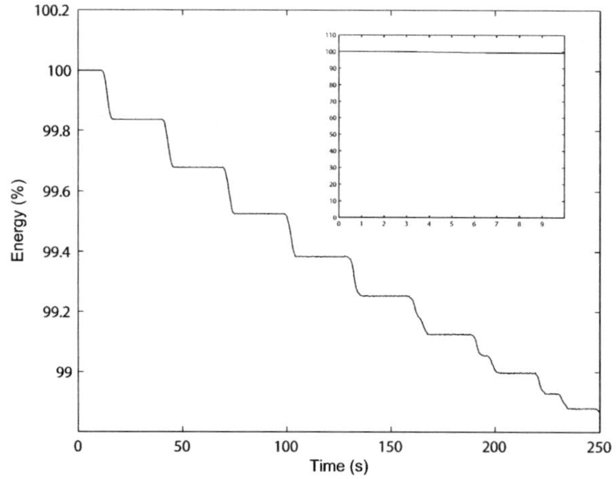

Fig. 12 Evolution of the mechanical energy of an elastic sphere bouncing inside a rigid cube, with a total of 16 collision events. Only 1% of the energy is lost throughout this simulation, somewhat negligible if seen in the full energy scale shown in the inset.

way. Since we consider only frictionless contact we need only track the evolution of the center of mass of the sphere. Its trajectory is integrated as evolving with a constant velocity between any two contact events with an element in the mesh, and its momentum changed to be the reaction to the impulse acting on each node of

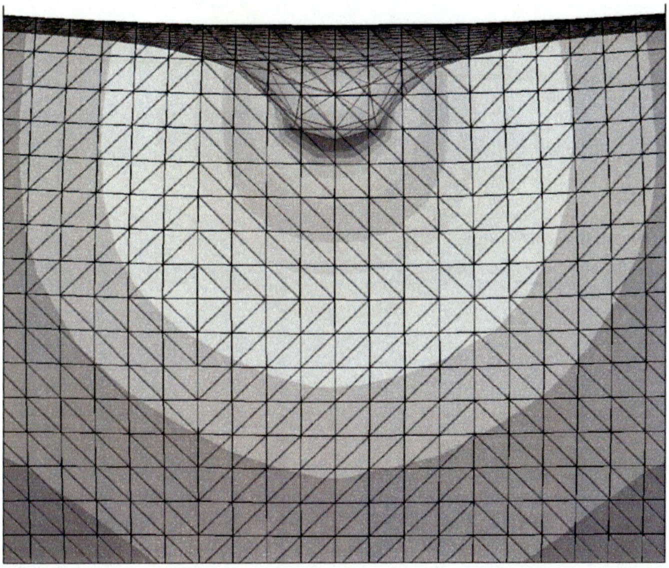

Fig. 13 Impact of a rigid sphere into an elastic block. The snapshot corresponds to the deepest penetration of the block. Contours are of the displacements parallel to the incoming direction of the sphere.

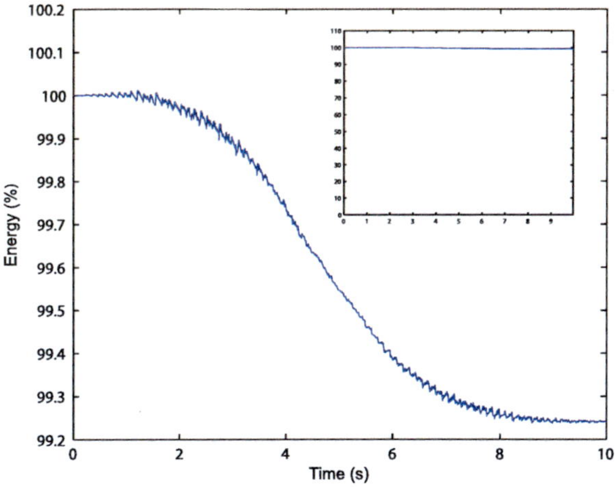

Fig. 14 Evolution of the mechanical energy for a rigid sphere impacting an elastic block. The simulations includes the initial impact until the maximum penetration in Figure 13, and full rebound out of the block. The simulation used linear tetrahedral elements, the bulk sound speed in the undeformed elastic block was 1, while the initial velocity of the sphere was 0.25.

the mesh any time the contact conditions are applied. A snapshot of the deformed elastic block when the sphere attains zero velocity is shown in Figure 13, while the evolution of the energy of the elastic block plus the sphere is displayed in Figure 14. Less than 1% of the total energy is lost during the entire contact event for the selection of time steps in this example.

Afterword At the time it was written, this conference paper was meant to be a qualitative description of the ideas behind the algorithm. In particular, we have not carefully dealt with many of the non-smooth features of the contact problem here, which we do in [15].

Acknowledgements This research was carried out with government support under and awarded by DoD, Air Force Office of Scientific Research, National Defense Science and Engineering Graduate (NDSEG) Fellowship, 32 CFR 168a; and the Department of the Army Research Grant, grant number W911NF-07-2-0027.

References

1. Armero, F., Petocz, E.: Formulation and analysis of conserving algorithms for frictionless dynamic contact/impact problems. Computer Methods in Applied Mechanics and Engineering 158, 269–300 (1998)
2. Bond, S.D., Leimkuhler, B.J.: Stabilized integration of Hamiltonian systems with hard-sphere inequality constraints. SIAM Journal on Scientific Computing 30(1), 134–147 (2007)
3. Cirak, F., West, M.: Decomposition contact response (DCR) for explicit finite element dynamics. International Journal for Numerical Methods in Engineering 64(8), 1078–1110 (2005)
4. Fetecau, R.C., Marsden, J.E., Ortiz, M., West, M.: Nonsmooth Lagrangian mechanics and variational collision integrators. SIAM Journal on Applied Dynamical Systems 2(3), 381–416 (2003)
5. Fetecau, R.C., Marsden, J.E., West, M.: Variational multisymplectic formulations of nonsmooth continuum mechanics. In: Kaplan, E., Marsden, J.E., Sreenivasan, R.K. (eds.) Perspectives and Problems in Nonlinear Science. Springer, Berlin (2001)
6. Fong, W., Darve, E., Lew, A.: Stability of asynchronous variational integrators. Journal of Computational Physics 227(18), 8367–8394 (2008)
7. Harmon, D., Vouga, E., Smith, B., Tamstorf, R., Grinspun, E.: Asynchronous contact mechanics. In: Proceedings of ACM SIGGRAPH 2009. ACM, New York (2009)
8. Kale, K., Lew, A.: An explicit asynchronous contact algorithm. In: Proceedings of Second Structured Integrators Workshop, Stanford, CA (2006)
9. Kale, K., Lew, A.: Parallel asynchronous variational integrators. International Journal for Numerical Methods in Engineering 70, 291–321 (2007)
10. Laursen, T.A.: Computational Contact and Impact Mechanics: Fundamentals of Modeling Interfacial Phenomena in Nonlinear Finite Element Analysis. Springer, Berlin (2002)
11. Lew, A., Marsden, J.E., Ortiz, M., West, M.: Asynchronous variational integrators. Archive for Rational Mechanics and Analysis 167(2), 85–146 (2003)

12. Lew, A., Marsden, J.E., Ortiz, M., West, M.: Variational time integrators. International Journal for Numerical Methods in Engineering 60(1), 153–212 (2004)
13. Lew, A., Marsden, J.E., Ortiz, M., West, M.: An overview of variational integrators. In: Franca, L.P., Tezduyar, T.E., Masud, A. (eds.) Finite Element Methods: 1970s and Beyond. CIMNE, Spain (2004)
14. Rangarajan, R., Ryckman, R., Lew, A.: Towards long-time simulation of soft tissue simulant penetration. In: Conference Proceedings for Army Science Conference (26th), DTIC, 2008 (2010), http://handle.dtic.mil/100.2/ADA505859
15. Ryckman, R., Lew, A.: An explicit asynchronous contact algorithm for elastic body-rigid wall interaction (2011) (submitted)
16. Wriggers, P.: Computational Contact Mechanics. Wiley, New York (2002)

Dynamics of a Soft Contractile Body on a Hard Support

A. Tatone, A. Di Egidio and A. Contento

Abstract The motion of a soft and contractile body on a hard support is described by fields of short range contact forces. Besides repulsion these forces are able to describe also viscous friction, damping and adhesion allowing the body to have complex motions which look rather realistic. The contractility is used to make the body behave like a living body with some basic locomotion capabilities. The simulated motions, showing jumping or crawling, are driven either by a contraction or by a contractile couple. Although only homogeneous deformations are allowed, the model arises from a general theory of remodeling in finite elasticity. The body is made of a viscoelastic incompressible neo-Hookean material.

1 Introduction

The aim of this paper is to use a contact model to describe the complex motions of stiff, soft and contractile bodies interacting with a rigid flat support. The body model, though restricted to homogeneous deformations, is derived from a general continuum theory of remodeling in finite elasticity, as set up in [3, 7]. A summary of that theory is given in Sections 2 and 3. The body is made of a viscoelastic incompressible neo-Hookean material.

Contractility is the ability of bodies, like muscle cells and fibres to contract or to extend in order to apply forces. More precisely, it can be defined as the ability to modify the zero stress, or relaxed, configuration [11]. Contractility can also give an elementary body some motility and locomotion capabilities allowing it to move over a substrate [14, 9]. The simulated motions described here, showing jumping or crawling, are driven either by a contraction or by a contractile couple.

A. Tatone · A. Di Egidio · A. Contento
Department of Structural, Hydraulic and Geotechnical Engineering, University of L'Aquila, Via G. Gronchi 18, 67100 L'Aquila (AQ), Italy; e-mail: amabile.tatone@univaq.it

G. Zavarise & P. Wriggers (Eds.): Trends in Computational Contact Mechanics, LNACM 58, pp. 193–210.
springerlink.com

We use a contact model described by constitutive laws for the contact forces arising from the interaction of the body boundary and the support surface, which get close to each other during a motion, without enforcing explicitly unilateral constraints, in the spirit of [15, chapter 5]. An interatomic potential was used in [8, 10, 6] to add adhesion to the Hertz model, and in contact modeling based on molecular dynamics, as in [4]. In macroscopic modeling the interatomic potential is replaced by a surface contact potential, as in [16, 1] and more recently discussed in [12, 13].

The contact tractions on the body boundary are given here by four short range force fields of different kind, potential or dissipative, all of them depending on the distance from the support: (i) a repulsive force field; (ii) an adhesive force field, both derived from a Lennard-Jones–like surface contact potential; (iii) a damping force field, describing the impact dissipation and depending on the normal velocity; (iv) a viscous frictional force field, depending on the sliding velocity. They decay very fast as the distance increases and grow to infinity as the body and the support get closer and closer. Hence the body will never *touch* the support. Instead, the true *contact distance* will depend on the motion, although some characteristic values can be related to the constitutive properties of each contact force field.

Although the law used for the frictional force field does not implement a Coulomb like friction, the two dissipation mechanisms (iii) and (iv) turn out to be suitable to describe the dynamical behavior of a body over a support.

Summarizing, this paper briefly shows how a simple model can be defined within the context of finite elasticity and contact mechanics, including all the details needed to set up numerical physics-based simulations, showing a body bouncing and rolling and even crawling over a flat rigid support.

Though the simulations do not make use of experimental data for the material properties of neither the body nor the substrate, they seem useful to provide some insight into the dynamics of a simple body over a substrate and the mechanisms on which locomotion could be based.

2 Affine Contractile Body

The motion of a body $\mathscr{B}$ is described by a mapping

$$\mathbf{p} : \mathscr{D} \times \mathscr{I} \rightarrow \mathscr{E}\,, \tag{1}$$

transforming the *reference shape* $\mathscr{D}$, at each time $t \in \mathscr{I}$, into the *current shape* $\mathbf{p}(\mathscr{D},t)$ in a three-dimensional Euclidean space $\mathscr{E}$. An *affine* or *homogeneous* motion is completely defined at any time t by the current position $\mathbf{p}_0(t)$ of a point occupying a position $\mathbf{x}_0$ in $\mathscr{D}$ and by the *deformation gradient* $\nabla\mathbf{p}(t)$, with $\det\nabla\mathbf{p}(t) > 0$, through the following representation:

$$\mathbf{p}(\mathbf{x},t) = \mathbf{p}_0(t) + \nabla\mathbf{p}(t)(\mathbf{x}-\mathbf{x}_0)\,, \quad \forall\mathbf{x} \in \mathscr{D}\,. \tag{2}$$

As a consequence, an affine velocity field $\mathbf{v}$ at any time t has the representation

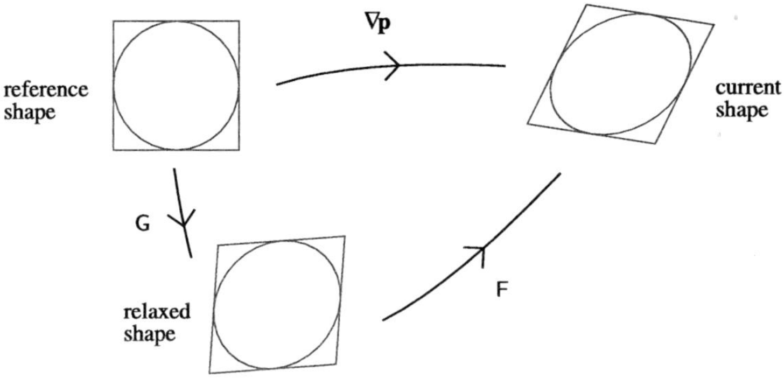

Fig. 1 Kröner–Lee decomposition of the deformation gradient $\nabla\mathbf{p}$.

$$\mathbf{v}(\mathbf{x}) = \mathbf{v}_0 + \nabla\mathbf{v}\,(\mathbf{x} - \mathbf{x}_0)\,, \tag{3}$$

where $\nabla\mathbf{v}$ is the velocity gradient. Along the motion (2) at time t

$$\nabla\mathbf{v} = \nabla\dot{\mathbf{p}}(t)\,, \quad \mathbf{v}_0 = \dot{\mathbf{p}}_0(t)\,, \tag{4}$$

where the dot denotes time derivatives.

In order to describe *contractility* we introduce a new tensor $\mathsf{G}(t)$, such that $\det\mathsf{G}(t) > 0$, transforming the reference shape $\mathscr{D}$ into a *relaxed shape* at time t, and will assume that the strain energy is a function of F as defined by the Kröner–Lee decomposition (see Figure 1)

$$\mathsf{F}(t) := \nabla\mathbf{p}(t)\,\mathsf{G}(t)^{-1}\,. \tag{5}$$

Let us denote by V the velocity corresponding to G which takes the value

$$\mathsf{V} = \dot{\mathsf{G}}(t)\,\mathsf{G}(t)^{-1} \tag{6}$$

at time t along a motion described by $(\mathbf{p},\mathsf{G})$. We assume as *balance principle* (following [5]) that at any time t for any test velocity field $(\mathbf{v},\mathsf{V})$

$$\begin{aligned} &\int_{\mathscr{D}} \mathbf{b}(\mathbf{x},t)\cdot\mathbf{v}\,dV + \int_{\partial\mathscr{D}} \mathbf{q}(\mathbf{x},t)\cdot\mathbf{v}\,dA \\ &\quad - \mathsf{S}(t)\cdot\nabla\mathbf{v}\,\mathrm{vol}(\mathscr{D}) + \big(\mathsf{Q}(t) - \mathsf{A}(t)\big)\cdot\mathsf{V}\,\mathrm{vol}(\mathscr{D}) = 0\,, \end{aligned} \tag{7}$$

where the bulk density force $\mathbf{b}$, denoting by ρ the reference mass density, is composed of the inertial force and the gravity force densities

$$\mathbf{b}(\mathbf{x},t) := -\rho(\mathbf{x})\,\big(\ddot{\mathbf{p}}(\mathbf{x},t) + \mathbf{g}\big)\,, \tag{8}$$

while the traction $\mathbf{q}$ on the boundary is assumed to be the sum of different contact force fields $\mathbf{q}_j$. The tensor $\mathsf{S}(t)$ is the Piola stress in the reference shape, related to the Cauchy stress T by

$$\mathsf{T} = \mathsf{S}\,\nabla\mathbf{p}^{\mathsf{T}}(\det\nabla\mathbf{p})^{-1}\,. \tag{9}$$

The tensors $\mathsf{A}(t)$ and $\mathsf{Q}(t)$ are the *internal* and *external contractile couples* per unit reference volume. The first one describes the material response while the second one describes an action from outside the mechanical system, which could be controlled by an electrical or biochemical signal.

The balance equations corresponding to (7) turn out to be

$$-m\ddot{\mathbf{p}}_0(t) - m\mathbf{g} + \mathbf{f}(t) = 0\,, \tag{10}$$

$$-\nabla\ddot{\mathbf{p}}(t)\,\mathsf{E} + \mathsf{M}(t) - \mathsf{S}(t)\,\mathrm{vol}\,(\mathscr{D}) = 0\,, \tag{11}$$

$$\mathsf{Q}(t) - \mathsf{A}(t) = 0\,. \tag{12}$$

The scalar quantity m is the total mass $\int_{\mathscr{D}} \rho\,\mathrm{d}V$, while E is the Euler tensor

$$\mathsf{E} := \int_{\mathscr{D}} \rho(\mathbf{x})(\mathbf{x}-\mathbf{x}_0)\otimes(\mathbf{x}-\mathbf{x}_0)\,\mathrm{d}V\,, \tag{13}$$

where $\mathbf{x}_0$ has been chosen as the barycenter of $\mathscr{D}$. The contact force fields $\mathbf{q}_j$ give rise to the total force

$$\mathbf{f}(t) := \sum_j \int_{\partial\mathscr{D}} \mathbf{q}_j(\mathbf{x},t)\mathrm{d}A \tag{14}$$

and to the total moment tensor[1]

$$\mathsf{M}(t) := \sum_j \int_{\partial\mathscr{D}} (\mathbf{x}-\mathbf{x}_0)\otimes\mathbf{q}_j(\mathbf{x},t)\mathrm{d}A\,. \tag{15}$$

The volume of the reference shape is denoted by $\mathrm{vol}\,(\mathscr{D})$.

3 Dissipation Inequality and Material Characterization

The material response can be characterized (as in [3] and [7]) by assuming that along any motion at any time t

$$\mathsf{A}\cdot\dot{\mathsf{G}}\mathsf{G}^{-1} + \mathsf{S}\cdot\nabla\dot{\mathbf{p}} - \frac{d}{dt}\big(J\,\varphi(\mathsf{F})\big) \geq 0\,, \tag{16}$$

[1] Here we use the following definition of tensor product:

$$(\mathbf{u}\otimes\mathbf{f})\,\mathbf{e} = (\mathbf{u}\cdot\mathbf{e})\mathbf{f} \quad \forall\mathbf{e}\,.$$

where φ is the *strain energy density* per unit relaxed volume and $J := \det \mathsf{G}$. Note that, because of the balance principle (7), the power $(\mathsf{A}\cdot\dot{\mathsf{G}}\mathsf{G}^{-1}+\mathsf{S}\cdot\nabla\dot{\mathbf{p}})$ equals the power, per unit reference volume, of the bulk forces, the contact forces and the external contractile couple. Thus the dissipation principle above states that the external power is not entirely balanced by a rate of change of the strain energy. By replacing $\nabla\dot{\mathbf{p}}$ with the time derivative of the Kröner–Lee decomposition of $\nabla\mathbf{p}$ defined in (5), we get

$$\mathsf{A}\cdot\dot{\mathsf{G}}\mathsf{G}^{-1}+\mathsf{S}\mathsf{G}^{\mathsf{T}}\cdot\dot{\mathsf{F}}+\mathsf{F}^{\mathsf{T}}\mathsf{S}\mathsf{G}^{\mathsf{T}}\cdot\dot{\mathsf{G}}\mathsf{G}^{-1}-J\frac{d}{dt}\varphi(\mathsf{F})-J\varphi(\mathsf{F})\mathsf{I}\cdot\dot{\mathsf{G}}\mathsf{G}^{-1}\geq 0. \tag{17}$$

Since $d\,\varphi(\mathsf{F})/dt$ is linear in $\dot{\mathsf{F}}$, we can define the *elastic response* for the Piola stress $\mathbb{S} := J^{-1}\mathsf{S}\mathsf{G}^{\mathsf{T}}$, *pull-back* of T to the relaxed shape, as the function $\widehat{\mathbb{S}}$ such that in any motion

$$\widehat{\mathbb{S}}(\mathsf{F})\cdot\dot{\mathsf{F}}=\frac{d}{dt}\varphi(\mathsf{F}). \tag{18}$$

By requiring φ to be frame-indifferent it turns out that the Cauchy stress (9) is a symmetric tensor. Going back to (17), after substituting (18) we can collect all terms in two groups

$$\big(\mathsf{S}-\widehat{\mathsf{S}}(\mathsf{F})\big)\nabla\mathbf{p}^{\mathsf{T}}\cdot\dot{\mathsf{F}}\mathsf{F}^{-1}+\big(\mathsf{A}+\mathsf{F}^{\mathsf{T}}\mathsf{S}\mathsf{G}^{\mathsf{T}}-J\varphi(\mathsf{F})\mathsf{I}\big)\cdot\dot{\mathsf{G}}\mathsf{G}^{-1}\geq 0 \tag{19}$$

with $\widehat{\mathbb{S}}(\mathsf{F})=J^{-1}\widehat{\mathsf{S}}(\mathsf{F})\mathsf{G}^{\mathsf{T}}$. Setting

$$\begin{aligned}\mathsf{S}^{+} &:= \mathsf{S}-\widehat{\mathsf{S}}(\mathsf{F}),\\ \mathsf{A}^{+} &:= \mathsf{A}+\mathsf{F}^{\mathsf{T}}\mathsf{S}\mathsf{G}^{\mathsf{T}}-J\varphi(\mathsf{F})\mathsf{I},\end{aligned} \tag{20}$$

the dissipation inequality (19) takes the form

$$\mathsf{S}^{+}\nabla\mathbf{p}^{\mathsf{T}}\cdot\dot{\mathsf{F}}\mathsf{F}^{-1}+\mathsf{A}^{+}\cdot\dot{\mathsf{G}}\mathsf{G}^{-1}\geq 0. \tag{21}$$

In order for S^{+} and A^{+} to satisfy a-priori the dissipation inequality (21) both of them have to depend on $\dot{\mathsf{F}}\mathsf{F}^{-1}$ and $\dot{\mathsf{G}}\mathsf{G}^{-1}$. A possible constitutive prescription, recovering the classical viscous stress, consists in assuming

$$\begin{aligned}\mathsf{S}^{+}\nabla\mathbf{p}^{\mathsf{T}} &= \mu\,\mathrm{sym}\,(\dot{\mathsf{F}}\mathsf{F}^{-1}),\\ \mathsf{A}^{+} &= \mu_{\gamma}\dot{\mathsf{G}}\mathsf{G}^{-1},\end{aligned} \tag{22}$$

with positive scalars μ, the *viscosity*, and μ_{γ}, the *resistance to contraction.*

Hence A and S are constitutively characterized by the expressions

$$\begin{aligned}\mathsf{S} &= \mathsf{S}^{+}+\widehat{\mathsf{S}}(\mathsf{F})=\mu\,\mathrm{sym}\,(\dot{\mathsf{F}}\mathsf{F}^{-1})\nabla\mathbf{p}^{-\mathsf{T}}+\widehat{\mathsf{S}}(\mathsf{F}),\\ \mathsf{A} &= \mathsf{A}^{+}-\mathsf{F}^{\mathsf{T}}\mathsf{S}\mathsf{G}^{\mathsf{T}}+J\varphi(\mathsf{F})\mathsf{I}=\mu_{\gamma}\dot{\mathsf{G}}\mathsf{G}^{-1}-\mathsf{F}^{\mathsf{T}}\mathsf{S}\mathsf{G}^{\mathsf{T}}+J\varphi(\mathsf{F})\mathsf{I}.\end{aligned} \tag{23}$$

Now the balance equation (12) takes the form of an evolution equation

$$\mu_\gamma \dot{\mathsf{G}}\mathsf{G}^{-1} = \mathsf{F}^\mathsf{T}\mathsf{S}\mathsf{G}^\mathsf{T} - J\,\varphi(\mathsf{F})\mathsf{I} + \mathsf{Q}\,, \tag{24}$$

where S is meant to be given by the first of (23). The external contractile couple Q is the driving force, which could be related to some other quantity like an electrical or biochemical signal. In some of the simulations shown in this paper the motion is driven by G instead. In those cases Q is just a reactive contractile couple given by (24). We will consider an incompressible material defined by the neo-Hookean strain energy function

$$\varphi(\mathsf{F}) := c_1(\iota_1(\mathsf{C}) - 3)\,, \tag{25}$$

where c_1 is the elastic moduli and $\iota_1(\mathsf{C})$ is the trace of the Cauchy–Green tensor $\mathsf{C} := \mathsf{F}^\mathsf{T}\mathsf{F}$. Because of the incompressibility constraint $\det \mathsf{F} = 1$, the velocity fields turn out to be isochoric, i.e. such that $\operatorname{tr} \dot{\mathsf{F}}\mathsf{F}^{-1} = 0$. Hence there exists a *reactive* spherical part $-\pi\mathsf{I}$ of S, while the inequality (19) characterizes only the deviatoric part S_0. Thus the first of (23) will be replaced by

$$\mathsf{S} = \mu \,\operatorname{sym}(\dot{\mathsf{F}}\mathsf{F}^{-1})\nabla\mathbf{p}^{-\mathsf{T}} + \widehat{\mathsf{S}}_0(\mathsf{F}) - \pi\nabla\mathbf{p}^{-\mathsf{T}}\,. \tag{26}$$

4 Surface Energy and Contact Force Characterization

The flat surface $\mathscr{S}$ of the rigid support can be defined by the position $\mathbf{o}$ of a point on it and by the exterior unit normal vector $\mathbf{n}$. The distance of a point $\mathbf{x}$ on $\partial\mathscr{D}$ from $\mathscr{S}$ at time t is

$$d(\mathbf{x},t) := (\mathbf{p}(\mathbf{x},t) - \mathbf{o})\cdot\mathbf{n}\,, \tag{27}$$

where $\mathbf{p}(\mathbf{x},t)$ is the position occupied by $\mathbf{x}$ at time t.

In order to describe the contact interaction it is convenient to consider the body and the support as a whole body and to define a *surface contact potential* ψ as a density per unit reference area on $\partial\mathscr{D}$. If we assume that this potential depends only on the distance d it turns out to be frame-indifferent. Accordingly, the dissipation inequality (16) should be changed into

$$\begin{aligned} &\operatorname{vol}(\mathscr{D})\big(\mathsf{A}\cdot\dot{\mathsf{G}}\mathsf{G}^{-1} + \mathsf{S}\cdot\nabla\dot{\mathbf{p}}\big) - \int_{\partial\mathscr{D}} \mathbf{q}\cdot\dot{\mathbf{p}}\,dA \\ &-\operatorname{vol}(\mathscr{D})\frac{d}{dt}\big(J\,\varphi(\mathsf{F})\big) - \frac{d}{dt}\int_{\partial\mathscr{D}} \psi(d)\,dA \geq 0\,, \end{aligned} \tag{28}$$

where the terms in the first row, through (7), equal the power of the external bulk forces and the external contractile couple. This condition can be replaced by the stronger requirement that (16) be satisfied together with the following condition:

$$-\mathbf{q}(\mathbf{x},t)\cdot\dot{\mathbf{p}}(\mathbf{x},t) - \frac{d}{dt}\psi(d(\mathbf{x},t)) \geq 0 \quad \forall \mathbf{x}\in\partial\mathscr{D} \tag{29}$$

Since the rate of change of the contact potential is linear in $\dot{d} = \mathbf{n} \cdot \dot{\mathbf{p}}$, we can define the *potential contact force* through a scalar field $\hat{q}$ such that in any motion

$$\hat{q}(\mathbf{x},t)\,\mathbf{n} \cdot \dot{\mathbf{p}}(\mathbf{x},t) = -\frac{d}{dt}\psi(d(\mathbf{x},t)), \tag{30}$$

which allows (29) to be rewritten as

$$-\big(\mathbf{q}(\mathbf{x},t) - \hat{q}(\mathbf{x},t)\,\mathbf{n}\big) \cdot \dot{\mathbf{p}}(\mathbf{x},t) \geq 0 \quad \forall \mathbf{x} \in \partial\mathscr{D}, \tag{31}$$

or simply

$$-\mathbf{q}^{+}(\mathbf{x},t) \cdot \dot{\mathbf{p}}(\mathbf{x},t) \geq 0, \tag{32}$$

with $\mathbf{q}^{+}(\mathbf{x},t) := (\mathbf{q}(\mathbf{x},t) - \hat{q}(\mathbf{x},t)\,\mathbf{n})$ *the dissipative contact force*. The requirement above is a restriction on constitutive laws for contact forces.

5 Contact Constitutive Laws

The *repulsive* traction field is assumed to be defined on $\partial\mathscr{D}$ by a surface potential

$$\psi_{\mathfrak{r}}(d(\mathbf{x},t)) := \frac{\alpha_{\mathfrak{r}}}{\nu_{\mathfrak{r}} - 1}\, d(\mathbf{x},t)^{-\nu_{\mathfrak{r}}+1}. \tag{33}$$

where the coefficient $\alpha_{\mathfrak{r}}$ is a positive real number and the exponent $\nu_r > 1$ is an integer. The corresponding constitutive law, through (30), turns out to be

$$\mathbf{q}_{\mathfrak{r}}(\mathbf{x},t) = \alpha_{\mathfrak{r}}\, d(\mathbf{x},t)^{-\nu_{\mathfrak{r}}}\, \mathbf{n}, \tag{34}$$

A value for $\alpha_{\mathfrak{r}}$ can be obtained by requiring the repulsive force to balance the gravity force when the body stays at rest at an equilibrium distance d_0 from a horizontal surface, thus relating $\alpha_{\mathfrak{r}}$ to a characteristic distance.

An impact dissipation can be described by the following *damping* traction on $\partial\mathscr{D}$

$$\mathbf{q}_{\mathfrak{d}}(\mathbf{x},t) = -\beta_{\mathfrak{d}}\, d(\mathbf{x},t)^{-\nu_{\mathfrak{d}}}\, (\mathbf{n} \otimes \mathbf{n})\, \dot{\mathbf{p}}(\mathbf{x},t), \tag{35}$$

where the damping factor $\beta_{\mathfrak{d}}$ is a positive real number and $\nu_{\mathfrak{d}}$ a positive integer. The tensor $(\mathbf{n} \otimes \mathbf{n})$ is the projector onto the direction orthogonal to $\mathscr{S}$.

The simulations shown in this paper make use of a viscous *friction* defined by the following constitutive law

$$\mathbf{q}_{\mathfrak{f}}(\mathbf{x},t) = -\beta_{\mathfrak{f}}\, d(\mathbf{x},t)^{-\nu_{\mathfrak{f}}}\, \dot{\mathbf{p}}_{\tau}(\mathbf{x},t), \tag{36}$$

which is linear in the tangent velocity projection $\dot{\mathbf{p}}_{\tau}(\mathbf{x},t) := (\mathsf{I} - \mathbf{n} \otimes \mathbf{n})\, \dot{\mathbf{p}}(\mathbf{x},t)$ and depends also on d. The coefficient $\beta_{\mathfrak{f}}$ is a positive real number and $\nu_{\mathfrak{f}}$ is a positive integer. If we allow $\beta_{\mathfrak{f}}$ not to be a constant we could also use the following law:

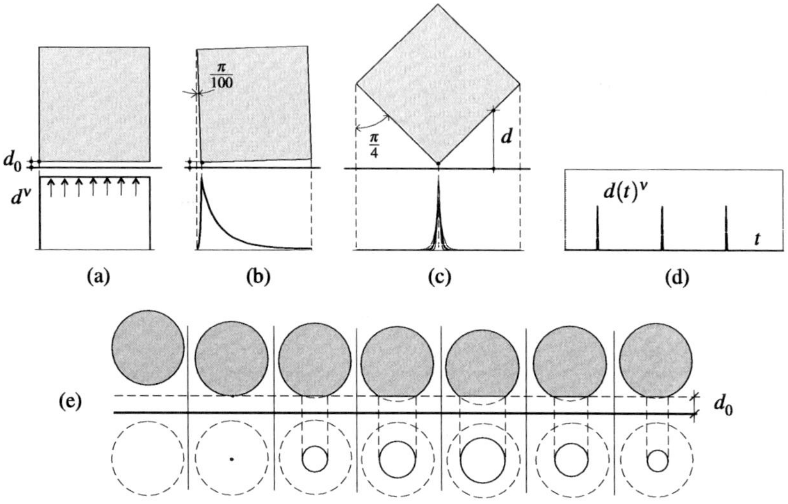

Fig. 2 (a) Spatial distribution of the function $d^{-\nu}$, with $d_0 = 0.01$ and lower face parallel to the support surface; (b) body rotated by $\vartheta = \pi/100$ about the left lower edge; (c) body rotated by $\vartheta = \pi/4$. (d) Time evolution of the function $d^{-\nu}$ at the bottom edge during an undamped vertical bouncing motion, with $\vartheta = \pi/4$. (e) Different frames of a bouncing rigid sphere and sections at distance d_0.

$$\mathbf{q}_{\mathfrak{f}_C}(\mathbf{x},t) = -\alpha_\tau\, d(\mathbf{x},t)^{-\nu_\tau}\, \mu_\mathfrak{f}(\|\dot{\mathbf{p}}_\tau(\mathbf{x},t)\|)\, \frac{\dot{\mathbf{p}}_\tau(\mathbf{x},t)}{\|\dot{\mathbf{p}}_\tau(\mathbf{x},t)\|}\,. \tag{37}$$

This expression could be given the form of a regularized Coulomb law [15, chapter 5], depending on the repulsive traction (34) and on a regular positive function $\mu_\mathfrak{f}$ of the tangent velocity. In some of the following simulations adhesion will be introduced through an additional potential contact force which is given by the following law:

$$\mathbf{q}_\mathfrak{a}(\mathbf{x},t) = \left(\beta_\mathfrak{r}\, d(\mathbf{x},t)^{-\nu_\mathfrak{r}} - \beta_\mathfrak{a}\, d(\mathbf{x},t)^{-\nu_\mathfrak{a}}\right) \mathbf{n}\,, \tag{38}$$

where $\beta_\mathfrak{r}$ and $\beta_\mathfrak{a}$ are positive real numbers and $\nu_\mathfrak{a}$ is a positive integer.

It is worth noting that the condition that $\beta_\mathfrak{d}$, $\beta_\mathfrak{f}$, $\mu_\mathfrak{f}$ is positive makes each of the above constitutive laws for the dissipative traction fields $\mathbf{q}_\mathfrak{d}$, $\mathbf{q}_\mathfrak{f}$, $\mathbf{q}_{\mathfrak{f}_C}$ fulfill separately the requirement (32).

6 Contact Force Distributions

In order to illustrate the role of the parameters on which the contact tractions depend, it could useful to recall some elementary properties. All of the traction fields given by (34)-(38) depend on the function $d^{-\nu}$. Figures 2 (a), (b) and (c) show how rapidly the graph of $d^{-\nu}$ changes when rotating a body in the shape of a cube around an edge: when the lower face of the body is parallel to the support $\mathscr{S}$ (Figure 2 a), the graph is flat; as soon as the body rotates by a very small angle (Figure 2 b) the graph rapidly decreases from a maximum value attained at the edge; when the body is in the unstable equilibrium configuration (Figure 2 c) the graph becomes very sharp: the higher the value of the exponent ν the sharper the graph. Finally, Figure 2(d) shows the time evolution of the function $d^{-\nu}$ at the edge close to the support, when the body bounces vertically.

Even though the support does not deform and the body never touches its surface, we can consider the cross section of the body shape at a characteristic distance d_0 from the support as a "contact area". In Figure 2(d), different frames of a bouncing rigid sphere are shown together with the corresponding contact cross sections.

7 Numerical Simulations

7.1 Parameter Choice and Computational Details

Several numerical simulations have been performed using different constitutive parameters and different initial conditions. The whole boundary of the body was supposed to be able to interact with the support surface, with uniform properties. The body has a nonzero uniform mass density and is subjected to a downward gravity field, while the support is rigid, flat and usually horizontal.

All the simulations were aimed at testing the ability of both the body model and the contact model to exhibit a somewhat "qualitatively realistic" behavior. By this we mean the ability of bouncing and rolling, and also jumping and crawling, within a time interval of few seconds, with a length scale of 1 m, a mass density of about 10^3 kg/m^3, an elastic modulus around 1 MPa. Calibration of the parameters was done to this end. No comparison was made with experimental data. That is why the presented simulations do not constitute a quantitative benchmark set. Some simulations of a three-dimensional motion of a rigid body can be found in [2].

The computational scheme can shortly be described as follows: the main procedure consists in the numerical integration of the equations of motion (10), (11), (12) starting from given initial conditions. For plane motions, the number of corresponding scalar time differential equations is 2 for eqn. (10), 4 for eqn. (11), and 4 for eqn. (24) which is the explicit form of eqn. (12). At a lower level, for each time step, the main task consists in computing the integrals (14) and (15) over the boundary $\partial\mathscr{D}$. The whole procedure has been implemented in *Mathematica*®, which has also been used to derive the general expressions for each of the traction fields in sect. 4, starting from the motion description (2) and making use of (7). The time integration, as well as the integration on the boundary, has been performed by using the Mathematica built-in functions with only some parameter tweaking.

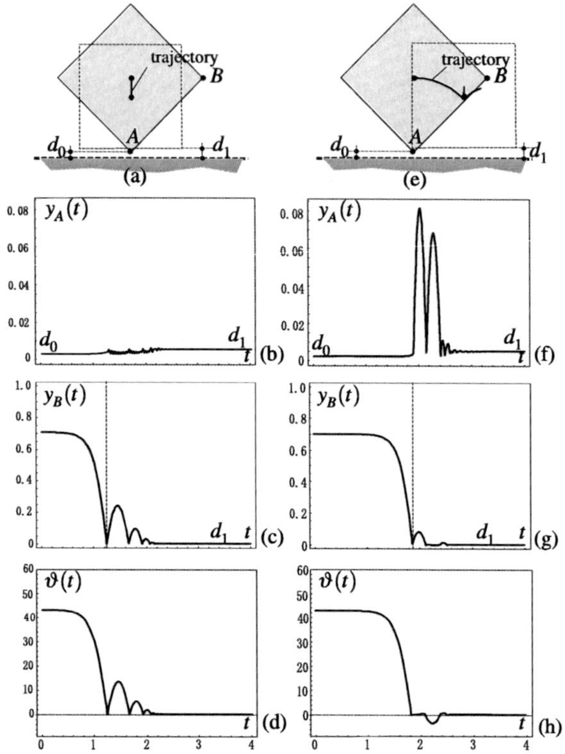

Fig. 3 Plane motion of a rigid body in the shape of a cube. Left column: contact without friction ($L = 1$ m, $d_0 = 0.002$ m, $v_0 = 0$, $\vartheta_0 = 0.99\pi/4$, $\nu_\mathfrak{r} = 8$, $\rho = 10^3$ kg/m^3, $\nu_\mathfrak{d} = 2$, $\beta_\mathfrak{d}/d_0^{\nu_\mathfrak{d}} = 2.5\times10^5$ Pa·s/m, $\beta_\mathfrak{f} = 0$). Right column: contact with friction ($\nu_\mathfrak{f} = 2$, $\beta_\mathfrak{f}/d_0^{\nu_\mathfrak{f}} = 2.5\times10^7$ Pa·s/m); (a)–(e) initial and final configurations and trajectory of the center; (b)–(f) distance of A from the support; (c)–(g) distance of B from the support; (d)–(h) rotation amplitude.

The outcome of each time integration, after a dump of the Mathematica session, was routinely processed producing graphs and movies.

7.2 Sliding, Bouncing and Rocking

Figure 3 shows the plane motion of a rigid body in the shape of a cube, with edge length L, starting from a slightly perturbed unstable equilibrium configuration. The left column (a)–(d) refers to a contact without friction, while the right column (e)–(h) refers to a contact with friction. The graphs show the time evolution of the distances $y_A(t)$ and $y_B(t)$ of the A and B edges from the support, together with the time evolution of the rotation amplitude $\vartheta(t)$. Both the initial configurations (gray) and the limit configurations (dashed) can be seen at the top of the figure. Looking at the left column (Figures 3b–c) we can see how the edge *A* changes only slightly its distance from the support, while the edge *B* falls down in a clockwise rotation of the body (Figure 3d) until it starts bouncing. Both *A* and *B* edges reach, in a long enough time span, the same distance from the support, slightly greater than the initial one because the body ends up lying on a flat face instead of an edge. This fact, together with the small oscillations the edge *A* exhibits in the transition (Figure 3b), reveals the absence of a real contact surface. When frictional forces are added, the system exhibits a richer dynamics.

As can be seen from the bouncing of both edges *A* and *B* (Figures 3f–g), the rigid motion resembles a rocking motion until it fades out. In the frictionless case (Figure 3a) the trajectory of the center of the rigid body turns out to be vertical. That means that while the body rotates, the edge *A* slides leftward. Instead, if the friction coefficient is large enough the edge *A* does not slide any more, though it bounces for a while (Figures 3e–g), thus making the trajectory of the center very different from the previous case and even longer. These differences could be caught also by comparing the first impact time (Figures 3c–g).

7.3 Bouncing and Rolling

Figure 4 shows the plane motion of a rigid body in the shape of a circular cylinder, with diameter *L*, which drops on the support from a distance v_0 with initial horizontal velocity $\dot{u}_0$. The left column (b)–(d) refers to a contact without friction, while the right column (e)–(g) refers to a contact with friction. Graphs (b)–(e) and (c)–(f) show the time evolution of the vertical and horizontal coordinates of the center, $y_C(t)$ and $x_C(t)$, while graphs (d)–(g) show the time evolution of the angular velocity. The trajectory of the center drawn in the top panel (a) has been rescaled to make the bouncing more visible. Although friction does not affect the time evolution of the distance of the center from the support (the vertical motion of the body), as can be seen comparing Figures 4(b) and 4(e), it makes the motion quite different: at the first impact the angular velocity rises suddenly (Figure 4g), as a consequence of the initial value of the horizontal velocity. That means that the friction makes the cylinder roll while, at the same time, lowering the horizontal velocity (compare Figures 4(f) and 4(c)). Finally, it is worth to note how the angular velocity decreases after the bouncing has faded out. This is a consequence of the damping forces (35) acting on points close to the contact, whose vertical velocity is different from zero because of rolling.

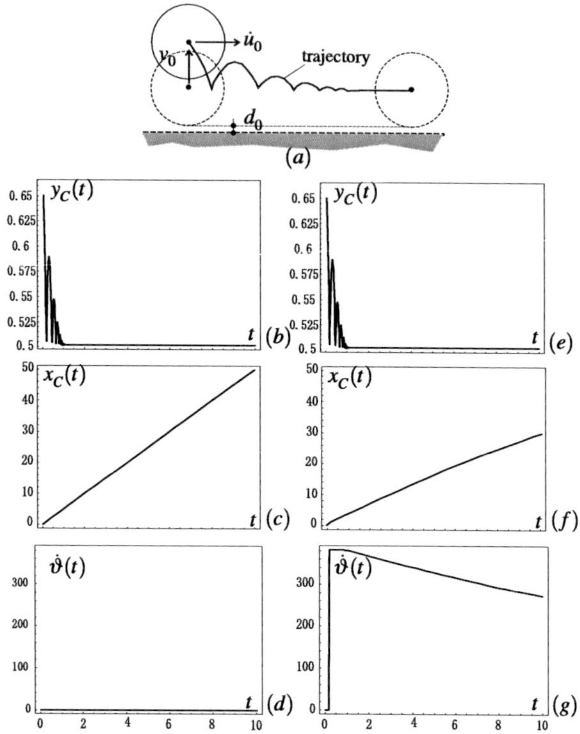

Fig. 4 Plane motion of a rigid circular cylinder. Left column: contact without friction ($L = 1$ m, $d_0 = 0.002$m, $v_0 = 0.15$ m, $\dot{u}_0 = 5$ m/s, $\nu_\tau = 8$, $\rho = 10^3$ kg/m^3, $\nu_\partial = 4$, $\beta_\partial / d_0^{\nu_\partial} = 6.25 \times 10^5$ Pa·s/m, $\beta_f = 0$). Right column: contact with friction ($\nu_f = 6$, $\beta_f / d_0^{\nu_f} = 1.6 \times 10^{19}$ Pa·s/m).

7.4 Adhesion and Detachment

Figure 5 shows the outcome of simulations where adhesive contact forces (38) have been added. To better understand the influence of adhesive forces these simulations consist in computing the motion generated by throwing the body against either a horizontal support (like a ceiling, Figure 5a) or a vertical support (like a wall, Figure 5b), in order not to confuse the adhesive forces with the gravity force. The initial velocity has been calibrated in such a way to let the body touch the support without bouncing back. Once the body has got stuck to the support, the mass density of the body is increased gradually as a trick to make the bond brake. And that is exactly what happens: the body detaches from the support and falls down.

The role of the friction is very different in the two simulations. While in case (a) the friction just slows down the body until it stops sliding on the ceiling, in case (b) the friction prevents the body from sliding down the wall until it suddenly

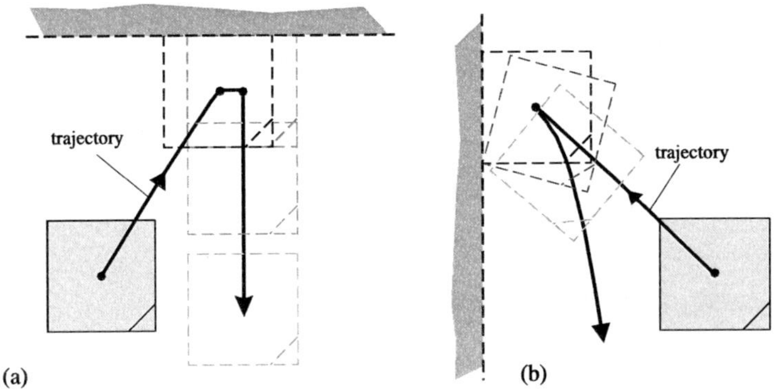

(a) (b)

Fig. 5 Effect of the adhesive forces: (a) adhesion to a ceiling; (b) adhesion to a vertical wall; ($L = 1$ m, $d_0 = 0.002$ m, $\nu_{\mathfrak{t}} = 8$, $\nu_{\mathfrak{a}} = 6$, $\nu_{\mathfrak{d}} = 3$, $\nu_{\mathfrak{f}} = 6$, $\beta_{\mathfrak{a}} = 4 \times 10^6\, \alpha_{\mathfrak{t}}$ $\beta_{\mathfrak{d}} = 4 \times 10^6\, \alpha_{\mathfrak{t}}$, $\beta_{\mathfrak{f}} = 4 \times 10^6\, \alpha_{\mathfrak{t}}$).

starts detaching. The trajectory of the center of the body and a few frames help to understand the motion.

7.5 Bouncing and Vibrations of a Soft Body

For plane deformations it is convenient to enforce *a-priori* the incompressibility constraint $\det \mathsf{F} = (\det \mathsf{R})(\det \mathsf{U}) = 1$ by giving the matrix of the stretch U the following parameterized form

$$\begin{pmatrix} \frac{1+\kappa^2}{\chi} & \kappa \\ \kappa & \chi \end{pmatrix}. \tag{39}$$

The principal stretches λ and $1/\lambda$, turn out to be given by the expression

$$\lambda := \left(1 + \kappa^2 + \chi^2 + \sqrt{(1+\kappa^2)^2 + 2(\kappa^2 - 1)\chi^2 + \chi^4}\right) / (2\chi). \tag{40}$$

Denoting by θ the amplitude of the rotation R, the tensor F will be described by the three parameters: θ, κ, χ.

Figure 6 shows a cylinder, with diameter L, bouncing in a plane vertical motion, after dropping on the support from a distance v_0. In this case λ denotes the vertical stretch while $1/\lambda$ denotes the horizontal stretch, given by the ratio $l(t)/L$. Both friction and impact damping have been neglected. The motion described by the graphs in Figures 6(b)–(c) is only slightly damped by a dissipative stress with a low value for the coefficient μ. It is worth noticing how the deformation of the body reflects on the bouncing. Comparing the time-histories of the center C and the bottom B (Figure 6b), we can see a sequence of bounces, due to the motion of the center,

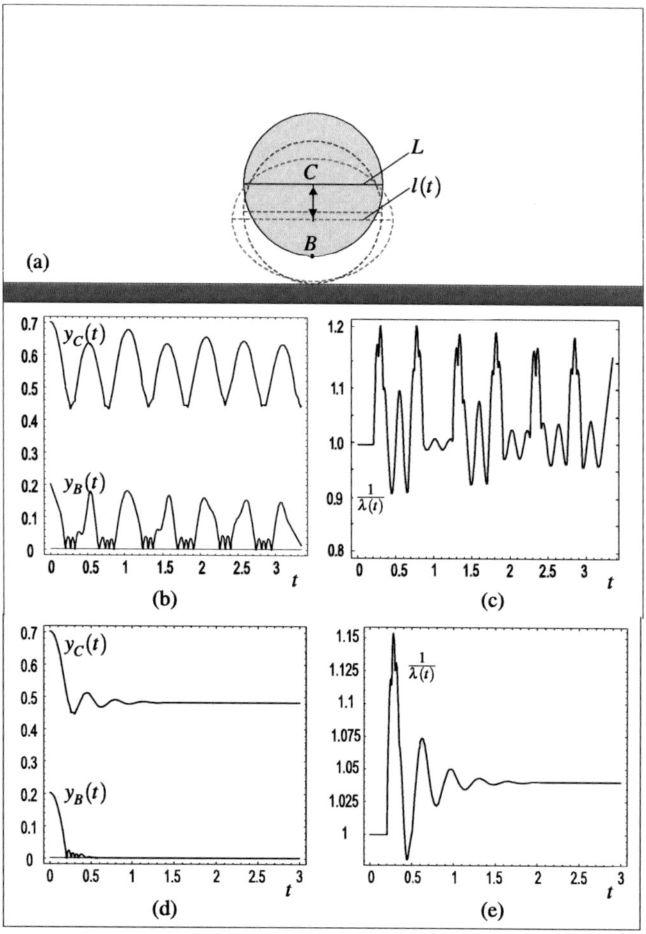

Fig. 6 Motion of an elastic cylinder ($L = 1$ m, $d_0 = 0.002$ m, $v_0 = 100 d_0$, $c_1 = 6 \times 10^4$ Pa, $\rho = 4 \times 10^3$ kg/m^2, $\nu_\tau = 8$, $\nu_\partial = 2$, $\nu_f = 2$, $\beta_\partial = 0$, $\beta_f = 10^9 \alpha_\tau$, $\mu = 10^2$ Pa·s): (a) selected frames; (b) distance of B and C from the support; (c) principal stretch; (d)–(e) effect of a higher dissipative stress ($\mu = 5 \times 10^3$ Pa·s).

together with other bounces with a lower amplitude and a higher frequency, due to the stretching. The frequencies of the two kinds of bouncing seem to be far enough not to interact significantly with each other. The graphs in Figures 6(d)–(e) show the effects of a higher value for the coefficient μ, to be compared with the graphs above. After a short while the dissipation is able to slow down both the bouncing and the vibrations.

The body in Figure 7 falls down on the support from a distance v_0 with an initial leftward horizontal velocity $\dot{u}_0$. The friction makes it start rolling while the impact is followed by a few bounces. The body is stiffer and heavier than the body in

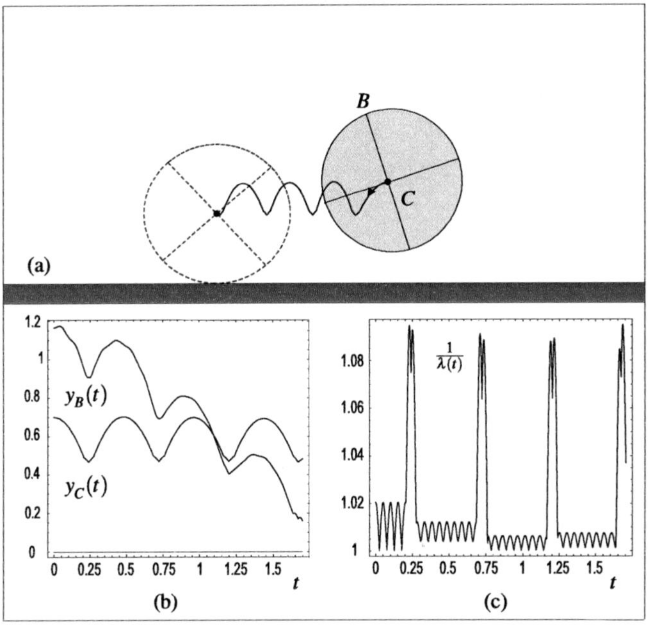

Fig. 7 Motion of an elastic cylinder ($L = 1$ m, $d_0 = 0.002$ m, $c_1 = 6 \times 10^6$ Pa, $\rho = 10^5$ kg/m^2, $\nu_\tau = 8$, $\nu_\eth = 2$, $\nu_f = 2$, $\beta_\eth = 0$, $\beta_f = 10^7\,\alpha_\tau$, $\mu = 10^2$ Pa·s, $v_0 = 100d_0$, $\dot{u}_0 = -1$ m/s).

Figure 6. The resulting vibration frequency is higher while it bounces almost at the same frequency.

7.6 Driven Motion of a Soft Contractile Body

Figure 8 shows the vertical motion of a contractile body driven by an oscillating external contractile couple Q. The contraction is assumed to be isochoric and with a fixed eigenvector **n**, the external unit normal to the support. In such a motion G is described by a scalar time function γ, which is one of its eigenvalues together with $1/\gamma$. The couple Q is described by a scalar function as well, which has been assigned the law $Q(t) = Q_0 \sin(2\pi t/T)$, with $T = 0.4$ s, $Q_0/\mu_\gamma = 0.9$ s^{-1}. The resistance to contaction μ_γ was set to a very high value in order to prevent plastic deformation or relaxation induced by the stress and the energy terms in (24). The body initially lies at rest on the support, slightly deformed by its weight. As soon as the signal is activated the body starts oscillating and, as can be noticed in the selected frames in Figures 8(a)–(b), and from the graph (d), it contracts and jumps upward. The evolution of the relaxed shape is shown in Figure 8(c). The graph in Figure 8(e) describes

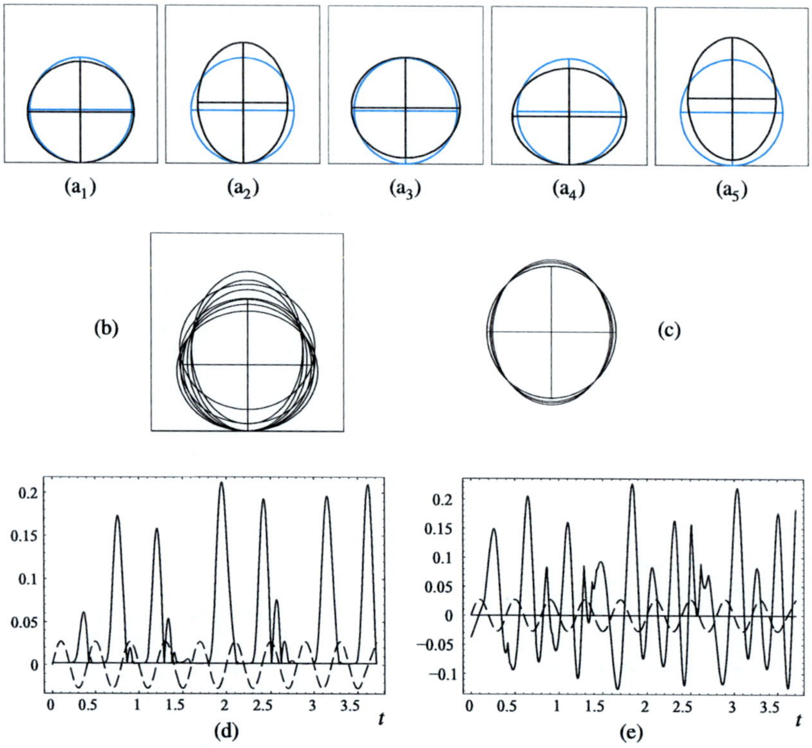

Fig. 8 Motion generated by an oscillating contractile couple with fixed principal axes: (a) selected frames showing how the body jumps upward; (b) overlapped body shapes during the motion; (c) oscillating relaxed shape; (d) bouncing of the bottom over the support, and driving contractile couple Q (dashed line, rescaled amplitude); (e) vertical elongation (solid line) and driving contractile couple Q (dashed line, rescaled amplitude).

the time evolution of the vertical elongation $(\lambda(t)\gamma(t)-1)$, which is compared with $Q(t)$.

Figure 9 shows a motion driven directly by an oscillating tensor G. The contraction is assumed again to be isochoric and is assigned by the eigenvalue law $\gamma(t) = 1 + (\gamma_0 - 1)\sin(2\pi t/T)$, with $T = 1.1$ s, $\gamma_0 = 1.2$, and by a rotating eigenvector $\mathbf{a}(t)$, with a constant angular velocity $(0.8\pi/T)$. It is worth noting that $\mathsf{G}(t)$ at any time t is a symmetric tensor with positive eigenvalues. Hence G does not generate a rotating relaxed shape but just a pulsing relaxed shape with varying pulse axis (Figure 9c). The body starts oscillating and soon, from an initial configuration on the right side of Figure 9(a), it moves leftward rolling, almost crawling, and even jumping a little. In Figure 9(d) the increasing horizontal displacement of the center C and of the bottom B in the starting configuration are showed together with the oscillating driving contraction.

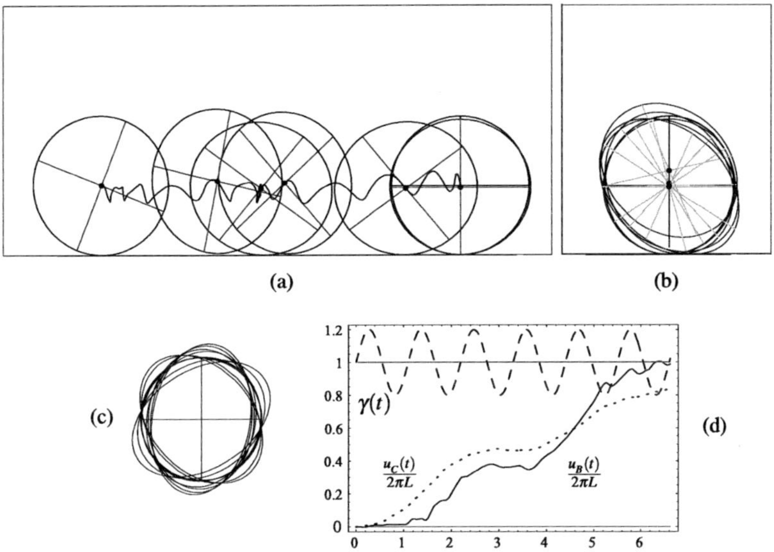

Fig. 9 Motion driven by an oscillatory contraction with rotating principal axes: (a) trajectory of the center and some frames from right to left; (b) frictionless motion; (c) the oscillating relaxed shape; (d) normalized leftward displacement of the initial bottom point B (solid line) and of the center C (dotted line), and driving contraction amplitude (dashed line).

As expected, locomotion on a support relies on the frictional traction: removing the friction the body cannot move forward any more while the center follows a vertical trajectory (Figure 9b).

The contractile couple Q, which was a given quantity in the previous case, can now be computed through (24), as a reactive couple. In both cases the power expended per unit volume to sustain the body motion is $\mathsf{Q} \cdot \dot{\mathsf{G}}\mathsf{G}^{-1}$.

8 Conclusions

The aim of this paper was to study the motion of a soft contractile body over a rigid substrate. To this end a non linear elastic model has been used together with a contact model based on constitutive laws for different kind of interactions. The body model, though restricted to homogeneous deformations, accounts for large deformations and also for an evolving relaxed shape. This makes it possible to give a precise meaning to contractility. The constitutive characterization of both the material and the contact is based on a purely mechanical dissipation principle which enlighten the role of energy functions for both stress and contact forces. The presented simulations are meant to illustrate the realizable motions, in the presence of contact interactions, like repulsion, adhesion, impact damping and friction, of a body with different material properties, showing the interplay between contact, vibrations and contractions. In particular it is shown how the contractility endows a body with motility capabilities which can be exploited for locomotion. All the computations, both symbolic and numerical, have been performed using *Mathematica*®, starting from the very basic expressions in Sections 2, 3 and 4. Further work should be done for gaining better physical interpretations of numerical simulations by using parameters based on experimental data and by comparing results with other models.

References

1. Argento, C., Jagota, A., Carter, W.C.: Surface formulation for molecular interactions of macroscopic bodies. J. Mech. Phys. Solids 45, 1161–1183 (1997)
2. Contento, A., Di Egidio, A., Dziedzic, J., Tatone, A.: Modeling the contact of stiff and soft bodies with a rigid support by short range force fields. TASK Quarterly 13, 1001–1027 (2009)
3. Di Carlo, A., Quiligotti, S.: Growth and balance. Mech. Res. Comm. 29, 449–456 (2002)
4. Gao, J., Luedtke, W.D., Gourdon, D., Ruths, M., Israelachvili, J.N., Landman, U.: Frictional forces and Amontons' law: From the molecular to the macroscopic scale. J. Phys. Chem. B 108, 3410–3425 (2004), doi:10.1021/jp036362l.
5. Germain, P.: The method of virtual power in continuum mechanics. Part 2: Microstructure. SIAM J. Appl. Math. 25, 556–575 (1973)
6. Greenwood, J.A.: Adhesion of elastic spheres. Proc. R. Soc. Lond. A 453, 1277–1297 (1997), doi:10.1098/rspa.1997.0070
7. Gurtin, M.E.: A gradient theory of single-crystal viscoplasticity that accounts for geometrically necessary dislocations. J. Mech. Phys. Solids 50, 5–32 (2002)
8. Johnson, K.L., Kendall, K., Roberts, A.D.: Surface energy and the contact of elastic solids. Proc. R. Soc. Lond. A 324, 301–313 (1971), `http://www.jstor.org/pss/78058`
9. Mogilner, A.: Mathematics of cell motility: Have we got its number? J. Math. Biol. 58, 105–134 (2009)
10. Muller, V.M., Yushchenko, S.V., Derjaguin, B.V.: On the influence of molecular forces on the deformation of an elastic sphere and its sticking to a rigid plane. J. Colloid Interface Sci. 77, 91–101 (1980), doi:10.1016/0021-9797(80)90419-1
11. Nardinocchi, P., Teresi, L.: On the active response of soft living tissues. J. Elasticity 88, 27–39 (2007)
12. Sauer, R.A., Li, S.: A contact mechanics model for quasi-continua. Int. J. Numer. Meth. Engng. 71, 931–962 (2007)
13. Sauer, R.A., Wriggers, P.: Formulation and analysis of a three-dimensional finite element implementation for adhesive contact at the nanoscale. Comp. Meth. Appl. Mech. Engng. 198, 3871–3883 (2009)
14. Spolenak, R., Gorb, S., Gao, H., Arzt, E.: Effects of contact shape on the scaling of biological attachments. Proc. R. Soc. Lond. A 461, 305–319 (2005)
15. Wriggers, P.: Computational Contact Mechanics. John Wiley & Sons, Chichester (2006)
16. Yu, N., Polycarpou, A.A.: Adhesive contact based on the Lennard–Jones potential: A correction to the value of the equilibrium distance as used in the potential. J. Colloid Interface Sci. 278, 428–435 (2004)

Two-Level Block Preconditioners for Contact Problems

C. Janna, M. Ferronato and G. Gambolati

Abstract Contact mechanics can be addressed numerically by Finite Elements using either a penalty formulation or the Lagrange multipliers. The penalty approach leads to a linearized symmetric positive definite system which can prove severely ill-conditioned, with the iterative solution to large 3D problems requiring expensive preconditioners to accelerate, or even to allow for, convergence. If the nodal unknowns are numbered properly, the system matrix takes on a two-level block structure that may be efficiently preconditioned by matrices having the same block structure. The present study addresses two different approaches, the Mixed Constraint Preconditioner (MCP) and the Multilevel Incomplete Factorization (MIF). It is shown that both MCP and MIF can prove very effective in the solution of large size 3D contact problems discretized by a penalty formulation, where classical algebraic preconditioners, such as the incomplete Cholesky decomposition, may exhibit poor performances.

1 Introduction

The increasing capabilities of the modern computers enhance a continuous improvement in the numerical treatment of contact problems, allowing for very sophisticated and detailed three-dimensional (3D) analyses. The numerical discretization by Finite Elements (FE) gives rise to highly non-linear systems of equations whose solution is usually obtained with a Newton-like method. This basically relies on the solution of a sequence of $n \times n$ linear systems:

$$A\mathbf{u} = \mathbf{f} \tag{1}$$

C. Janna · M. Ferronato · G. Gambolati
Department of Mathematical Methods and Models for Scientific Applications, University of Padova, via Trieste 63, 35121 Padova, Italy; e-mail: {janna, ferronat, gambo}@dmsa.unipd.it

G. Zavarise & P. Wriggers (Eds.): Trends in Computational Contact Mechanics, LNACM 58, pp. 211–226.
springerlink.com

where the number of the unknowns, depending on the size of the domain and the required numerical accuracy, may easily grow up to several hundreds of thousands. As the memory occupation and processing time for a direct solver can often be prohibitive for a 3D simulation, the use of iterative Krylov subspace techniques, such as Preconditioned Conjugate Gradient (PCG), may become virtually mandatory [19]. The penalty formulation of contact, however, which is still widely used for its ease of implementation, generally produces ill-conditioned Symmetric Positive Definite (SPD) matrices needing expensive preconditioners to accelerate, or even to allow for, convergence. Two-level preconditioners can be developed that take advantage of the matrix structure arising from the penalty approach and can significantly outperform popular algebraic preconditioners such as incomplete Cholesky factorizations (ILLT). If the unknowns are ordered properly, by numbering first those related to the standard FE nodes $\mathbf{u}_m$ and last those related to the penalty contact elements $\mathbf{u}_c$, the tangential stiffness matrix takes on the following 2×2 block form [10]:

$$A\mathbf{u} = \mathbf{f} \rightarrow \begin{bmatrix} K & B \\ B^T & C \end{bmatrix} \begin{Bmatrix} \mathbf{u}_m \\ \mathbf{u}_c \end{Bmatrix} = \begin{Bmatrix} \mathbf{f}_m \\ \mathbf{f}_c \end{Bmatrix} \tag{2}$$

where K and B are structural blocks, C is the penalty block and $\mathbf{f}_m$, $\mathbf{f}_c$ are the residual forces corresponding to standard and penalty nodes, respectively. The main source of ill-conditioning of system (2) is the several orders of magnitude difference between the spectral radius of K and C. An efficient preconditioner for A can be a block matrix:

$$M^{-1} = \begin{bmatrix} G & B \\ B^T & \tilde{C} \end{bmatrix}^{-1} \tag{3}$$

with G and $\tilde{C}$ proper approximations of K and C, respectively. The application of M^{-1} basically requires the sequential solution of two inner systems that involves G and $S = \tilde{C} - B^T G B$, i.e. the Schur complement of (3). It goes without saying that the effectiveness of M^{-1} relies on a most appropriate choice of G and $\tilde{C}$ that have to be as cheap and effective to compute and apply as possible. In the present paper we consider two novel approaches:

- the Mixed Constraint Preconditioner (MCP), originally developed for coupled consolidation problems [7, 8] and later successfully applied to contact mechanics [10], and
- the Multilevel Incomplete Factorization (MIF), originally developed for multilevel non-linear FE problems [15].

The use of either MCP or MIF as a preconditioner in the PCG scheme, according to the specific problem at hand, allows for a significant saving in both the computational memory and CPU time required to solve numerically non-linear contact problems.

The paper is organized as follows. First, the MCP and MIF construction and application algorithms are introduced with a description of the user-specified parameters needed for their setup. Then, MCP and MIF computational performance is compared to that of ILLT in two large size contact test problems borrowed from the geomechanical simulation of faulted reservoirs. Finally, a few concluding remarks close the paper.

2 Mixed Constraint Preconditioner

The preconditioner in (3) can be regarded as a SPD variant of the Constraint Preconditioner first proposed in [17] for indefinite matrices. This class of block preconditioners, originally developed for constrained optimization problems, was later successfully used in the FE solution to Navier–Stokes [25] and coupled consolidation equations [7, 8], where the $(2,2)$ block of A is negative definite. If $\tilde{C}$ is set equal to C, all the eigenvalues of the preconditioned matrix $M^{-1}A$ corresponding to those of the penalty block C are unitary [9], thus removing the main source of ill-conditioning in the whole system.

Set $\tilde{C} = C$. The application of (3) in a PCG iteration consists of the computation of $\mathbf{y} = M^{-1}\mathbf{r}$ with $\mathbf{r}$ the residual vector, i.e. the solution to the system:

$$\begin{bmatrix} G & B \\ B^T & C \end{bmatrix} \begin{bmatrix} \mathbf{y}_m \\ \mathbf{y}_c \end{bmatrix} = \begin{bmatrix} \mathbf{r}_m \\ \mathbf{r}_c \end{bmatrix} \tag{4}$$

This can be carried out by solving for $\mathbf{y}_m$ in the upper set of equation (4):

$$\mathbf{y}_m = G^{-1}(\mathbf{r}_m - B\mathbf{y}_c) \tag{5}$$

and then substituting equation (5) into the lower set, giving rise to an inner reduced system:

$$S\mathbf{y}_c = \mathbf{r}_c - B^T G^{-1}\mathbf{r}_m \tag{6}$$

where S is the Schur complement of M:

$$S = C - B^T G^{-1} B \tag{7}$$

At variance with Navier–Stokes and coupled consolidation problems, S in equation (7) is a difference between SPD matrices, hence is not guaranteed to be definite for any choice of G and might even be singular. In practice, S is positive definite if G is a "sufficiently" good approximation of K, as it tends to the exact SPD Schur complement of A in the limit when $G \to K$.

Define G as the incomplete Cholesky factorization of K with user-specified fill-in degree:

$$G = L_K L_K^T \tag{8}$$

The solution to the inner system (5) can be readily obtained from a forward and backward substitution. On the other hand, the solution to the second inner system (6) is much more expensive as the explicit computation of S in (7), i.e. the explicit inversion of L_K, is required. To avoid this computational cost, a proper approximation of S can be used. Following Bergamaschi et al. [8], we resort to a so-called mixed approach where S is approximated by the inexact Schur complement $\widehat{S}$:

$$\widehat{S} = C - B^T Z Z^T B \tag{9}$$

with ZZ^T the factorized approximate inverse AINV [4, 6] of K:

$$G^{-1} \simeq K^{-1} \simeq ZZ^T \tag{10}$$

Z is an upper triangular matrix. To save memory and CPU time, a dropping strategy is enforced in the $\widehat{S}$ computation aimed at neglecting its smallest entries:

$$\tilde{S} = \text{ drop}[\widehat{S}] \tag{11}$$

The approximation above is equivalent to implementing the preconditioner M^{-1} in equation (3) where $\tilde{C}$ is the following approximation of C:

$$\widehat{C} = \tilde{S} + B^T L_K^{-T} L_K^{-1} B \tag{12}$$

This is denoted as MCP in the following. Using equation (8) for G, it can be readily verified that MCP can be factorized as follows:

$$M^{-1} = \left(\begin{bmatrix} L_K & 0 \\ B^T L_K^{-T} & I \end{bmatrix} \begin{bmatrix} I & 0 \\ 0 & \tilde{S} \end{bmatrix} \begin{bmatrix} L_K^T & L_K^{-1} B \\ 0 & I \end{bmatrix} \right)^{-1} \tag{13}$$

where I is the identity matrix. Equation (13) shows right away that M^{-1} is SPD only if $\tilde{S}$ is so. For more details on the constraint preconditioning theory, see [8, 10, 17].

Since the main cost of the preconditioner application rests on the solution of the inner system (6), to reduce the computational effort a further simplification is introduced by computing the incomplete Cholesky factorization of $\tilde{S}$ with user-specified fill-in degree:

$$\tilde{S} \simeq L_{\tilde{S}} L_{\tilde{S}}^T \tag{14}$$

thus replacing the solution to (6) with a forward and backward substitution. Recalling equations (13) and (14), the actual factorized form of M^{-1} finally reads:

$$M^{-1} = \left(\begin{bmatrix} L_K & 0 \\ B^T L_K^{-T} & L_{\tilde{S}} \end{bmatrix} \begin{bmatrix} L_K^T & L_K^{-1} B \\ 0 & L_{\tilde{S}}^T \end{bmatrix} \right)^{-1} = \left(LL^T \right)^{-1} \tag{15}$$

The construction of M^{-1} requires two symmetric incomplete factorizations (L_K and $L_{\tilde{S}}$), one approximate inverse factor computation (Z), two sparse matrix-matrix products ($W = B^T Z$ and WW^T) and one sparse merge of matrices ($C - WW^T$). Despite the high computational complexity for the preconditioner setup, a significant part of M^{-1}, i.e. the computation of L_K, Z, W and WW^T, can be actually made just once at the beginning of a complete simulation in all the problems where contact is the only source of non-linearity, with the penalty block C alone being the varying part of system (2).

The user-specified parameters needed for the MCP setup are the following:

- ρ_K, i.e. the number of terms computed and stored in each row of L_K in excess to the non-zeroes of K;
- the AINV tolerance τ_Z, i.e. the fraction to the Z diagonal term below which an extra-diagonal coefficient in the same row is dropped;
- γ_S, i.e. the number of terms computed and stored in each row of $\tilde{S}$ in excess to the non-zeroes of C;
- ρ_S, i.e. the number of terms computed and stored in each row of $L_{\tilde{S}}$ in excess to the non-zeroes of $\tilde{S}$.

3 Multi-Level Incomplete Factorization

The partial factorization of a real symmetric $(n+m) \times (n+m)$ matrix A is defined as

$$A = \begin{bmatrix} A_{11} & A_{12} \\ A_{12}^T & A_{22} \end{bmatrix} = \begin{bmatrix} L_1 & 0 \\ A_{12}^T L_1^{-T} D_1^{-1} & I \end{bmatrix} \begin{bmatrix} D_1 & 0 \\ 0 & S_1 \end{bmatrix} \begin{bmatrix} L_1^T & D_1^{-1} L_1^{-1} A_{12} \\ 0 & I \end{bmatrix} \tag{16}$$

where L_1 and D_1 are a $n \times n$ lower unitary triangular and diagonal matrix, respectively, such that

$$A_{11} = L_1 D_1 L_1^T \tag{17}$$

and S_1 is the $m \times m$ Schur Complement of A:

$$S_1 = A_{22} - A_{12}^T L_1^{-T} D_1^{-1} L_1^{-1} A_{12} \tag{18}$$

A preconditioner for A can be obtained by approximating the exact factorization (16). First, the exact A_{11} decomposition (17) is replaced by an incomplete one based on an appropriate dropping strategy, e.g. [20, 18]

$$\tilde{L}_1 \tilde{D}_1 \tilde{L}_1^T \simeq A_{11} \tag{19}$$

As a consequence, the rectangular block $D_1^{-1} L_1^{-1} A_{12}$ and the exact Schur complement S_1 cannot be computed. Hence, the factorization (16) becomes a partial incomplete factorization

$$A \simeq M = L_1 D_1 L_1^T = \begin{bmatrix} \tilde{L}_1 & 0 \\ H_1^T & I \end{bmatrix} \begin{bmatrix} \tilde{D}_1 & 0 \\ 0 & \tilde{S}_1 \end{bmatrix} \begin{bmatrix} \tilde{L}_1^T & H_1 \\ 0 & I \end{bmatrix} \tag{20}$$

where H_1 and $\tilde{S}_1$ are chosen so that the four blocks of M resemble as much as possible those of A, i.e.

$$\tilde{L}_1 \tilde{D}_1 H_1 \simeq A_{12} \tag{21}$$

$$\tilde{S}_1 + H_1^T \tilde{D}_1 H_1 \simeq A_{22} \tag{22}$$

For instance, H_1 can be computed as the rectangular matrix arising naturally from the factorization process of A_{11} extended so as to include the corresponding rows (columns) of A_{12} (A_{12}^T), and $\tilde{S}_1$ as $A_{22} - H_1^T \tilde{D}_1 H_1$.

The application of M^{-1} in (20) to $\mathbf{r}$ can be carried out by a three-stage procedure: (1) the forward substitution $L_1 \mathbf{z} = \mathbf{r}$; (2) the inversion of the block diagonal system $D_1 \mathbf{x} = \mathbf{z}$; and (3) the backward substitution $L_1^T \mathbf{y} = \mathbf{x}$. In the second stage the solution of the block diagonal system is found as

$$\begin{bmatrix} \tilde{D}_1 & 0 \\ 0 & \tilde{S}_1 \end{bmatrix} \begin{Bmatrix} \mathbf{x}_1 \\ \mathbf{x}_2 \end{Bmatrix} = \begin{Bmatrix} \mathbf{z}_1 \\ \mathbf{z}_2 \end{Bmatrix} \rightarrow \begin{cases} \mathbf{x}_1 = \tilde{D}_1^{-1} \mathbf{z}_1 \\ \mathbf{x}_2 = \tilde{S}_1^{-1} \mathbf{z}_2 \end{cases} \tag{23}$$

While the computation of $\mathbf{x}_1$ is straightforward, a full system with matrix $\tilde{S}_1$ has to be solved to get $\mathbf{x}_2$. This system can be solved in an approximate way by performing again a partial incomplete factorization of $\tilde{S}_1$.

The basic idea of some multi-level preconditioners, e.g. [23, 22, 21], which underlies also the present one, is to use recursively a partial incomplete factorization of the Schur complement of each level. Matrix $\tilde{S}_1$ in (20) is the first-level approximate Schur complement which can be regarded as a four blocks matrix and factorized as:

$$\tilde{S}_1 \simeq \begin{bmatrix} \tilde{L}_2 & 0 \\ H_2^T & I \end{bmatrix} \begin{bmatrix} \tilde{D}_2 & 0 \\ 0 & \tilde{S}_2 \end{bmatrix} \begin{bmatrix} \tilde{L}_2^T & H_2 \\ 0 & I \end{bmatrix} \tag{24}$$

Therefore A in (20) can be viewed as a zero-level Schur complement $\tilde{S}_0$. At the i-th level, $\tilde{S}_i$ is factorized as

$$\tilde{S}_i = \begin{bmatrix} \tilde{S}_{i,11} & \tilde{S}_{i,12} \\ \tilde{S}_{i,12}^T & \tilde{S}_{i,22} \end{bmatrix} \simeq \begin{bmatrix} \tilde{L}_{i+1} & 0 \\ H_{i+1}^T & I \end{bmatrix} \begin{bmatrix} \tilde{D}_{i+1} & 0 \\ 0 & \tilde{S}_{i+1} \end{bmatrix} \begin{bmatrix} \tilde{L}_{i+1}^T & H_{i+1} \\ 0 & I \end{bmatrix} \tag{25}$$

thus giving rise to an $(i+1)$-th level Schur complement such that

$$\tilde{S}_{i+1} + H_{i+1}^T \tilde{D}_{i+1} H_{i+1} \simeq \tilde{S}_{i,22} \tag{26}$$

with $\tilde{L}_{i+1} \tilde{D}_{i+1} \tilde{L}_{i+1}^T \simeq \tilde{S}_{i,11}$ and $\tilde{L}_{i+1} \tilde{D}_{i+1} H_{i+1} \simeq \tilde{S}_{i,12}$. The algorithm stops when the size of $\tilde{S}_{i+1}$ equals 0, i.e. when $i+1$ equals the user specified number of levels ℓ. For more details on the multilevel preconditioning, see [15, 21, 22]. The fill-in degree is controlled at each level i by two parameters:

- $\rho_1^{(i)}$, i.e. the maximum allowable number of nonzeroes for each row of L_{i+1} in excess to those of the $(1,1)$ block of $\tilde{S}_i$;
- $\rho_2^{(i)}$, i.e. the maximum allowable number of nonzeroes for each row of $\tilde{S}_{i+1}$ in excess to those of the $(2,2)$ block of $\tilde{S}_i$.

The efficiency of a multi-level preconditioner basically relies on the level subdivision and the algorithm used to compute $\tilde{L}_{i+1}$, $\tilde{D}_{i+1}$, H_{i+1} and $\tilde{S}_{i+1}$. Generally speaking, the $(i+1)$-th level blocks should be computed as inexpensively as possible giving at the same time good approximations of $\tilde{S}_{i,11}$, $\tilde{S}_{i,12}$ and $\tilde{S}_{i,22}$. Decreasing the number of the retained entries, the cost for the preconditioner setup and application reduces but conversely the total number of iterations required for convergence increases. Moreover, M in (20) may become indefinite because of dropping, thus losing the key SPD property of the native matrix A needed for PCG to converge. In these cases a breakdown of PCG preconditioned with MIF can occur.

There are two major sources for such breakdown:

1. negative terms may arise in $\tilde{D}_{i+1}$ during the incomplete factorization of $\tilde{S}_{i,11}$;
2. even though $\tilde{D}_{i+1}$ has all positive diagonal terms, $\tilde{S}_{i+1}$ can be indefinite if its computation is not accurate enough because it arises from the difference between two SPD matrices.

These issues are addressed in the following subsections with the aim at preserving the MIF positive definiteness. For simplicity of notation, in the sequel the $\sim$ symbol above the i-th level Schur complement will be dropped.

3.1 Positive Definiteness of $S_{i,11}$

Consider the i-th level Schur complement:

$$S_i = \begin{bmatrix} S_{i,11} & S_{i,12} \\ S_{i,12}^T & S_{i,22} \end{bmatrix} \tag{27}$$

A breakdown of PCG preconditioned with MIF may be caused by the occurrence of negative entries in $\tilde{D}_{i+1}$ during the factorization of $S_{i,11}$. Although the (1,1) block of a SPD matrix is also SPD, this is not sufficient to guarantee that its incomplete factorization is SPD too. A few strategies have been advanced in the literature to address this point. For instance, in [2] a diagonal compensated reduction of the positive off-diagonal entries is suggested in order to transform the native matrix into a M-matrix before the incomplete factorization process is performed. A different algorithm is proposed in [5] where the incomplete factor is obtained through a matrix-orthogonalization instead of a Cholesky decomposition. In the present paper we use the strategy discussed in [1], that has proved quite promising in structural mechanics applications [14, 24]. Following this approach, we can write

$$S_{i,11} = L_{i+1} D_{i+1} L_{i+1}^T = \tilde{L}_{i+1} \tilde{D}_{i+1} \tilde{L}_{i+1}^T + E_{i+1} \tag{28}$$

where $L_{i+1} D_{i+1} L_{i+1}^T$ is the exact root-free factorization of $S_{i,11}$, $\tilde{L}_{i+1} \tilde{D}_{i+1} \tilde{L}_{i+1}^T$ the incomplete factorization and E_{i+1} an error matrix with all the discarded entries. Note that $\tilde{L}_{i+1} \tilde{D}_{i+1} \tilde{L}_{i+1}^T$ in (28) is the exact factorization of $S_{i,11} - E_{i+1}$. As $S_{i,11}$ is SPD, it follows immediately that a sufficient condition for $\tilde{L}_{i+1} \tilde{D}_{i+1} \tilde{L}_{i+1}^T$ to be SPD is that E_{i+1} is negative semidefinite.

Suppose that we have computed the first column of L_{i+1} and we want to drop the term l_{j1} and the corresponding l_{1j} term in L_{i+1}^T. Denoting the entry in position

ij of $S_{i,11}$ by s_{ij}, the dropped term l_{j1} is equal to s_{j1}/s_{11}. Hence, the incomplete factorization is equivalent to the exact factorization of $\bar{S}_{i,11}$ where the entries s_{1j} and s_{j1} are replaced by zero, and the error matrix E_{i+1} contains s_{j1} in position $1j$ and $j1$. For example, in a 3×3 matrix with $j = 2$ we have

$$S_{i,11} = \begin{bmatrix} s_{11} & s_{12} & s_{13} \\ s_{21} & s_{22} & s_{23} \\ s_{31} & s_{32} & s_{33} \end{bmatrix} = \begin{bmatrix} s_{11} & 0 & s_{13} \\ 0 & s_{22} & s_{23} \\ s_{31} & s_{32} & s_{33} \end{bmatrix} + \begin{bmatrix} 0 & s_{12} & 0 \\ s_{21} & 0 & 0 \\ 0 & 0 & 0 \end{bmatrix} = \bar{S}_{i,11} + E_{i+1} \tag{29}$$

E_{i+1} can be forced to be negative semidefinite by adding two coefficients α and β in position $(1,1)$ and (j,j), such that the submatrix

$$\begin{bmatrix} \alpha & s_{1j} \\ s_{j1} & \beta \end{bmatrix} \tag{30}$$

be negative semidefinite. The matrix $\bar{S}_{i,11}$ must be modified accordingly by subtracting α and β from the corresponding diagonal terms $(1,1)$ and (j,j). Though different choiches are possible, e.g. [1, 24], the option $\alpha = \beta = -|s_{1j}|$ is particularly attractive because in such a way the sum of the absolute value of the arbitrarily introduced entries $|\alpha| + |\beta|$ is minimal.

Applying the same procedure to the other dropped entries of L_{i+1} gives rise to a factorization which is guaranteed to be the exact factorization of a SPD matrix:

$$\tilde{L}_{i+1}\tilde{D}_{i+1}\tilde{L}_{i+1}^T = \bar{S}_{i,11} \tag{31}$$

Obviously, the quality of the above procedure depends on the quality of $\bar{S}_{i,11}$ as an approximation of $S_{i,11}$.

3.2 Positive Definiteness of S_{i+1}

A factorization procedure is presented in [26] that proceeds by rows and updates the Schur complement with a quadratic formula ensuring its positive definiteness. The theoretical robustness of this algorithm has been proved in [16]. Here the above idea is extended to levels in place of rows, thus implying matrix by matrix operations instead of vector by vector.

Using equations (27) and (28), the matrix S_i can be additively decomposed as:

$$S_i = \begin{bmatrix} S_{i,11} & S_{i,12} \\ S_{i,12}^T & S_{i,22} \end{bmatrix} = \bar{S}_i + R_i = \begin{bmatrix} \bar{S}_{i,11} & S_{i,12} \\ S_{i,12}^T & S_{i,22} \end{bmatrix} + \begin{bmatrix} E_{i+1} & 0 \\ 0 & 0 \end{bmatrix} \tag{32}$$

Assume that the procedure described in the previous section has been used, so that E_{i+1}, hence R_i, is negative semidefinite. It follows that $\bar{S}_i$ is positive definite. Recalling (31), the partial factorization of $\bar{S}_i$ is formally equal to equation (25):

$$\bar{S}_i = \begin{bmatrix} \tilde{L}_{i+1} & 0 \\ H_{i+1}^T & I \end{bmatrix} \begin{bmatrix} \tilde{D}_{i+1} & 0 \\ 0 & S_{i+1} \end{bmatrix} \begin{bmatrix} \tilde{L}_{i+1}^T & H_{i+1} \\ 0 & I \end{bmatrix} \tag{33}$$

with the $(i+1)$-th level Schur complement given by

$$S_{i+1} = S_{i,22} - H_{i+1}^T \tilde{D}_{i+1} H_{i+1} \tag{34}$$

Theoretically, S_{i+1} is the Schur complement of a SPD matrix, hence it is SPD. Unfortunately, it is expensive to be computed. In fact, though $\tilde{L}_{i+1}$ is usually characterized by a high degree of sparsity, the matrix $H_{i+1} = \tilde{D}_{i+1}^{-1} \tilde{L}_{i+1}^{-1} S_{i,12}$ may have a very large number of non-zeroes and hence some of them are worth dropping. Collecting all the dropped entries in the error matrix E_H, we can write $H_{i+1} = \tilde{H}_{i+1} + E_H$ with the Schur complement rewritten as

$$S_{i+1} = S_{i,22} - \tilde{H}_{i+1}^T \tilde{D}_{i+1} \tilde{H}_{i+1} - E_H^T \tilde{D}_{i+1} \tilde{H}_{i+1} - \tilde{H}_{i+1}^T \tilde{D}_{i+1} E_H - E_H^T \tilde{D}_{i+1} E_H \tag{35}$$

Ignoring the last three terms of (35) gives

$$S_{i+1}^{(1)} = S_{i,22} - \tilde{H}_{i+1}^T \tilde{D}_{i+1} \tilde{H}_{i+1} \tag{36}$$

that is the standard Schur complement usually computed in a partial incomplete factorization. As $E_H^T \tilde{D}_{i+1} \tilde{H}_{i+1}$ is generally indefinite, $S_{i+1}^{(1)}$ may be indefinite too, thus causing a potential breakdown in a PCG iteration. As a remedy, add and subtract $\tilde{H}_{i+1}^T \tilde{D}_{i+1} \tilde{H}_{i+1}$ to the right-hand side of (35):

$$\begin{aligned} S_{i+1} &= S_{i,22} + \tilde{H}_{i+1}^T \tilde{D}_{i+1} \tilde{H}_{i+1} - (\tilde{H}_{i+1} + E_H)^T \tilde{D}_{i+1} \tilde{H}_{i+1} \\ &\quad - \tilde{H}_{i+1}^T \tilde{D}_{i+1} (\tilde{H}_{i+1} + E_H) - E_H^T \tilde{D}_{i+1} E_H \end{aligned} \tag{37}$$

Recalling that

$$\tilde{H}_{i+1} + E_H = H_{i+1} = \tilde{D}_{i+1}^{-1} \tilde{L}_{i+1}^{-1} S_{i,12} \tag{38}$$

equation (37) becomes

$$S_{i+1} = S_{i,22} + \tilde{H}_{i+1}^T \tilde{D}_{i+1} \tilde{H}_{i+1} - S_{i,12}^T \tilde{L}_{i+1}^{-T} \tilde{H}_{i+1} - \tilde{H}_{i+1}^T \tilde{L}_{i+1}^{-1} S_{i,12} - E_H^T \tilde{D}_{i+1} E_H \tag{39}$$

Neglecting the last term in (39) we obtain another expression for the Schur complement:

$$S_{i+1}^{(2)} = S_{i,22} + \tilde{H}_{i+1}^T \tilde{D}_{i+1} \tilde{H}_{i+1} - S_{i,12}^T \tilde{L}_{i+1}^{-T} \tilde{H}_{i+1} - \tilde{H}_{i+1}^T \tilde{L}_{i+1}^{-1} S_{i,12} \tag{40}$$

Note that $S_{i+1}^{(2)}$ is always SPD. In fact, equation (39) yields

$$S_{i+1}^{(2)} = S_{i+1} + E_H^T \tilde{D}_{i+1} E_H \tag{41}$$

which is SPD independently of the degree of dropping enforced on $\tilde{H}_{i+1}$.

It is important to remark that to guarantee the positive definiteness of $S^{(2)}_{i+1}$, S_{i+1} must be SPD. To this aim, a SPD incomplete factorization of $S_{i,11}$ might not suffice. By distinction, if $\bar{S}_i$ in (33) is SPD, then S_{i+1} is also. Such a condition can be ensured by using the procedure described in the previous subsection. This is why combining both procedures introduced above can contribute to improve MIF.

4 Numerical Results

The MCP and MIF performance is evaluated in 3D large size contact problems addressing the geomechanical simulation of faulted reservoirs where penalty interface elements [11] are used to model the friction between the two sides of a fracture. As reference we use the efficient ILLT implementation of [14] where the fill-in degree is controlled by the user-specified parameter ρ_A representing the maximum allowable number of entries retained in each row of A in excess to the original non zeroes. The geomechanical model, discretizing a faulted porous volume with a horizontal section of 35×50 km^2 down to a depth of 10,000 m, reproduces a real faulted reservoir located in the Po river plain, Italy [12]. Two test cases are used for the comparison. Sixteen and thirteen faults are considered in test case A and B, respectively. While in test case A the faults have a relatively small size, in test case B they extend up to the ground surface from a 9,000 m depth. A plane view of the computational grids is shown in Figure 1 along with the trace of the faults, while an axonometric view is provided in Figure 2. The Young modulus of the porous medium in the reservoir varies according to the non-linear constitutive law developed in [3], in the surrounding medium is constant depending on depth, while the penalty parameter used to discretize the faults is set to 10^9 MPa/m. Three levels can be easily recognized, the first for the unknowns connected to the linear elements, the second for the unknowns connected to non-linear elements within the reservoir, and the last one for the unknowns connected to the non-linear interface elements on the faults

$$A = \begin{bmatrix} K_1 & B_{12} & B_{13} \\ B_{12}^T & K_2 & B_{23} \\ B_{13}^T & B_{23}^T & C \end{bmatrix} \tag{42}$$

The number of unknowns belonging to each level is summarized in Table 1, while the block structure and sparsity pattern of the global matrix A is shown in Figure 3. When using MCP, the block K of equation (2) is taken equal to

$$K = \begin{bmatrix} K_1 & B_{12} \\ B_{12}^T & K_2 \end{bmatrix} \tag{43}$$

with $B = [B_{13}\ B_{23}]^T$. Note that test case A and B basically differ by the size of C relative to K.

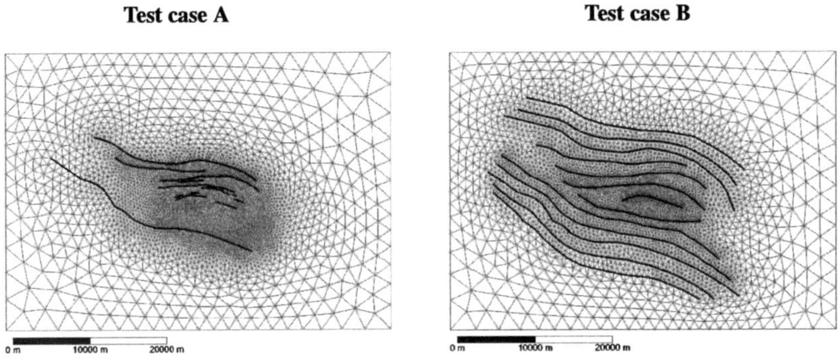

Fig. 1 Plane view of the computational grids.

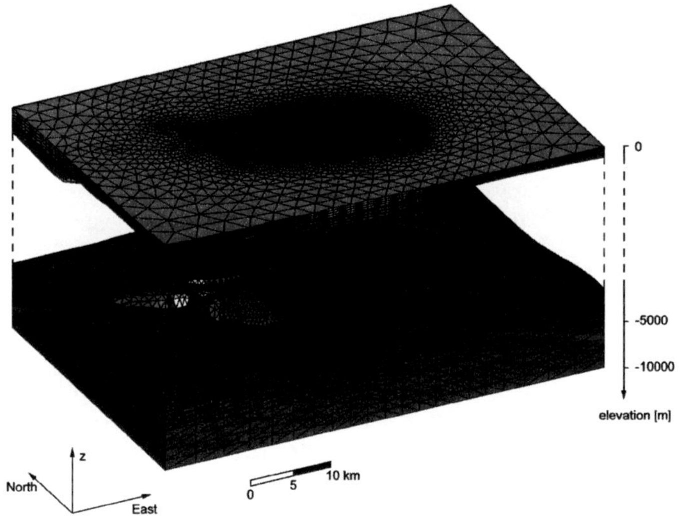

Fig. 2 Axonometric view of the FE grid used in test case A.

The computational performance of MCP and MIF is compared using the following criteria:

- number of iterations for the solver to converge;
- CPU time needed for the preconditioner setup (T_p) and the solver iterations (T_s);
- preconditioner density (μ) defined as the ratio between the number of double precision words required to store the preconditioner and the system matrix.

Convergence is achieved whenever the following test on the relative residual r_r is met:

$$r_r = \frac{\|\mathbf{f} - A\mathbf{u}\|_2}{\|\mathbf{f}\|_2} \leq 10^{-10} \tag{44}$$

Table 1 Size and number of non-zeroes of sub-matrices K_1, B_{12}, B_{13}, K_2, B_{23} and C.

	Test case A		Test case B	
	size	# non-zeroes	size	# non-zeroes
K_1	435,207 × 435,207	18,685,287	143,082 × 143,082	5,362,722
B_{12}	435,207 × 163,581	122,613	143,082 × 87,921	19,251
B_{13}	435,207 × 40,014	340,005	143,082 × 89,490	779,895
K_2	163,581 × 163,581	6,766,623	87,921 × 87,921	3,227,103
B_{23}	163,581 × 40,014	320,994	87,921 × 89,490	525,186
C	40,014 × 40,014	1,595,430	89,490 × 89,490	3,610,908
A	638,802 × 638,802	28,614,564	320,493 × 320,493	14,849,397

Table 2 Memory occupation, preconditioner density, user-defined parameters, number of iterations and CPU time to convergence for the best perfomance of PCG preconditioned with ILU, MCP and MIF.

	ILU (ρ_A)		MCP (ρ_K, γ_S)		MIF ($\rho_1^{(1)}, \rho_1^{(2)}, \rho_1^{(3)}$)	
test case	A	B	A	B	A[(a)]	B[(b)]
μ	3.90	2.96	2.03	3.09	1.83	2.15
parameters	70	50	50, 10	10, 10	10, 10, 110	0, 10, 60
# iter.	102	67	119	71	103	103
T_p [s]	111.4	39.6	34.1	29.9	14.8	15.4
T_s [s]	54.2	15.6	54.6	16.6	31.2	17.3
T_t [s]	165.6	55.2	88.7	46.5	46.0	32.7

[(a)] pre-processing time: 10.7 s.
[(b)] pre-processing time: 2.0 s.

All tests have been performed on a scalar machine equipped with an Intel Core2 Duo processor at 2.13 GHz, 2 GB of core memory and 2 MB of secondary cache.

Table 2 provides the best performance from ILLT, MCP and MIF. Especially in test case A, ILLT turns out to require quite a dense preconditioner with $\mu \simeq 4$. MCP generally allows for a better performance than ILLT with a lower memory occupation, at least for test case A. The MCP optimization is done by acting on ρ_K and γ_S only. The other user-specified parameters have been set as follows: $\tau_Z = 0.03$ and 0.02 in test cases A and B, respectively, and $\rho_S = 40$. The selected values of τ_Z and ρ_S are those producing the sparsest converging preconditioner. Due to the penalty coefficients, $\tilde{S}$ typically turns out to be quite ill-conditioned, hence dense Z and $L_{\tilde{S}}$ are required. In these test cases, K is not constant during a multi-step simulation, so the MCP setup cost includes the computation of L_K, Z and $B^T ZZ^T B$ at each solution of the linearized procedure. As a consequence, whenever material non-linearities are also considered, MCP typically outperforms ILLT but MIF appears to be the most convenient option (Table 2). A different fill-in for each matrix level can be used with MIF, yielding a sparser preconditioner and a lower total CPU time. Note that the number of MIF iterations may be even larger than that of ILLT and MCP, the lower setup and application costs being the actual key factor for its superiority. Moreover, in the previous test cases the factorization of level 1 can be

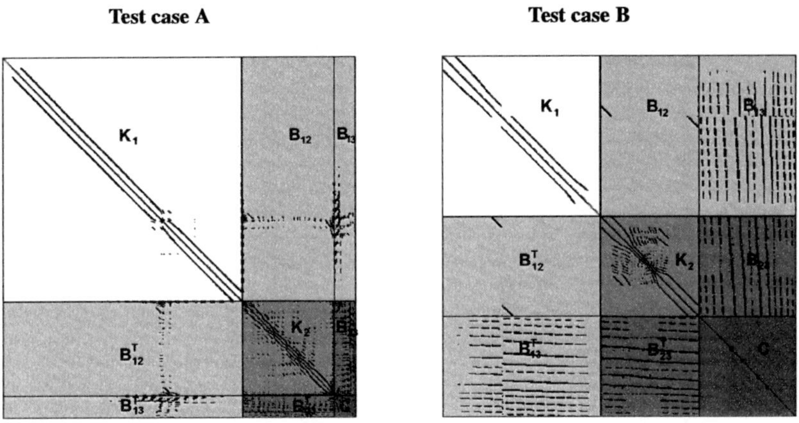

Fig. 3 Sparsity pattern and block structure of A. The overall matrix size in test case A is twice as large as in test case B.

performed just once in a multi-step simulation, thus representing a pre-processing time. In return, MIF requires a larger number of user-defined parameters than ILLT and MCP.

The robustness of ILLT, MCP and MIF to the user-specified parameters is shown in Figure 4. ILLT appears to be quite sensitive to ρ_A and PCG converges only over a limited upper ρ_A range with significantly different final CPU times. MCP converges over a much wider ρ_K and γ_S range, so their selection is not a difficult task. MIF is very stable with a good performance over a wide $\rho_1^{(i)}$ range. Only $\rho_1^{(i)}$, $i = 1,2,3$, were changed keeping $\rho_2^{(i)} = \rho_1^{(i)}$, with no significant difference being observed for other selections. The most influential parameter appears to be $\rho_1^{(\ell)}$, i.e. the last level fill-in, which should be taken large enough to allow for convergence. Therefore, though the number of user-specified parameters for MIF is high, their practical selection does not prove overly difficult.

In non-linear problems, where the high set-up cost of MCP may prevent its use, a modified version, denoted as ModMCP, can be efficiently implemented. The Schur complement of M is directly approximated by C and exactly factorized with an efficient direct solvers, e.g. MA57 [13]. The resulting preconditioner is theoretically poorer, as the penalty block C by itself is generally a bad approximation of S, however it might take advantage from its simple setup (one user-specified parameter only, ρ_K) and cheap application. The best ModMCP performance in test cases A and B is provided in Table 3. ModMCP has a smaller density than and can compete with MIF in test case B, i.e. if the size of the penalty block C is large relative to K. Figure 4d shows that ModMCP is not as stable as MIF to ρ_K, however the need for setting one parameter only can make it attractive in some problems.

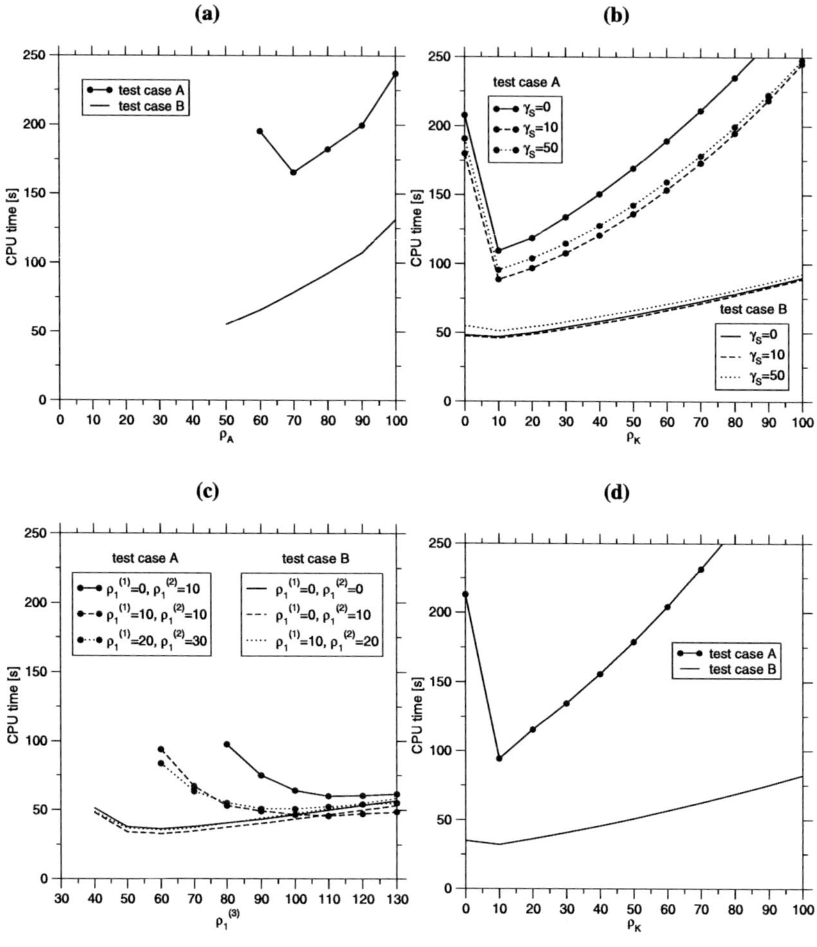

Fig. 4 Total CPU time for PCG preconditioned with (a) ILU; (b) MCP; (c) MIF; (d) ModMCP. PCG does not converge where a profile is missing.

5 Concluding Remarks

Two novel block preconditioners have been developed for 3D contact problems addressed by a penalty approach with the aim at yielding a fast convergence of the linear solver (PCG) using a limited amount of computational resources. Their performance has been tested on large size geomechanical simulations of faulted rocks. The comparison, including also the popular incomplete Cholesky decomposition ILLT, shows that:

- MCP, though requiring slightly more memory in some cases, outperforms ILLT up to a factor 2 in the most favourable condition;

Table 3 Memory occupation, preconditioner density, user-defined parameters, number of iterations and CPU time to convergence for the best perfomance of PCG preconditioned with ModMCP.

	ModMCP (ρ_K)	
test case	A	B
μ	1.50	1.98
parameters	10	10
# iter.	180	116
T_p [s]	14.9	6.5
T_s [s]	79.1	25.5
T_t [s]	94.0	32.0

- if the continuous body is elastic, the most expensive stage of the preconditioner setup can be carried out only once at the beginning of the simulation making the computational gain of MCP over ILLT even larger;
- with material non-linearities, MIF appears to be the most appropriate choice requiring a low memory occupation and providing the best performance;
- if the penalty block C is relatively large, ModMCP, offering a simpler setup, can be a valuable alternative to MIF also with material non-linearities.

Acknowledgements This study has been partially funded by the Italian MIUR project (PRIN) "Advanced Numerical Methods and Models for Enviromental Fluid-Dynamics and Geomechanics".

References

1. Ajiz, M.A., Jennings, A.: A robust incomplete Cholesky-conjugate gradient algorithm. Int. J. Numer. Meth. Eng. 20, 949–966 (1984)
2. Axelsson, O., Kolotilina, L.Y.: Diagonally compensated reduction and related preconditioning methods. Numer. Linear Algebra Appl. 1, 155–177 (1994)
3. Baù, D., Ferronato, M., Gambolati, G., Teatini, P.: Basin scale compressibility of the Northern Adriatic by the radioactive marker technique. Geotechnique 52, 605–616 (2002)
4. Benzi, M., Tůma, M.: A sparse approximate inverse preconditioner for nonsymmetric linear systems. SIAM J. Sci. Comput. 19, 968–994 (1998)
5. Benzi, M., Tůma, M.: A robust incomplete factorization preconditioner for positive definite matrices. Numer. Linear Algebra Appl. 10, 385–400 (2003)
6. Benzi, M., Cullum, K., Tůma, M.: Robust approximate inverse preconditioning for the conjugate gradient method. SIAM J. Sci. Comput. 22, 1318–1332 (2000)
7. Bergamaschi, L., Ferronato, M., Gambolati, G.: Novel preconditioners for the iterative solution to FE-discretized coupled consolidation equations. Comput. Method. Appl. Mech. 196, 2647–2656 (2007)
8. Bergamaschi, L., Ferronato, M., Gambolati, G.: Mixed constraint preconditioners for the iterative solution of FE coupled consolidation equations. J. Comput. Phys. 227, 9885–9897 (2008)

9. Durazzi, C., Ruggiero, V.: Indefinitely preconditioned conjugate gradient method for large sparse equality and inequality constrained quadratic problems. Numer. Linear Algebra Appl. 10, 673–688 (2003)
10. Ferronato, M., Janna, C., Gambolati, G.: Mixed constraint preconditioning in computational contact mechanics. Comput. Method. Appl. Mech. 197, 3922–3931 (2008)
11. Ferronato, M., Gambolati, G., Janna, C., Teatini, P.: Numerical modelling of regional faults in land subsidence prediction above gas/oil reservoirs. Int. J. Numer. Anal. Meth. Geomech. 32, 633–657 (2008)
12. Ferronato, M., Gambolati, G., Janna, C., Teatini, P.: Geomechanical issues of anthropogenic CO_2 sequestration in exploited gas fields. Energ. Convers. Manage. 51, 1918–1928 (2010)
13. HSL Archive: A catalogue of subroutines, Aea Technology, Engineering Software. In: CCLRC (2004), http://www.cse.clrc.ac.uk/nag/hsl
14. Janna, C., Comerlati, A., Gambolati, G.: A comparison of projective and direct solvers for finite elements in elastostatics. Adv. Eng. Softw. 40, 675–685 (2009)
15. Janna, C., Ferronato, M., Gambolati, G.: Multilevel incomplete factorizations for the iterative solution of non-linear FE problems. Int. J. Numer. Meth. Eng. 80, 651–670 (2009)
16. Kaporin, I.E.: High quality preconditioning of a general symmetric positive definite matrix based on its $U^TU+U^TR+R^TU$-decomposition. Numer. Linear Algebra Appl. 5, 483–509 (1998)
17. Keller, C., Gould, N.I.M., Wathen, A.J.: Constraint preconditioning for indefinite linear systems. SIAM J. Matrix Anal. A 21, 1300–1317 (2000)
18. Lin, C., Moré, J.J.: Incomplete Cholesky factorizations with limited memory. SIAM J. Sci. Comput. 21, 24–45 (1999)
19. Saad, Y.: Iterative Methods for Sparse Linear Systems. SIAM, Philadelphia (2003)
20. Saad, Y.: ILUT: A dual threshold incomlete LU factorization. Numer. Linear Algebra Appl. 1(4), 387–402 (1994)
21. Saad, Y.: ILUM: A multi-elimination ILU preconditioner for general sparse matrices. SIAM J. Sci. Comput. 17(4), 830–847 (1996)
22. Saad, Y., Suchomel, B.: ARMS: An algebraic recursive multilevel solver for general sparse linear systems. Numer. Linear Algebra Appl. 9, 359–378 (2002)
23. Saad, Y., Zhang, J.: BILUM: Block version of multi-elimination and multi-level ILU preconditioner for general sparse linear systems. SIAM J. Sci. Comput. 20, 2103–2121 (1999)
24. Saint-George, P., Warzee, G., Notay, Y., Beauwens, R.: Problem-dependent preconditioners for iterative solvers in FE elastostatics. Comput. Struct. 73, 33–43 (1999)
25. Silvester, D.J., Elman, H.C., Kay, D., Wathen, A.J.: Efficient preconditioning of the linearized Navier-Stokes equations for incompressible flow. J. Comput. Appl. Math. 128, 261–279 (2001)
26. Tismenetsky, M.: A new preconditioning technique for solving large sparse linear systems. Linear Algebra Appl. 156, 331–356 (1991)

A Local Contact Detection Technique for Very Large Contact and Self-Contact Problems: Sequential and Parallel Implementations

V.A. Yastrebov, G. Cailletaud and F. Feyel

Abstract The local contact detection step can be very time consuming for large contact problems reaching the order of time required for their resolution. At the same time, even the most time consuming technique all-to-all does not guarantee the correct establishment of contact elements needed for further contact problem resolution. Nowadays the limits on mesh size in the Finite Element Analysis are largely extended by powerful parallelization methods and affordable parallel computers. In the light of such changes an improvement of existing contact detection techniques is necessary. The aim of our contribution is to elaborate a very general, simple and fast method for sequential and parallel detection for contact problems with known a priori and unknown master-slave discretizations. In the proposed method the strong connections between the FE mesh, the maximal detection distance and the optimal dimension of detection cells are established. Two approaches to parallel treatment of contact problems are developed and compared: SDMR/MDMR – Single/Multiple Detection, Multiple Resolution. Both approaches have been successfully applied to very large contact problems with more than 2 million nodes in contact.

1 Introduction

By local contact detection we mean a procedure which detects elements (nodes, surfaces) of one part of a finite element mesh which potentially come in contact with another part of the mesh on a current computational step. In the context of the node-to-segment discretization (NTS) [15] the contact detection is an establishment of the

V.A. Yastrebov · G. Cailletaud
Centre des Matériaux, Mines ParisTech, CNRS UMR 7633, BP 87, 10 rue Henri Desbruères, 91003 Evry, France; e-mail: {vladislav.yastrebov, georges.cailletaud}@mines-paristech.fr

F. Feyel
ONERA, BP 72, 29 avenue de la Division Leclerc, 92322, Chatillon, France;
e-mail: frederic.feyel@onera.fr

G. Zavarise & P. Wriggers (Eds.): Trends in Computational Contact Mechanics, LNACM 58, pp. 227–251.
springerlink.com

closest opposing master segment for each node of the slave surface which will potentially come in contact. Further these nodes and the corresponding surfaces form abstract contact elements which contribute consequently to the weak form associated with the finite element problem. Therefore, wrong contact detection results in incorrect solution or even in its failure.

It should be distinguished two contact search phases [11]: spatial search and contact detection. The first notion is used for search between separate solids coming in contact, i.e. rather between separate geometries than discretizations. Contact spatial search methods are of big importance in multibody systems and the discrete element method where interaction between more or less identical particles such as crashed stone, sand, snow is considered to analyse mud flows, opencast mines, avalanches, etc. It is worth mentioning that previously the particular attention of the scientific community has been paid mostly to this phase of contact search, because the local discretization of solids remained rather moderate and the bucket detection method [1] in its very general form or even the simplest all-to-all approach remained rather efficient and fast enough techniques; especially in case of small slip when only one execution of the detection procedure is required. Recent progress in parallel computing makes possible extremely large implicit and explicit contact simulations between just a few but very finely meshed solids. The phase of local contact detection becomes crucial and time consuming part of the computational process, especially in case of finite slip and large deformation.

The detection phase consists in establishment of contact elements which by-turn in NTS discretization consist of one slave node and a master surface of another element. The simplest and straightforward detection method is all-to-all: each master segment is checked for proximity to each slave node. The growth rate of the method is $O(N_s \times N_m)$, where N_s and N_m are numbers of slave nodes and master segments respectively. If one considers second order master surface each check of projection requires the solution of nonlinear equation which takes several iterations. For example, let us estimate the time needed to perform the simplest contact detection procedure within two surfaces consisting of 1024×1024 elements each. If each search of projection requires 5 iteration and the computer performs the detection with 10^6 flops, then the detection will be achieved in more than two months(!). This time surpasses considerably the time needed to a parallel resolution of a FE problem possibly associated with such a FE mesh.

The first simplification of the detection is to start from node-to-node detection in stead of node-to-segment. For further improvement let us imagine a set of spatially distributed points. The problem is to detect for a given point the closest one from this set. Human vision accomplishes this task easily analyzing just few close points. It does not need any analysis of the whole point set while the simple detection algorithm does, because it is "blind" and needs to "touch" all the points one by one and compare distances between them. The techniques which have been worked out for contact detection are aimed at the reduction of quantity of points to "touch". Among these methods there are the bucket method [1,4], the heapsort and the octree method [10] and others. The two last algorithms have been developed mainly for the multibody simulations, i.e. for spatial contact search, however they can be adapted for the local contact detection as well. But these methods lack for generality (self-contact, parallel detection) and more elaborated analysis (optimal parameters), moreover several of them require facilities to handle tree data structures. Recently a new powerfull detection algorithm [13] has been proposed. It accounts the links between nodes and that is why it is more adapted for local contact detection. The method is based on the bounding volume trees associated with sets of segments. However it is designed for mortar based contact formulations and apparently for the NTS discretization its rapidity is inferior to the method proposed here.

Even if a fast enough node-to-node detection procedure is worked out, many problems remain, such as optimal bounding box construction, challenging detection of nodes in blind spots and as a particular case – detection of passing by nodes. These difficulties appear from the fact that the finite element discretization of contacting surfaces is continuous but not smooth. Finally it can be affirmed that the robustness of the detection and its rapidity depends strongly on the way the mentioned difficulties are overcame as well as on the carefulness of coding.

In this contribution the bucket or grid detection method [1] will be improved and adapted for a very general case. Its sequential and parallel implementations will be discussed in detail for very large finite element simulations in the framework of the node-to-segment discretization.

First, the principal notions are introduced and all the steps of the method are considered in details. Optimal detection parameters (proximity criterion, detection distance and cell size) are derived based on the numerous large scale detection tests. Efficient procedures for bounding box construction, neighbouring cell detection and verification of "passing by" nodes are proposed. The performance of the method is demonstrated on several extremely large contact problems containing up to 2 million nodes in contact. Further the method is extended for a very general case of unknown a priori master-slave contact surfaces, which is of a great importance for self-contact treatment. The last part is devoted to the detection phase in case of parallel treatment of contact problems, different approaches are considered.

2 Method Description

The grid (bucket) detection method [1, 4] is natural and simple. But its implementation and choice of the internal parameters should be discussed in more details. First, a short description of the method is given, then each stage of the procedure is discussed in detail, the optimality of parameters is analyzed and finally some numerical examples both artificial (contact between two curved surfaces) and real engineering (tyre on road contact) are given to demonstrate the performance of the method. The ultimate aim is an improvement of the grid method and determination of detection parameters which would reduce the required CPU time.

First of all the master-slave approach and the associated node-to-segment (NTS) discretization have to be shortly explained. Two contacting surfaces in the master-slave approach are distinguished: one is called slave or impactor and the other is called master or target. For clarity the slave-master notation will be used. Such distinction comes both from the node-to-segment discretization and from geometrical description of contact, precisely from the asymmetry of the closest distance definition between contacting surfaces. The slave surface manages nodes of the first surface and neglects their connections, i.e. all the interpolations between slave nodes are not taken into account. The master manages segments of the second contacting surface and its description is closely connected with the order of the elements and consequently with the interpolation functions. It is worth mentioning that the use of nonlinear interpolations with many discretization techniques (including NTS which is considering here) leads to incorrect results of contact analysis (see e.g. [11]).

"Contact element" (here NTS contact element) is an abstract (not structural) element consisting of a slave node and several master nodes united by a master surface segment. All the geometrical quantities such as normal gap g_n and tangential velocity $\dot{\mathbf{g}}_t$ are evaluated in the master reference frame, i.e. the geometry of contact is described by an interaction of a slave node with a master surface segment. Such contact elements take care of the local contact interaction between two bodies in the resolution phase and hence they have to be created before the slave node slides on or penetrates the master surface. Thus the slave nodes which are close enough to the master surface have to be detected and included in consideration before the resolution step.

Before discussing particular details let us derive a short description of the grid detection method. Two phases can be distinguished: preliminary phase and detection phase. In the preliminary phase the optimal size of the grid is evaluated, further a potential contact area is determined and divided with an enumerated regular grid. That allows to reduce locally the area of closest nodes search. Finally all slave and master nodes situated in the detection area are distributed in the cells of the grid. In the detection phase for each slave node we check for the closest master node in the current cell. And if necessary we check one or several neighbouring cells for possible proximal master nodes. As the closest master node is found the existence of the slave node projection onto all segments attached to the master node is verified. If at least one projection exists then a contact element will be established otherwise the verification if the node is in a blind spot or is a "passing by node" is needed.

2.1 Preliminary Stage of Contact Detection

First of all a key parameter for the contact detection procedure has to be introduced – maximal detection distance $d_{\max}$. In case of node-to-segment detection $d_{\max}$ determines the following: if a slave node is closer to the master surface than $d_{\max}$, then it is supposed that this node can come in contact during the following time step, otherwise not. The method considered here is based on the node-to-node detection, so the meaning of the maximal detection distance is different. If the distance between two nodes $d_{ij} = \text{dist}(\mathbf{r}_i, \mathbf{r}_j)$ is smaller than the maximal detection distance, then the corresponding slave node $\mathbf{r}_i$ and one of master surfaces containing the mentioned

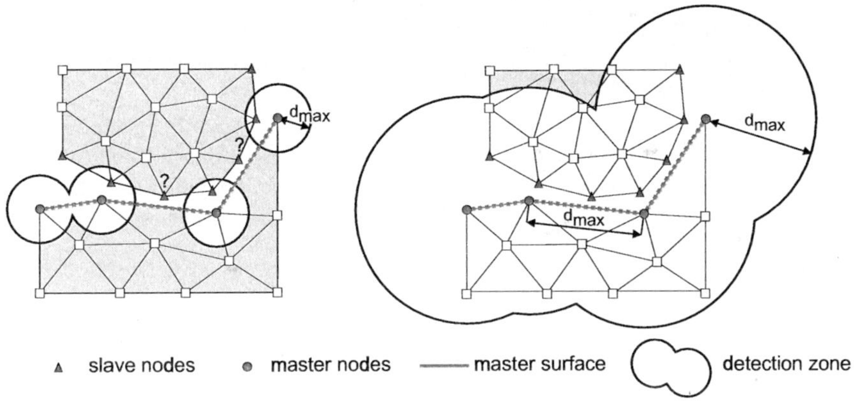

Fig. 1 Maximal detection distance $d_{\max}$: on the left not correct on the right correct choice.

node $\mathbf{r}_j$ are considered to be potentially in contact during the following time step otherwise not. This difference naturally results in a limitation on the minimal value of the $d_{\max}$. Here the $\mathrm{dist}(\mathbf{r}_i, \mathbf{r}_j)$ denotes Euclidian metric in the global reference frame $\mathrm{dist}(\mathbf{r}_i, \mathbf{r}_j) = |\mathbf{r}_i - \mathbf{r}_j|$. The value of the key parameter $d_{\max}$ for the detection procedure can be determined automatically accordingly to the discretization of the master or self-contact surface, to the loading and deformation rate. First, it should be mentioned that the maximal detection distance in proposed method has to be unique for the entire contact area and greater than one half of the maximal distance between master nodes attached to one segment

$$d_{\max} > \frac{1}{2} \max_{i=1,\, j=1,\, k=j+1}^{i=N_m,\, j=N_n^i-1,\, k=N_n^i} \mathrm{dist}(\mathbf{r}_j^i, \mathbf{r}_k^i), \tag{1}$$

where N_m is a total number of master segments, N_n^i is a total number of master nodes attached to the i-th master segment, $\mathbf{r}_j^i$ is a coordinate of the j-th node of the i-th master segment. If the condition (1) is not fulfilled, then some slave nodes coming in contact with master surface can be lost (see Figure 1, here and further for the sake of simplicity and clarity all figures represent two dimensional cases but can be easily extended to three dimensions).

For a reasonable number of time steps for geometrically or physically nonlinear problem the maximal detection distance can be determined as dimension of the biggest master segment, i.e. accordingly to the discretization of the geometry

$$d_{\max} = \max_{i=1,\, j=1,\, k=j+1}^{i=N_m,\, j=N_n^i-1,\, k=N_n^i} |r_j^i - r_k^i|. \tag{2}$$

Such estimation is reasonable in case of a regular discretization of the master surface. On the other hand if the distribution of the master nodes is very heterogeneous, i.e. fine surface mesh in one contact region and rough in another, the value of $d_{\max}$ appears to be highly overestimated for certain regions. This fact decreases the efficiency of the method, but in general for an adequate finite element mesh the increase of the detection time is not so high. The influence of the maximal detection distance on detection time will be discussed later.

In the case of linearly elastic material and frictionless contact, the geometry can change significantly during one time step. So the analysis of discretization can give only lower bound for $d_{\max}$ and that is why its value should be augmented manually or automatically accordingly to the deformation and/or displacement rate, for example in the following way:

$$d_{\max} = \max\left\{ \max_{i=1,\, j=1,\, k=j+1}^{i=N_m,\, j=N_n^i-1,\, k=N_n^i} |\mathbf{r}_j^i - \mathbf{r}_k^i|;\ 2\max_{i=1}^{N_c} |\Delta\mathbf{r}_i| \right\}, \tag{3}$$

where N_c is a total number of slave and master nodes and $\Delta\mathbf{r}_i$ is an estimation of the maximal displacement of the i-th node, 2 takes care of possible opposite translations of master and slave nodes. In case of remeshing or sufficiently large deformations of the master, the detection parameter $d_{\max}$ should be recomputed at each remeshing or at each N-th time step.

Before carrying out any detection the spatial area where contact can take place during the following time step has to be chosen. It has to contain as few master and slave nodes as possible but on the other hand it has to include all the nodes potentially coming in contact on the following step. If needed this area has to be frequently updated. We propose to confine this area by a bounding box (parallelepiped) defined in the global reference frame.

The determination of the bounding box differs for known a priori and unknown master-slave discretizations. In case of unknown master-slave the bounding box should include all possible contacting surfaces. But frequently the discretization is known a priori even if contact occurs within one body (self contact). In this case the construction of an optimal bounding box allows to exclude from consideration some nodes which cannot come in contact during the following time step (Figure 2) and consequently it results in acceleration of the detection procedure. It is worth mentioning that here the very general case is considered: any slave node can potentially come in contact with any master segment during the loading. Often it is not the case and for each slave node the set of possible master segments is limited and partly predefined. But in order to take it into account the detection technique should be tuned for each particular case. The consideration of such techniques is out of the scope of this contribution.

First of all the dimensions of master and slave surfaces are estimated. Note than even if the master surface consists of several independent zones in the grid detection method it can be considered as one set of master nodes with associated segments. It is proposed to construct two independent bounding boxes $B_s : \{\mathbf{r}_s^1, \mathbf{r}_s^2\}$ and $B_m : \{\mathbf{r}_m^1, \mathbf{r}_m^2\}$ containing all slave and master nodes respectively, where $\mathbf{r}^1$ and $\mathbf{r}^2$ are the vectors in the global reference frame of two opposite corners determining the bounding boxes. Note that each bounding box confining master and slave nodes includes also a node free margin zone of the size of maximal detection distance at each side.

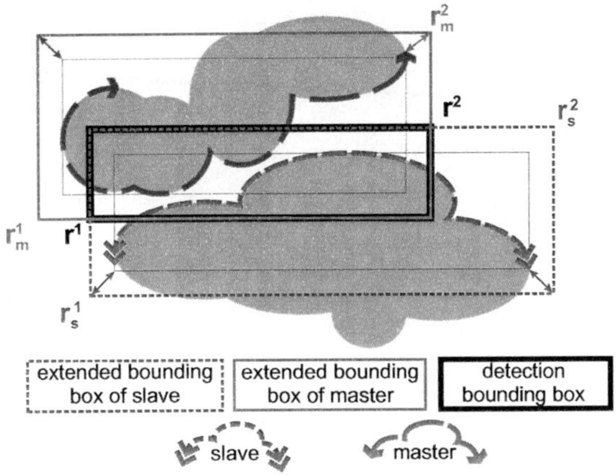

Fig. 2 Determination of the bounding box for the contact detection procedure in case of simple master-slave contact.

$$\mathbf{r}^1 : r^1_{\{x,y,z\}} = \min_{i=1}^{N_b}\{\mathbf{e}_{\{x,y,z\}} \cdot \mathbf{r}_i\} - d_{\max}\mathbf{e}_{\{x,y,z\}},$$

$$\mathbf{r}^2 : r^2_{\{x,y,z\}} = \max_{i=1}^{N_b}\{\mathbf{e}_{\{x,y,z\}} \cdot \mathbf{r}_i\} + d_{\max}\mathbf{e}_{\{x,y,z\}}, \tag{4}$$

where N_b is a number and $\mathbf{r}_i$ is a vector of nodes to be included in the bounding box and $\mathbf{e}_{\{x,y,z\}}$ are orthonormal basis vectors in the global reference frame. The margin of $\pm d_{\max}$ is introduced to avoid any loss of possible contact elements. Some improvements can be introduced in order to reduce the time of bounding box construction. The user can precise that one or several contact surfaces are rigid and do not move, then permanent bounding boxes can be assigned to these surfaces and there is no need to update them. Another possible feature is the prediction by the user that the deformation and displacement of a contact surface is connected to the displacement of certain nodes. It allows to avoid the verification of all nodes in (4). Since the nodal coordinates are kept in memory in the global reference frame it is much more faster to work directly with these coordinates so no rotation to the bounding boxes must be applied. The resultant bounding box $B : \{\mathbf{r}^1, \mathbf{r}^2\}$ is taken as the intersection of master and slave bounding boxes $B = B^m \bigcap B^s$. The practice shows that a further contraction of the bounding box does not reduce significantly the detection time.

When a bounding box is determined, an internal grid should be constructed in a proper way. In the grid detection method this grid should be regular and the cell size d_c should be optimum: not too large in order to keep the number of slave and master nodes in the cell as small as possible and not too small at least greater than the maximal detection distance $d_c \geq d_{\max}$. For smaller cells, the determination of the neighbouring cells which should be investigated is not evident; moreover, the growth rate of their maximal number N_c is cubical

$$\text{if } d_c = \frac{d_{\max}}{n}, n > 1 \Rightarrow N_c = (3 + 2n)^3 \tag{5}$$

Fig. 3 Example of finite element meshes used to determine the optimal cell size. Proximal meshes with homogeneous (left top) and heterogeneous (left bottom) spatial node distribution and convex meshes (right).

The smaller the cell size, the higher the total number of cells and consequently the smaller the number of contact nodes per cell. But on the other hand small cell size increases the necessity to carry out the detection in neighbouring cells. It can be shown analytically by means of probability methods that for homogeneous node distribution both in 2D and 3D cases the minimal detection time is unique and corresponds to the minimal cell size. Such simple analysis predicts quadratic growth of the detection time in 2D case and cubic in 3D. To demonstrate it for real cases let us analyse the dependence of the detection CPU time t on the cell size d_c. Several finite element problems have been considered, each problem consists of two separate finite element meshes curved in a different way. Slave and master surfaces consist of over 10200 nodes. Three sets have been considered: proximal meshes with homogeneous (Figure 3, left top) and heterogeneous (Figure 3, left bottom) node distribution and convex mesh with heterogeneous node distribution (Figure 3, right). Each set is represented by 5 different realizations of curved surfaces. By homogeneous node distribution we mean that the maximal segment dimension does not exceed 200% of the minimal one, otherwise the node distribution is considered to be heterogeneous.

In Figure 4 the dependence of the average detection CPU time and the average number of investigated neighbouring cells on the normalized cell size $d_c/d_{\max}$ is

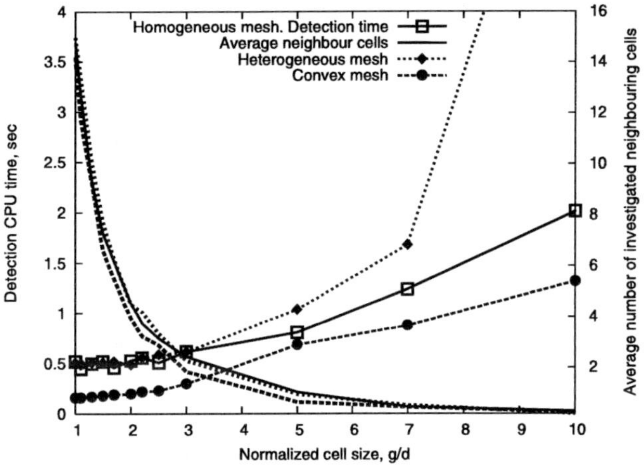

Fig. 4 The dependence of the detection time and the average number of neighbouring cell investigated during the detection on the normalized cell size.

represented for different sets. As shown in the figure, the detection time increases nonlinearly with higher slope for heterogeneous mesh than for homogeneous, because of initially higher maximal detection distance for such type of mesh. As expected, the detection time for convex meshes is smaller because of the smaller associated bounding boxes. Different discretizations (256×256, 512×512) have been tested and in all the cases the same dependence takes place. Accordingly to the analytical estimation and carried out tests, the optimal grid size is the minimal one, i.e. equal to the maximal detection distance

$$d_c = d_{\max}. \tag{6}$$

For such a choice, each grid cell contains the minimal number of nodes, but on the other hand it is necessary to carry out the detection procedure in many neighbouring cells: on average 12–16 cells (of 26 surrounding cells in 3D) (Figure 4).

When the maximal detection distance is determined and the bounding box is constructed, the internal grid has to be established in the bounding box and all the slave and master nodes have to be distributed in the cells of the grid. Since the optimal cell size is $d_{\max}$ the number of cells in each dimension of the grid is defined as

$$N_{x,y,z} = \max\left\{\left[\frac{r^2_{x,y,z} - r^1_{x,y,z}}{d_{\max}}\right];\, 1\right\}, \tag{7}$$

where $[x]$ stands for the integer part of x. Such choice of cell numbers provides the grid sizes Δx, Δy and Δz not smaller than the maximal detection distance in the case of $N > 1$

$$\Delta\{x,y,z\} = \frac{r^2_{x,y,z} - r^1_{x,y,z}}{N_{x,y,z}} \geq d_{\max}. \tag{8}$$

Each cell of the grid has to be enumerated, the unique integer number $N \in [0;\ N_x \times N_y \times N_z - 1]$ is given to each cell with spatial "coordinates" i_x, i_y and i_z, where $i_{x,y,z} \in [0;\ N_{x,y,z} - 1]$

$$N = i_x + i_y N_x + i_z N_x N_y. \tag{9}$$

Now the growth rate of the method can be estimated roughly as $O(N_s N_m / N_x N_y N_z)$. If number of master and slave nodes per cell is supposed to be constant $\rho = N/N_c$, where $N_c = N_x N_y N_z$ and N is an average number of master and slave nodes, then the growth rate of the method can be rewritten as $O(N)$. However, in practice the distribution of nodes is not homogeneous and this value appears to be underestimated.

Slave and master nodes situated in the bounding box have to be distributed in the cells. For this purpose two arrays A^s and A^m corresponding to slave and master nodes respectively are to be created. They contain slave and master node identification numbers (ID). For example element A^s_{ij} keeps the ID of the j-th slave node in the i-th cell of the grid, $i \in [0; N_x N_y N_z - 1]$, $j \in [0; N^s_i]$, N^s_i being the number of slave nodes in the i-th cell. In average the number of integer (32 bits) elements in array does not exceed the number of contact nodes and so even for extremely large problems it makes just a minor contribution in memory requirement. However, the arrays can be replaced by linked-list storages as in [4].

For each node with coordinates $\mathbf{r} : \{r_x, r_y, r_z\}$ inside the bounding box, the corresponding cell number is easily determined as

$$N_{\text{cell}} = \left[\frac{r_x - r^1_x}{\Delta x}\right] + \left[\frac{r_y - r^1_y}{\Delta y}\right] N_x + \left[\frac{r_z - r^1_z}{\Delta z}\right] N_x N_y. \tag{10}$$

2.2 Contact Detection

All steps described previously represent the preliminary part of the detection algorithm which demands in general 7–10% of the total detection time. The next steps correspond to the closest node detection, the determination of projections for the slave nodes onto the corresponding master segments and the establishment of contact elements. Let us discuss this stage in details.

For each grid cell c_i and for each slave node $\mathbf{r}^s_{ij}$ in this cell, i.e. for each node A^s_{ij} we look for the closest master node $\mathbf{r}^m_{ik}$ in the current cell, i.e. the closest node among A^m_{ik} if A^m_i is not empty. Among all master nodes in the cell the distance to the closest one is $d^s_{ij} \le d_{\max}$:

$$d^s_{ij} = \min\left\{ d_{\max}, \min_k\{|r^s_{ij} - r^m_{ik}|\}\right\}. \tag{11}$$

It is obvious that master nodes situated in neighbouring cells (maximum 8 cells in 2D, 26 in 3D) have to be checked as well. Not all the cells are considered, but only those that are sufficiently close to the slave node. The criterion of the proximity is the

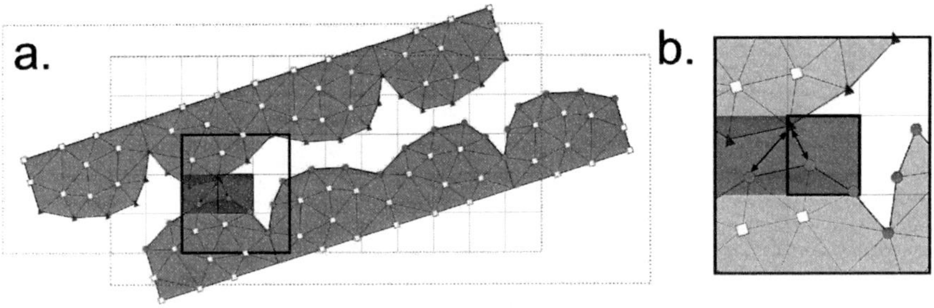

Fig. 5 Detection of the closest master node in current and neighbouring cell. Slave and master nodes represented by triangles and circles respectively.

following: if any boundary of the current cell (face, edge or vertex) is closer than the closest master node found up to the moment, i.e. closer than d^s_{ij}, then the detection procedure has to be carried out in neighbouring cells attached to this boundary one by one (Figure 5).

For example, let us consider a vertex of the i-th cell $\mathbf{r}^v_i$. For instance, after checking all master nodes in the current cell it was determined that the closest master node is situated at the distance of d^s_{ij} from the slave. If the considered slave node is closer to the vertex than this distance $\mathbf{r}^v_i$: $|\mathbf{r}^s_{ij} - \mathbf{r}^v_i| < d^s_{ij}$, then all master nodes in one of neighbouring cells attached to the vertex $\mathbf{r}^v_i$ have to be checked and consequently d^s_{ij} has to be decreased or kept the same (if no closer master node was found in this cell). And so on for other cells attached to this corner. In general the same procedure has to be performed for all 8 vertices, 12 edges and 6 faces of the i-th cell. To get more optimal algorithm such an investigation of neighbouring cells is better to start from the closest faces, further edges and finish the verification with vertices. Note that each verified cell may decrease the d^s_{ij} and consequently can decrease the number of cells to be checked. In such a manner all possibly proximal slave and master nodes are detected cell by cell. The average number of verified neighbouring cells for different meshes is represented in Figure 4. This number decreases with increasing normalized grid size $d_c/d_{\max}$ but as the optimal ratio $d_c/d_{\max} = 1$ the average number of verified neighbouring cells remains quite high (12–16 cells).

Every slave node in the bounding box has been considered and for certain of them $\mathbf{r}^{s*}_j$ corresponding proximal master nodes $\mathbf{r}^{m*}_j$ have been detected. To construct contact elements it is necessary to project each of such slave nodes $\mathbf{r}^{s*}_j$ onto surfaces containing its homologue master node $\mathbf{r}^{m*}_j$. The case when only one projection is found is trivial. There remains only to create the corresponding contact element spanned on the slave node and master surface possessing this projection. If several projections are found we choose the closest one and create the contact element. The case when no projection is found is not as trivial as the preceding ones and has to be considered in details. There are two possibilities:

1. the slave node is situated in a "blind spot" of the discretized master surface;
2. the slave node does not come in contact but just passes by close to the boundary of master surface.

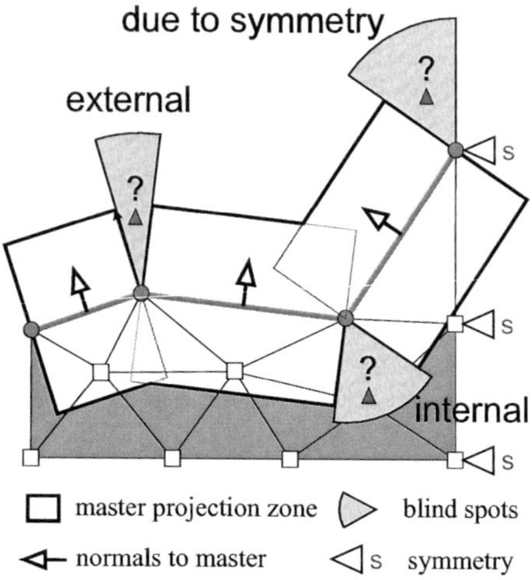

Fig. 6 Examples of blind spots: external, internal and due to the symmetry boundary conditions.

Case 1. Since the finite element method requires only continuity of the discretization ($\Gamma_c \in C^0$) so the contacting surface may be not smooth ($\Gamma_c \notin C^1$). Each master segment has its "projection" zone (Figure 6), each point in this zone has at least one projection onto the master surface [7, 8]. But often in the junction zone of master segments (at common edges and nodes) the intersections of "projection" zones does not fill the surrounding space entirely but with some gaps of form of prisms and pyramids in 3D or of form of sectors in 2D. This problem exists not only for linear master elements but for any order. Three types of blind spots can be distinguished: internal, external or blind spot due to boundary conditions (see Figure 6). If a slave node in a blind spot is overlooked, different consequences depending on the type of blind spot are possible.

- External blind spot. Slave nodes situated in this kind of spot are not detected before they penetrate under the master surface. After such penetration during the next time step it can be detected and brought back onto the surface, but the solution has been already slightly changed. In certain cases especially in force driven problems such penetration can lead to a failure of solution.
- Internal blind spot. Contact is predicted correctly, but if slave node penetrates just a little under the master surface and appears in its internal blind spot this node will be lost for the contact detection at least during the next time step. Such little penetrations take place if the penalty method for contact resolution is used or just due to the limited precision of the iterative solution.
- Blind spot due to boundary conditions. This type of blind spot is situated at the boundary and can be either internal or external. It appears due to the presence of symmetric or periodic boundary conditions on the master surface, for example, the basic Hertz contact problem with axisymmetrical 2D finite element mesh.

Obviously if the detection procedure which does not consider blind spots is repeated at every iteration then some problems can be avoided but on the other hand it is very

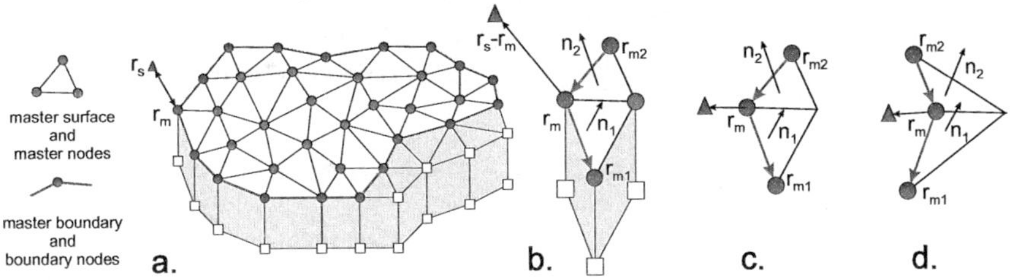

Fig. 7 Detection of the passing by nodes. (a) Master surface and its boundary. (b) Zoom on the geometry close to the passing by node. (c) Convex master boundary. (D) Concave master boundary.

time consuming and not always efficient (for example, if a convex edge with an external blind spot becomes a concave with an internal blind spot the penetration of the slave node situated in this spot can be irreversible).

There are different possibilities to avoid the loss of contact in blind spots:

- Artificial smoothing of master surface for large sliding contact problems [8, 12]. There are no more gaps in "projection" zones except gaps due to symmetry, i.e. there are almost no more blind spots and the problem of passing by nodes (Case 2) does not exist. However most of these methods have some inherent drawbacks: and derive sometimes not correct deformation close to the edge of the master surface boundary.
- A "proximal volume" can be constructed by an extrusion of the master surface in the normal direction and in the opposite one which fills both projection zones and blind spots. If a slave node is situated in this volume then it is considered as node in contact and the master surface is further detected. "Passing by" nodes can be easily detected as they do not appear in the "proximal volume".

The first group of methods in general is too "expensive" if one uses them only for the detection purpose and are not applicable for arbitrary meshes, the second one is quite time consuming as well.

We use rather rough but quite simple and robust treatment of blind spots. If a detected slave node has no projection and is not a passing by node then the corresponding contact element is constructed with the closest [15] or randomly chosen master surface attached to the closest master node. For sufficiently small time step such approach is quite reliable. There remains only to determine if the node is passing by or not. One possible technique is represented in Figure 7.

Case 2. First of all, in the preliminary phase the boundary master nodes surrounding the master contact surface have to be marked. Let us assume the situation when one of such marked nodes $\mathbf{r}_m$ is found to be the closest to slave node $\mathbf{r}_s$. If it has no projection onto master segments attached to the marked master node, then two alternatives are possible: either the slave node is situated in a blind spot or it passes by the master surface. To choose between these possibilities it is possible either to verify if the slave node is situated in one of blind spots attached to the master node or to check if the slave node is situated in the local proximal volume of the master surface. The second possibility seems to be more simple and natural. Note that such a verification is slightly different for locally convex and concave master surface boundaries. The convexity can be known as nodes of each master segment are ordered. The condition of convexity is

$$(\mathbf{r}_m - \mathbf{r}_m 2) \times (\mathbf{r}_{m1} - \mathbf{r}_m) \cdot (\mathbf{n}_1 + \mathbf{n}_2) \geq 0, \tag{12}$$

where $\mathbf{n}_1$ and $\mathbf{n}_2$ denote average normals to master segments possessing the edges $\{\mathbf{r}_m, \mathbf{r}_{m1}\}$ and $\{\mathbf{r}_m, \mathbf{r}_{m2}\}$ respectively. Then the criterion of the slave node being in the proximal volume is

$$\mathbf{n}_2 \times (\mathbf{r}_m - \mathbf{r}_{m2}) \cdot (\mathbf{r}_s - \mathbf{r}_m) \geq 0 \text{ AND } \mathbf{n}_1 \times (\mathbf{r}_{m1} - \mathbf{r}_m) \cdot (\mathbf{r}_s - \mathbf{r}_m) \geq 0. \tag{13}$$

If this condition is fulfilled then the slave node is taken into account and the contact element is established with the closest master segment. For the concave surface AND should be replaced by OR in (13).

More elaborated approaches (see, for example, [7]) take into account node-to-node and node-to-edge contacts in case of blind spot detection, average normals can be established at edges and vertices. But for most of the contact problems, consideration of only node-to-segment discretization provides the correct results.

As one can see the proposed algorithm is quite simple and natural except may be the verification of passing by nodes. The proposed method does not require any special data storage nor particular code structure and consequently can be easily implemented in any finite element code.

2.3 Validation and Performance

The preliminary validation of the grid detection method is easy to carry out on simple meshes. Normally a visual analysis of the constructed contact elements is sufficient. The further validation consists in comparison with the all-to-all detection method which is trivial to implement.

To demonstrate the performance of the grid detection method we consider a tyre-road contact problem. Such simulation can be rather helpful for example for an improvement of tread patterns (stick increase and noise reduction). We are particularly interested in this problem because the contact elements change intensively at each time step and consequently a fast detection procedure is highly desirable.

A finely and regularly meshed tyre wheel is translated over an artificially rough road surface and its FE mesh is deformed manually accordingly to the road roughness and next the contact detection procedure is executed. The finite element mesh of the tyre (Figure 8) consists of about 550,000 nodes with contact zone of about 105,000 nodes. The finite element mesh approximating the road roughness (Figure 9) consists of about 400,000 nodes one half of them being included in the master contact zone. Established contact elements are demonstrated in Figure 9 for different tyre-road dispositions and imprint deep. It can be noted that the choice of the bounding box as an intersection of master and slave bounding boxes reduces sig-

Fig. 8 A part of the tyre finite element mesh consisting of about 550,000 nodes.

Fig. 9 Tyre-road contact problem: general view, three tyre-road dispositions and corresponding contact elements on the bottom of the tyre for different imprint deep.

nificantly the number of contact nodes to be considered. The bounding box of the road is kept constant, whereas the bounding box of the tyre is updated at each step. The contact detection time at each time step in average is just 1.5–2 seconds on a laptop, i.e. the contact detection time can be neglected in comparison to the system resolution time. The analysis of the detection time shows that the estimation of the maximal detection distance takes about 30% of the time, preliminary stage takes about 20% and the detection procedure requires just 50% of the time.

Table 1 Detection of contact between rough surfaces (2 millions of master and slave nodes).

Geometry	Nodes in BB*	Contact elements	Detection time	Gain, $T_{\text{all-to-all}}/T_{\text{grid}}$
Two close surfaces	2,100,000	75,300	35 minutes	>300 times
Two convex surfaces	340,000	15,800	1 minute	>10,500 times
Two close but not contacting surfaces	50,000	0	4 seconds	>160,000 times

*BB – Bounding box

Fig. 10 Rendered surfaces of two finite element meshes (each contains 2^{20} contact nodes).

Another example is an artificial contact between two rough surfaces, each consisting of 2^{20} contact nodes. Rendered surfaces corresponding to the meshes are represented in Figure 10. Such a kind of problems requires a longer time for contact detection because the bounding box includes all or almost all contact nodes and there are as many slave nodes as master ones. If one uses the modified all-to-all method (not node-to-segment but node-to-node), the reliable estimation of the needed detection time exceeds 180 hours (!) (almost 8 days) and 2^{40} distance verifications are needed. The proposed grid detection method requires much less time than all-to-all method. The time strongly depends on the geometry and discretization, consequently on the constructed bounding box and the number of contact nodes located in it, for example, for close enough rough surfaces (Figure 10) the detection time is much higher than for convex surfaces and it is almost negligible if two surfaces are close enough but not so close to come in contact. The results are summarized in Table 1. Let us note that in presented computations the quadrilateral master segments are supposed to remain flat for both considered methods.

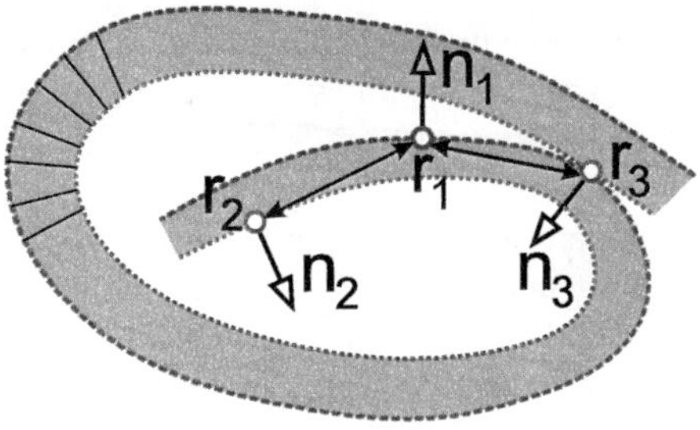

Fig. 11 Indistinguishable contact nodes in self-contact in case of the maximal detection distance higher than the minimal structure thickness.

3 Self-Contact Detection

There are mechanical problems for which determination of master and slave surfaces presents a big challenge or may be impossible. Among such problems there are multibody systems, problems with complicated geometries (for example highly porous media like metal foams), large deformation problems with not regular discretization and self-contact problems.

This class of contact problems needs a particular contact detection procedure. Here an adaptation of the grid detection method to problems with unknown a priori master-slave discretization is proposed, particular attention is paid to self-contact problems. Such adaptation demands considerable modifications in all stages of the grid detection procedure. Moreover an adapted finite element mesh can be required to make the detection possible. The growth rate of the method is the same as for the case of known a priori master-slave discretization. The method is straightforward and it does not need any complicated constructions and three data organization, as for example in recently proposed technique for mortar formulation of contact [14].

A self-contact is more probable for thin or oblong solids, for which one or two dimensions are much smaller than others, than for solids with all dimensions of the same order. But there is a challenge which reveals itself in Figure 11. For a thin solid with two sided contact zone in general case it is impossible to distinguish the contact with the reverse side node ($\mathbf{r}_1$ can be in contact with $\mathbf{r}_3$) from a simple neighbouring with it ($\mathbf{r}_1$ is close but cannot come in contact with $\mathbf{r}_2$). Even if in addition to node positions their normals $\mathbf{n}_1$, $\mathbf{n}_2$, $\mathbf{n}_3$ and corresponding surfaces are taken into account there is no way to distinguish $\mathbf{r}_2$ and $\mathbf{r}_3$. A possible solution to overcome this problem is to generate a finite element mesh with contact surfaces smaller than the minimal thickness of the structure (Figure 11). It provides the maximal detection distance higher than the distance to the back side and allows to avoid this confusion. But in a less general case the possibility of two sided contact can be omitted and two sides can be treated independently. In this case, the maximal detection distance should be limited by the doubled minimal thickness of the structure.

Let us enumerate the features of the implementation of the grid detection method in case of unknown a priori master-slave discretization. The main modification is that not only node coordinates but also associated normals have to be taken into account to determine potentially contacting elements as in [1].

1. The bounding box has to include all nodes of contact surface; it can be either constant if we know a priori a sufficiently small area where from the contact nodes do not escape or it can be a bounding box spanned all contact nodes.
2. In the beginning of every time step the normal have to be assigned to each contact node. An average of average normals of attached contact surfaces can be used.
3. Only one array A^c is created and filled with contact nodes. The logic is the same as in the case of simple contact.
4. Since we cannot distinguish master and slave nodes the detection of the closest node has to be carried out for each contact node $\mathbf{r}^c_{ij}$ against all other nodes $\mathbf{r}^c_{il}$, $l \neq j$ in the cell i. To be sure that the closest nodes can come in contact and are not attached to a common segment, the normals associated with nodes are checked to form an obtuse angle $\mathbf{n}_{ij} \cdot \mathbf{n}_{ik} \leq 0$. Obviously some neighbouring cells have to be verified as in case of simple contact.
5. When two proximal contact nodes j and k are detected then in order to determine the NTS contact element the local mesh density has to be analyzed. If the surface mesh surrounding j node is found to be more rough than the surface mesh of the second node k, then the node j is considered as master node, otherwise as slave. If one local mesh is as fine as another one, then the node j has to be checked against each surface attached to the node k and vice versa. If the projection exists then the contact element is created otherwise arbitrary one of nodes is considered as a slave and its opponent as a master and if the slave node is not passing by then the contact element is created.

Being adapted for the case of unknown master-surface, the detection procedure has been verified on the challenging artificial problem of the self-contact within a snail-operculum-like structure containing over 130,000 nodes on the surface, all nodes with attached segments are included in the contact detection step (see Figure 12). The detection time is higher than for the contact of the same order with known a priori master-slave discretization, because the preliminary stage requires the assignation of normals to every node and also because the main detection stage requires significantly more verifications of distances and normals than in master-slave conception. In practice the difference in detection time between known a priori and unknown master-slave depends significantly on the geometry and its evolution. For example, for the snail operculum problem for the known master-slave the detection time (≈ 9 sec) is only three times faster than for unknown a priori master-slave (≈ 30 sec).

In conclusion we affirm that the grid detection method can be adapted to the class of contact problems with unknown a priori master-slave discretization. The required detection time is of the same order of magnitude as the time needed for simple contact detection for the same problem. Availability of such powerful method extends significantly the capacities of the finite element analysis of contact problems.

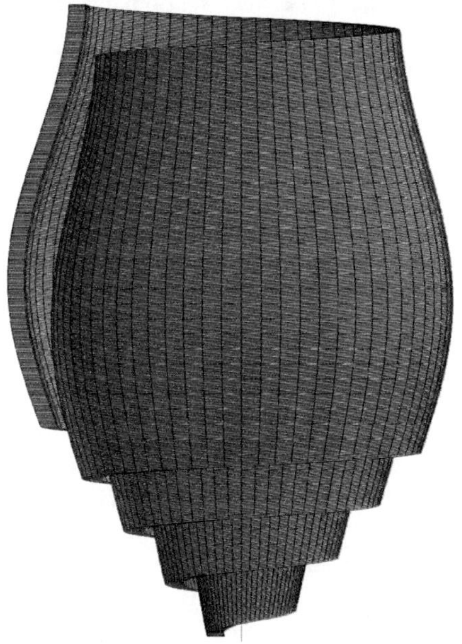

Fig. 12 Finite element mesh used to test the detection procedure for self-contact problems.

4 Parallelization

Sequential treatment of the problems presented above requires either too long computational time or even impossible due to the great amount of memory needed. The use of the parallelization paradigm is a good way out. Many parallelization techniques are available nowadays, the class of non-overlapping domain decomposition, also called iterative substructuring methods, is successfully and widely used in computational mechanics, see [3, 5, 9]. It implies a splitting of an entire finite element mesh into subdomains which intersect only on their interfaces. Each subdomain is treated by one or several associated processors and further the continuity of the solution across subdomain interfaces is enforced by displacement and force equality. The use of these techniques with affordable and powerful parallel computers allows to solve very large mechanical problems in reasonable terms. Since the resolution follows the detection procedure so the last one is very important for the efficiency of parallel computations [2]. It should not present a bottleneck in the whole process and if it is possible it has to use all available capacities of parallel computers.

The essential thing for the contact detection procedure in parallel treatment is the fact that the finite element mesh and possibly the contact surface is divided into some parts associated with different processors and in the case of distributed memory it is not available entirely on a particular processor. Since in principle we need the entire contact surface(s) to perform the detection procedure this repartition implies the data exchange between subdomains containing different parts of this surface(s). The smaller the amount of data transfer between subdomains on distributed memory, the faster the algorithm computers. It will be demonstrated below how this data transfer can be reduced significantly in the framework of the grid contact detection method.

Two ways of parallel treatment of contact problems are proposed and analyzed: Single processor Detection, Multiple processor Resolution (SDMR) and Multiple Detection, Multiple Resolution (MDMR). As it is evident from the notations SDMR carries out the contact detection on a single processor whereas MDMR uses all available resources. The last implies a parallelization of the detection procedure which will be discussed in details and tested.

First, let us consider the SDMR approach. The main idea is that all necessary information is collected by one processor which carries out the contact detection and distributes consequently the created contact elements among all concerned subdomains. This method can be efficiently applied to any contact problem and is easy to implement. On the other hand this method does not use efficiently all available resources, i.e. all except one processors are idle and inactive during the main detection phase however all processors possessing contact surface are active during the preliminary stage.

At first, the bounding box for the contact detection has to be defined. This task is easily performed in parallel. Each subdomain $i \in [1;N^c]$ possessing a part of contacting surfaces examines it and derives the corresponding bounding boxes ${}^m\mathbf{r}_i^1$, ${}^s\mathbf{r}_i^1$, ${}^m\mathbf{r}_i^2$, ${}^s\mathbf{r}_i^2$ and the maximal dimension of master segment $d^i_{\max}$. Further by means of data transfer the global maximal detection distance $d_{\max} = \max_{i=1}^{N^c}\{d^i_{\max}\}$ and master and slave bounding boxes are determined

$$ {}^{m,s}r^1_{\{x,y,z\}} = \min_{i=1}^{N^c}\{{}^{m,s}r^1_{i\{x,y,z\}}\} - d_{\max}, \quad {}^{m,s}r^2_{\{x,y,z\}} = \max_{i=1}^{N^c}\{{}^{m,s}r^2_{i\{x,y,z\}}\} + d_{\max}. \tag{14} $$

Finally, the resultant bounding box $\{\mathbf{r}^1, \mathbf{r}^2\}$ is constructed as the intersection of master and slave bounding boxes, exactly as in the sequential procedure. The data transfer consists in maximum $3N^c$ sends but the load is not uniformly distributed between processors, because not all of them contain the contact surface and among possessing it, the size of this surface can be quite different. In all cases this operation is quite fast even for huge meshes.

Next step consist in the union of all necessary parts of the contact surface at one processor-detector. First, the information about the global bounding box is distributed among the subdomains possessing the contact surface, each of them counts the number of master and slave nodes located in the bounding box, further the subdomain with the maximal number of master and slave nodes is chosen as the detector. Another possibility is that this choice can be made in concordance with processors network topology to accelerate the data transfer on the next detection step. As one can see, at this stage the data exchange between subdomains remains very limited.

It remains to transfer all master and slave nodes from the bounding box (global IDs, hosting subdomain ID, coordinates, and attached surfaces) to the detector subdomain, to carry out the detection as it is described in Section 2 and to distribute the

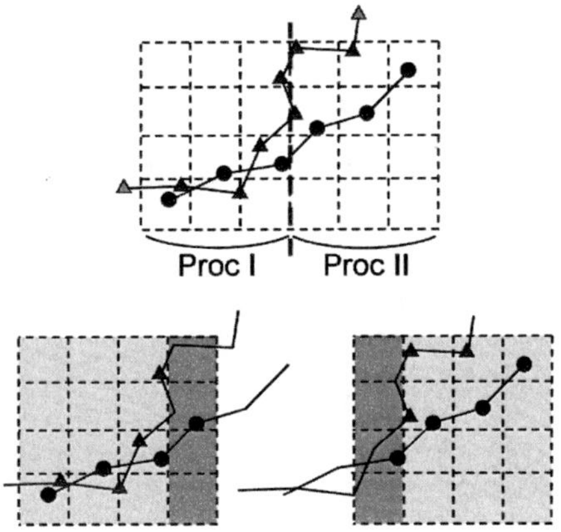

Fig. 13 Example of cells partition between two processors. Each one gets one half of the total number of cells (with slave and master nodes – represented by triangles and circles respectively) as well as one boundary layer from another half which contain only master nodes.

constructed contact elements between the corresponding subdomains. If a contact element unions slave node and master nodes from different subdomains, the interface between them has to be created or updated as well as duplicated slave or master nodes have to be formed.

In MDMR (Multiple Detection, Multiple Resolution) the preliminary part of a bounding box construction is exactly the same as in SDMR approach. The key difference between MDMR and SDMR consists in the following step. Instead of transferring all the necessary information to the detector, in MDMR this information is distributed between all subdomains in a special way. As it was shown above the grid is constructed in a way that for each slave node only one surrounding layer of neighbouring cells has to be verified to find the closest master node. If the self-contact is excluded from the consideration we do not care about slave nodes in neighbouring cells. That is why the bounding box can be divided into N non-overlapping parts, each part consists of integer number of cells. Further, each part is extended in all directions (not exceeding the bounding box) by a one-cell-overlapping layer; the extended part is filled only with master nodes (see an example for two subdomains in Figure 13). In other words each part consists of internal cells (non-overlapping with other parts) including both master and slave nodes and external cells (shared with neighbouring parts) including only master nodes. Each part is associated with a processor and all necessary data: nodes and surfaces located in the part (global IDs, hosting subdomain ID, coordinates, and attached surfaces) is collected from different subdomains and transferred to the considered one. Consequently the detection can be carried out absolutely independently, i.e. in parallel in each part. No additional data exchange is needed and it increase significantly the performance and scalability of the MDMR approach. The advantage of the method is that the total number of operations per processor during the main phase of detection does not increase with proportional increasing number of processors and contacting nodes. However, during the main detection phase the number of operations is not distributed homogeneously between processors.

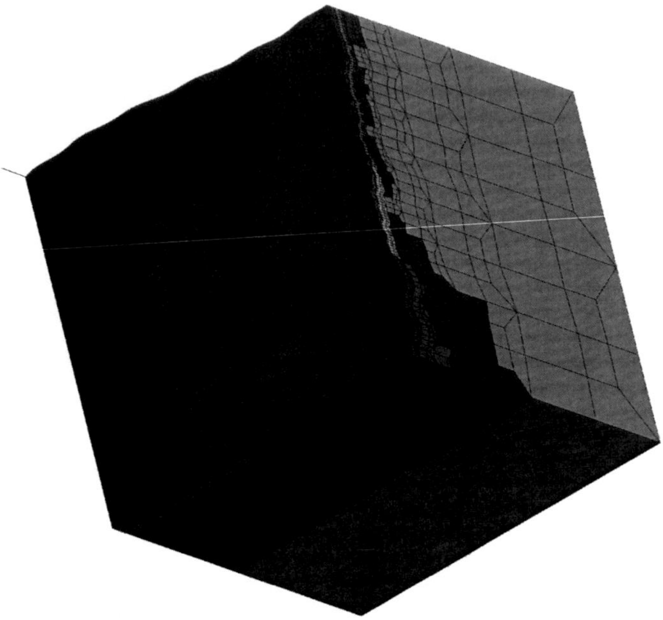

Fig. 14 The split of the FE mesh split into 16 sub-domains for parallel computations.

The same parallel procedure can be used for self-contact problems. The only difference is that master and slave nodes are not distinguished and hence all contact nodes have to be included in the overlapping cells. The described method is very similar to the parallelization of the Linked Cell Method widely used in molecular dynamic simulations for short-range interactions [6].

In Figure 14 the finite element mesh of a rough surface is presented, in the figure different tones of grey correspond to subdomains. The scalability test for MDMR approach has been performed between two such meshes containing over 560,000 nodes and over 66,000 contact nodes each. The scalability test for such meshes with slightly different surface roughness is represented in Figure 15. The heterogeneous distribution of active contact zones means that the parts of bounding box associated with different processors have quite different number of potential contact elements; the homogeneous distribution means that this number is more or less similar for different parts. And the average gain stands for averaged time of processors work. The difference between linear gain and the average gain makes evident the time necessary for the data exchange between subdomains. The pronounced difference between the gain for heterogeneous and homogeneous active contact zones distributions can be explained by the following observation. If there is no master node in the cell of the slave node, nor in the neighbouring cells, the time needed to conclude that is very small. On contrary if the considered cells are not empty and contain several master nodes it takes a longer time to derive the coordinates of these nodes, to

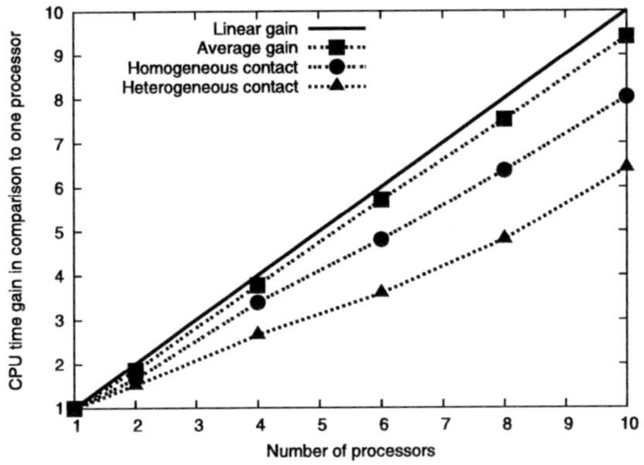

Fig. 15 Time gain for parallel contact detection procedure.

compare them with the slave node and to verify a projection availability. Nevertheless the gain is quite high and its rate does not decrease with increasing number of detecting processors (for reasonable ratio of contact nodes to number of processors).

The SDMR and MDMR approaches can be efficiently applied to parallel contact treatment. The second approach requires a larger amount of programming but its performance allows to neglect the detection time for large and extremely large contact problems.

5 Conclusion

The very fast local detection method has been elaborated on the base of the bucket method. Sequential and parallel implementations of the method have been discussed in details for known a priori and unknown master-slave discretizations.

The strong connections between the finite element mesh of the master surface, the maximal detection distance and the optimal dimension of detection cells are established. Analytical estimation and numerous tests demonstrate that the optimal cell size is equal to the maximal detection distance which by-turn is equal to the dimension of the biggest master segment. The particular attention in the article has been paid to the bounding box construction, optimal choice of the neighbouring cells to be verified, "passing by node" and blind spot analysis, master–slave surfaces definition in self-contact and especially to an efficient implementation of the method on distributed memory parallel computers.

The method is very flexible but it is not well adapted neither for very heterogeneous distribution of master segment dimensions nor for very different mesh densities of master and slave surfaces. In the method, the dimension of the biggest master segment is strongly connected with the maximal detection distance and consequently with the cell size. Therefore if the master surface has at least one segment which dimension is 10–100 times larger than the dimension of an average segment the detection time can be rather high, but always less than in all-to-all approach.

The validation of the method has been performed on different contact problems in sequential and parallel cases: contact between rough surfaces with different geometries, tyre-road contact, self-contact of a snail operculum and on the extremely large contact problem between two rough meshes including more than 1,000,000 segments at master surface against 1,000,000 slave nodes. In the latter problem, the detection time varies significantly for different geometries from several seconds to 30–40 minutes in comparison to almost 8 days needed for all-to-all detection techniques.

Acknowledgements The implementation of the algorithm and all demonstrated analyses have been carried out in the FEA package ZéBuLoN developed by the Centre des Matériaux (MINES ParisTech, France), ONERA (France) and Northwest Numerics (USA). This work has been performed with the support of the CNRS-SNECMA grant 920 78122.

References

1. Benson, D.J., Hallquist, J.O.: A single surface contact algorithm for the post-buckling analysis of shell structures. Computer Methods in Applied Mechanics and Engineering 78, 141–163 (1990)
2. Brown, K., Attaway, S., Plimpton, S., Hendrickson, B.: Parallel strategies for crash and impact simulations. Computer Methods in Applied Mechanics and Engineering 184, 375–390 (2000)
3. Farhat, C., Roux, F.-X.: Implicit parallel processing in structural mechanics. Computational Mechanics Advances 2(1), 1–24 (1994)
4. Fujun, W., Jiangang, C., Zhenhan, Y.: A contact searching algorithm for contact-impactor problems. Acta Mechanica Sinica (English series) 16(4), 374–382 (2000)
5. Gosselet, P., Rey, C.: Non-overlapping domain decomposition methods in structural mechanics. Archives of Computational Methods in Engineering 13(4), 515–572 (2006)
6. Griebel, M., Knapek, S., Zumbusch, G.: Numerical Simulation in Molecular Dynamics. Springer, Berlin (2007)
7. Konyukhov, A., Schweizerhof, K.: On the solvability of closest point projection procedures in contact analysis: Analysis and solution strategy for surfaces of arbitrary geometry. Computer Methods in Applied Mechanics and Engineering 197(33/40), 3045–3056 (2008)
8. Pietrzak, G., Curnier, A.: Large deformation frictional contact mechanics: Continuum formulation and augmented Lagrangian treatment. Computer Methods in Applied Mechanics and Engineering 177, 351–381 (1999)
9. Toselli, A., Widlund, O.: Domain decomposition methods – Algorithms and theory. Springer, Berlin (2005)
10. Williams, J.R., O'Connor, R.: Discrete element simulation and the contact problem. Archives of Computational Methods in Engineering 6(4), 279–304 (1999)

11. Wriggers, P.: Computational Contact Mechanics, 2nd edn. Springer, Berlin (2006)
12. Wriggers, P., Krstulovic-Opara, L., Korelc, J.: Smooth c1-interpolations for two-dimensional frictional contact problems. International Journal for Numerical Methods in Engineering 51(12), 1469–1495 (2001)
13. Yang, B., Laursen, T.A.: A contact searching algorithm including bounding volume trees applied to finite sliding mortar formulations. Computational Mechanics 41(2), 189–205 (2008)
14. Yang, B., Laursen, T.A.: A large deformation mortar formulation of self contact with finite sliding. Computer Methods in Applied Mechanics and Engineering 197(6/8), 756–772 (2008)
15. Zavarise, G., De Lorenzis, L.: The node-to-segment algorithm for 2D frictionless contact: Classical formulation and special cases. Computer Methods in Applied Mechanics and Engineering 198, 3428–3451 (2009)

Cauchy and Cosserat Equivalent Continua for the Multiscale Analysis of Periodic Masonry Walls

Daniela Addessi and Elio Sacco

Abstract The present paper deals with the problem of the determination of the in-plane behavior of periodic masonry material. Masonry is considered a composite material obtained as a regular distribution of blocks connected by horizontal and vertical mortar joints. The macromechanical equivalent Cauchy and Cosserat models are derived by means of the homogenization procedure, which make use of the Transformation Field Analysis (FTA) in order to account for the nonlinear effects occurring in the components. The micromechanical analysis is developed considering a Cauchy model for the masonry components. In particular, the linear elastic constitutive relationship is considered for the blocks, while a nonlinear constitutive law is proposed for the mortar joints, accounting for the damage and friction phenomena occurring during the loading history. Numerical applications are performed in order to assess the performances of the proposed models in reproducing the mechanical behavior of the masonry material. In particular, two different masonry textures are considered, remarking their different behavior. Moreover, for one masonry texture, the response is derived considering two possible RVEs.

1 Introduction

Masonry buildings are relevant part of the historical and architectural heritage in the World. The protection and conservation of this heritage is based on a careful evaluation of the building safety. Thus, the development of reliable structural analyses of masonry constructions is a very important task for the engineers and architects,

Daniela Addessi
Department of Structural Engineering, University of Rome "Sapienza", Via Eudossiana 18, 00198 Rome, Italy; e-mail: daniela.addessi@uniroma1.it

Elio Sacco
Department of Mechanics, Structures and Environment, University of Cassino, Via G. Di Biasio 43, 03043 Cassino, Italy; e-mail: sacco@unicas.it

G. Zavarise & P. Wriggers (Eds.): Trends in Computational Contact Mechanics, LNACM 58, pp. 253–268.
springerlink.com

in order to evaluate the stability and to design strengthening interventions. A crucial point for the development of effective stress analyses of masonry constructions is the definition of suitable constitutive laws. Indeed, masonry is a heterogeneous material, which is composed by natural or artificial bricks connected by mortar head and bed joints. Moreover, it shows a very complex mechanical behavior, due to the nonlinearity that appears also for limited values of the strain both under tensile and compressive stresses.

In fact, due to the complex nature of heterogeneous masonry materials, the mechanical behavior of masonry structures is one of the most challenging matters of the structural engineering both from a scientific and professional point of view. In the last decades, the scientific community has demonstrated great interest in the development of sophisticated numerical tools as an opposition to the tradition of rules-of-thumb or empirical formulae adopted to evaluate the safety of masonry buildings. In particular, the nonlinear models implemented in suitable finite element analysis (FEA) codes currently represent the most common advanced strategy to simulate the structural behavior of masonry structures.

Taking into account the heterogeneity of masonry material, the models proposed in literature can be framed in four different approaches: micro-mechanical models, phenomenological models, multi-scale models and macro-element models.

The multi-scale approach appears very appealing for the analysis of masonry structures, as it allows to derive in a rational way the stress-strain relationship of the masonry, accounting in a suitable manner for the mechanical properties of each material component and considering the failure micro-mechanisms of the masonry, which play a fundamental role in the overall behavior of the material.

The multi-scale approach has been satisfactorily employed in the structural analysis of periodic masonry [1–3].

Limiting the interest to the case of periodic masonry, the main idea consists in analyzing at the micro-mechanical level a properly selected repetitive Representative Volume Element (RVE) containing all the information about the microstructure. Then, two different Boundary Value Problems (BVP) have to be solved: one at the micro-scale level and the other at the macro-scale level. At the macro-scale, an equivalent homogeneous continuum is considered, whose formulation is completely stated except for the constitutive laws, which are derived by solving the proper formulated BVP at the micro-scale. In fact, at the micro-scale all the constituents are modeled, taking into account geometrical arrangement, size and constitutive laws of bricks and mortar. Bridging conditions between the two levels are then needed. Most of the multi-scale models presented in literature are based on the so-called first order homogenization technique, making use at both macro and micro scale of the classical Cauchy continuum.

Nevertheless, enhanced models have also been employed in literature for the modeling of the masonry constructions, in order to overcome the classical limitations and disadvantages of the Cauchy model. In fact, Cosserat, higher order and non-local models have been developed to reproduce the behavior of masonry panels in presence of high deformation gradients as well as of relevant microstructural sizes [4].

In particular, the Cosserat continuum, containing additional kinematical rotation fields with respect to the Cauchy model and enriched description of stresses and strains, has been proposed at the macro-scale [5, 6]. For instance, the Cosserat model is adopted by Trovalusci and Masiani [7], assuming the bricks as rigid elements in a periodic microstructure.

The present paper deals with the modeling of the masonry walls adopting a micromechanical analysis based on the Cauchy and Cosserat continua. The Cauchy and the Cosserat strain and stress measures in two-dimensional framework are introduced. The adopted constitutive model for each masonry component is described. Then, a nonlinear homogenization procedure, based on the so-called Transformation Field Analysis (TFA) [8, 9], is proposed. The TFA has been recently presented by Sacco [10] to perform the nonlinear first order homogenization procedure for masonry, making use of a micro-macro modeling approach based on a micromechanical analysis of the damaging process of the mortar material, the superposition of the effects and the FEA method. The TFA has been, then, extended also for the Cosserat modeling of the in-plane behavior of masonry walls [11].

Once the homogenization technique is illustrated, a solution algorithm is proposed; finally, numerical results are presented.

2 Micropolar Masonry Model for Periodic Masonry

Heterogeneous masonry material can be replaced by an equivalent Cauchy or Cosserat homogeneous medium at the macro level, developing a suitable homogenization procedure. The first step is the definition of the Representative Volume Element (RVE), which contains all the material and geometrical information of the heterogeneous medium.

Herein, the Cauchy and Cosserat continua are adopted at the macro level in order to describe the microstructural interaction effects, related to the size and shape of the constituents. On the other hand, a classical Cauchy continuum is assumed for the constituents of the RVE, where their complex nonlinear constitutive behavior is suitably modeled.

Adopting the Voigt notations and denoting by $\mathbf{U} = \{U_1 \;\; U_2 \;\; \Phi\}^T$ the kinematic descriptors of the 2D Cosserat medium, where only the first two components are considered for the Cauchy continuum, the deformation field is described at the macroscopic level by the infinitesimal strain vector

$$\mathbf{E} = \left\{ \begin{matrix} \mathbf{E}^c \\ \mathbf{K} \end{matrix} \right\}, \quad \mathbf{E}^C = \{E_1 \;\; E_2 \;\; \Gamma_{12}\}^T, \quad \mathbf{K} = \{\Phi \;\; K_1 \;\; K_2\}^T \tag{1}$$

where $\mathbf{E}^C$ is the classical Cauchy in-plane strain vector, with

$$E_1 = U_{1,1}, \quad E_2 = U_{2,2}, \quad \Gamma_{12} = E_{12} + E_{21} = U_{1,2} + U_{2,1} \tag{2}$$

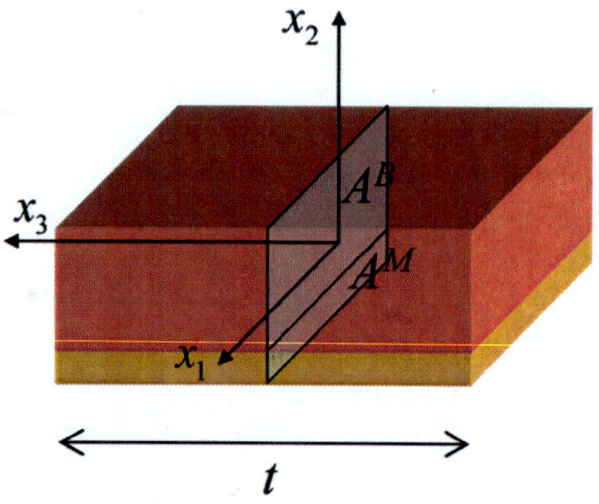

Fig. 1 Brick and mortar in the masonry thickness.

denoting the extensional strains and the classical Cauchy shear strain, respectively, and with $E_{12} = U_{2,1} - \Phi$ and $E_{21} = U_{1,2} + \Phi$ the Cosserat shear strains. As concerns the components of the vector $\mathbf{K}$, the rotational deformation and the flexural curvatures are defined as

$$\Phi = E_{12} - E_{21} = U_{2,1} - U_{1,2} - 2\Phi, \quad K_1 = \Phi_{,1}, \quad K_2 = \Phi_{,2} \tag{3}$$

Regarding the definition of the RVE for regular masonry, it is worth noting that the Cauchy model does not contain any information about a possible internal length of the material, so that for a given geometrical texture the overall response of the material does not depend on the RVE size. On the contrary, the Cosserat model accounts for an internal length of the material and, as consequence, the effective equivalent Cosserat continuum masonry for a given geometrical texture depends on the specific chosen RVE. As a consequence, the choice of the RVE in the framework of the Cosserat homogenized medium is an important issue, as it defines the Cosserat internal length.

Introducing a Cartesian coordinate system (O, x_1, x_2), the displacement vector at the typical point of the RVE is denoted as $\mathbf{u} = \{u_1 \ \ u_2\}^T$, while the strain vector, according to the Voigt notation, is given by $\boldsymbol{\varepsilon} = \{\varepsilon_1, \varepsilon_2, \gamma_{12}\}^T$, with $\varepsilon_1 = u_{1,1}$, $\varepsilon_2 = u_{2,2}$ and $\gamma_{12} = u_{1,2} + u_{2,1}$.

Although generalized plane state conditions should be adopted [12], in the following a simplified generalized plane state is considered. In fact, let t be the masonry thickness, and A^B and A^M the areas of the mid-surface of the brick and mortar, respectively, as schematically illustrated for the masonry RVE reported in Figure 1. Note that the superscripts B and M refer to the brick and mortar, respectively.

When the masonry is subjected to the vertical compressive stress σ_2, together with the vertical contraction, a transversal dilatation ε_3 is expected for the masonry. Note that if the wall is large enough along the x_1-direction and it is properly restrained to the soil, the strain ε_1 can be considered as negligible, i.e. $\varepsilon_1 = 0$.

Assuming the same values for the vertical normal stress σ_2 and for the transversal dilatation ε_3 in the brick and the mortar, under the in-plane strain condition $\varepsilon_1 = 0$, and considering both the brick and the mortar as elastic isotropic materials, the following constitutive equations can be written:

$$\begin{aligned} \sigma_2 &= 2\mu^B \varepsilon_2^B + \lambda^B(\varepsilon_2^B + \varepsilon_3) \\ \sigma_2 &= 2\mu^M \varepsilon_2^M + \lambda^M(\varepsilon_2^M + \varepsilon_3) \end{aligned} \tag{4}$$

$$\begin{aligned} \sigma_3^B &= 2\mu^B \varepsilon_3 + \lambda^B(\varepsilon_2^B + \varepsilon_3) \\ \sigma_3^M &= 2\mu^M \varepsilon_3 + \lambda^M(\varepsilon_2^M + \varepsilon_3) \end{aligned} \tag{5}$$

with μ^B, μ^M and λ^B, λ^M denoting the Lamé constants for the masonry constituents.

The average transversal stress is computed and enforced to be equal to zero:

$$\bar{\sigma}_3 = \frac{A^B}{A^B + A^M}\sigma_3^B + \frac{A^M}{A^B + A^M}\sigma_3^M = 0 \tag{6}$$

Equations (4), (5) and (6) can be regarded as a set of five equations in the five unknowns ε_2^B, ε_2^M, ε_3^B and ε_3^M. Once the system is solved, the effective Poisson ratios of the brick and mortar can be evaluated as

$$\bar{\nu}^B = -\frac{\varepsilon_3}{\varepsilon_2^B}, \quad \bar{\nu}^M = -\frac{\varepsilon_3}{\varepsilon_2^M} \tag{7}$$

Hence, developing the 2D plane stress analysis by assuming the Poisson ratios evaluated by formulas (7), the transversal deformation of the masonry is allowed with negligible average transversal stress, when the cell is mainly loaded in vertical compression, as it occurs often in the masonry.

The three RVEs shown in Figure 2 represent two textures of the periodic masonry, called running bond and stack bond. The first two RVEs are characterized by rectangular shape with dimensions $2a_1$ and $2a_2$, parallel to the coordinate axes x_1 and x_2. The mortar thickness is denoted by s and the brick sizes by b and h. The third RVE corresponds to the same masonry texture of the second one; it has dimensions $(b+s)\times(h+s)$. All three RVEs contains all the information regarding the geometrical and constitutive properties of the masonry microstructure.

The linear elastic constitutive law is considered for the brick. Thus, denoting by $\mathbf{C}^B$ the elastic matrix of the brick, the stress-strain relationship is written in the form:

$$\boldsymbol{\sigma}^B = \mathbf{C}^B \boldsymbol{\varepsilon} \tag{8}$$

where $\boldsymbol{\sigma}^B = \{\sigma_1^B, \sigma_1^B, \tau_{12}^B\}$ is the stress vector in the brick.

A constitutive law for the mortar material, coupling the cohesive damage mechanisms and friction phenomena, is developed. The stress-strain relationship proposed in [10] is adopted and is briefly introduced here. A local coordinate system is defined in the typical mortar joint, with T denoting the axis parallel to the mortar

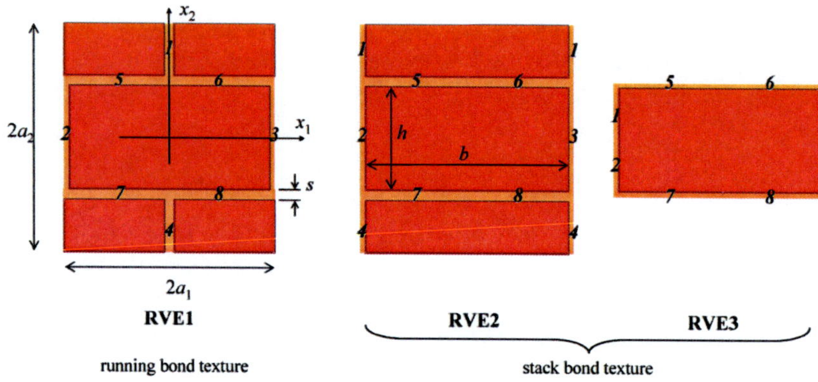

Fig. 2 Three RVEs for two different textures of periodic masonry.

joint and N the orthogonal direction. The stress $\boldsymbol{\sigma}^M$ at a typical point of the mortar, is obtained as suitable combination of two stresses, $\boldsymbol{\sigma}^u$ and $\boldsymbol{\sigma}^d$, according to

$$\boldsymbol{\sigma}^M = (1-D)\boldsymbol{\sigma}^u + D\boldsymbol{\sigma}^d \tag{9}$$

where D is the damage parameter. The two stress vectors $\boldsymbol{\sigma}^u$ and $\boldsymbol{\sigma}^d$ are related to the strain vector in the mortar $\boldsymbol{\varepsilon}$ by the constitutive equations

$$\boldsymbol{\sigma}^u = \mathbf{C}^M\boldsymbol{\varepsilon}, \quad \boldsymbol{\sigma}^d = \mathbf{C}^M(\boldsymbol{\varepsilon}-\boldsymbol{\varepsilon}^p) \tag{10}$$

where $\mathbf{C}^M$ represents the mortar elasticity matrix and $\boldsymbol{\varepsilon}^p$ is the vector of the inelastic strain due to the possible unilateral opening effect and to the friction sliding.

Taking into account the constitutive eq. (10), eq. (9) becomes

$$\boldsymbol{\sigma}^M = \mathbf{C}^M(\boldsymbol{\varepsilon}-\boldsymbol{\pi}) \tag{11}$$

where the vector $\boldsymbol{\pi}$ accounts for all the nonlinear effects and is defined as

$$\boldsymbol{\pi} = \left\{ \begin{array}{c} \pi_T \\ \pi_N \\ pi_{NT} \end{array} \right\} = D \left\{ \begin{array}{c} h(\varepsilon_N)\varepsilon_T \\ h(\varepsilon_N)\varepsilon_N \\ \gamma^p_{NT} \end{array} \right\} \tag{12}$$

with $h(\varepsilon_N)$ representing the Heaviside function, which assumes the following values: $h(\varepsilon_N) = 0$ if $\varepsilon_N < 0$ and $h(\varepsilon_N) = 1$ if $\varepsilon_N > 0$ and allows the modeling of the unilateral contact mechanism.

The friction is treated as a plasticity problem; in fact, the evolution of the inelastic slip strain component γ^p_{NT} is governed by the classical Coulomb yield function:

$$\varphi(\phi^d) = \mu\sigma^d_N + |\tau^d_{NT}| \tag{13}$$

where μ is the friction parameter. A non-associated flow rule is considered as

$$\dot{\gamma}^p_{NT} = \dot{\lambda} \frac{\tau^d_{NT}}{|\tau^d_{NT}|} \tag{14}$$

with the classical loading-unloading Kuhn–Tucker and consistency conditions:

$$\dot{\lambda} \geq 0, \quad \varphi(\boldsymbol{\sigma}^d) \leq 0, \quad \dot{\lambda}\varphi(\boldsymbol{\sigma}^d) = 0, \quad \dot{\lambda}\dot{\varphi}(\boldsymbol{\sigma}^d) = 0 \tag{15}$$

A model which accounts for the coupling of mode I and mode II of fracture is adopted for the damage parameter D evolution. The two quantities η_N and η_{NT}, which depend on the first cracking strains $\varepsilon_{N,0}$ and $\gamma_{NT,0}$, on the peak value of the stresses $\sigma_{N,0}$ and $\tau_{NT,0}$ and on the fracture energies G_{cl} and G_{ell}, respectively, are introduced

$$\eta_N = \frac{\varepsilon_{N,0}\sigma_{N,0}}{2G_{cl}}, \quad \eta_{NT} = \frac{\gamma_{NT,0}\tau_{NT,0}}{2G_{cll}} \tag{16}$$

The equivalent strain measures Y_N and Y_{NT} are defined as

$$Y_N = \langle \varepsilon_N \rangle^2, \quad Y_{NT} = (\gamma_{NT})^2 \tag{17}$$

where the bracket operator $\langle \cdot \rangle$ gives the positive part of its argument. Then the strain ratios are determined as

$$\eta = 1 - \frac{1}{\alpha^2}(Y_N\eta_N + Y_{NT}\eta_{NT}), \quad \beta = \sqrt{\frac{Y_N}{\varepsilon^2_{N,0}} + \frac{Y_{NT}}{\gamma^2_{NT,0}}} - 1 \tag{18}$$

with $\alpha = (Y_N + Y_{NT})^{1/2}$. Finally, the damage is evaluated by means of

$$D = \max_{\text{history}} \left\{ \min \left\{ 1, \frac{1}{\eta} \left(\frac{\beta}{1+\beta} \right) \right\} \right\} \tag{19}$$

According to the periodic homogenization technique, the displacement field in the generic point $\mathbf{x} = (x_1, x_2)^T$ of the RVE Cauchy medium is expressed in the following representation form:

$$\mathbf{u} = \bar{\mathbf{u}}(\mathbf{x}) + \tilde{\mathbf{u}}(\mathbf{x}) \tag{20}$$

where $\bar{\mathbf{u}}(\mathbf{x})$ is the assigned displacement field depending on the macroscopic deformation $\mathbf{E}$ and $\tilde{\mathbf{u}}(\mathbf{x})$ is the periodic perturbation satisfying periodicity boundary conditions.

In the spirit of the procedure proposed in [5] properly extended in order to consider rectangular RVEs, a third order polynomial expansion is adopted for $\bar{\mathbf{u}}(\mathbf{x})$, which results in

$$\bar{\mathbf{u}} = \underbrace{\underbrace{\begin{bmatrix} x_1 & 0 & \frac{1}{2}x_2 \\ 0 & x_2 & \frac{1}{2}x_1 \end{bmatrix}\mathbf{E}^C}_{\text{Cauchy}} + \begin{bmatrix} -\alpha(x_2^3 - 3\rho x_1^2 x_2) & -x_1 x_2 & -\frac{1}{2}x_2^2 \\ -\rho^2\alpha(\rho^2 x_1^3 - 3x_2^2 x_1) & \frac{1}{2}x_2^2 & x_1 x_2 \end{bmatrix}\mathbf{K}}_{\text{Cosserat}} \tag{21}$$

with $\alpha = 5(a_1^2 + a_2^2)/(4a_1^4)$ and $\rho = a_2/a_1$.

Following the TFA based nonlinear homogenization technique presented in [11], the total assigned Cosserat macroscopic strain $\mathbf{E}$ is additively decomposed into an elastic part $\mathbf{E}_e$ and an inelastic part $\mathbf{P}$. Firstly, the BVP on the RVE subjected to the elastic macroscopic strain $\mathbf{E}_e$ is numerically solved and the resulting micromechanical strain field can be expressed as follows:

$$\mathbf{e} = \mathbf{R}_e(\mathbf{x})\mathbf{E}_e \tag{22}$$

where $\mathbf{R}_e(\mathbf{x})$ represents the localization matrix, which allows to evaluate the Cauchy local strain $\mathbf{e}$ at any point of the RVE corresponding to the application of the Cosserat strain $\mathbf{E}_e$. By integrating the local strain $\mathbf{e}$ in each Ω^{M^i} of the eight mortar joints M^i, represented in Figure 2, the average strain in $\bar{\mathbf{e}}^{M^i}$ is obtained:

$$\bar{\mathbf{e}}^{M^i} = \frac{1}{\Omega^{M^i}} \int_{M^i} \mathbf{R}_e(\mathbf{x}) d\Omega \mathbf{E}_e = \bar{\mathbf{R}}_e^{M^i} \mathbf{E}_e, \quad i = 1,2,\ldots,8 \tag{23}$$

By applying the generalized Hill–Mandel principle [13, 14], the homogenized Cosserat stress in the RVE is evaluated as $\mathbf{\Sigma} = \mathbf{C}\mathbf{E}_e$, where $\mathbf{C}$ represents the overall elastic constitutive matrix. Similarly, the average stress in the mortar joint M^i may be evaluated as $\bar{\boldsymbol{\sigma}}_e^{M^i} = \mathbf{C}^M \bar{\mathbf{e}}^{M^i} = \mathbf{C}^M \bar{\mathbf{R}}_e^{M^i} \mathbf{E}_e$.

On the other hand, the solution of the BVP on the RVE subjected to an inelastic strain $\boldsymbol{\pi}^i$ prescribed in the mortar joint M^i, gives the resulting local strain field in the form:

$$\mathbf{p}^i = \mathbf{R}_{\pi^i}(\mathbf{x})\boldsymbol{\pi}^i \tag{24}$$

$\mathbf{R}_{\pi^i}(\mathbf{x})$ being the associated localization matrix. The elastic strain in the mortar joint M^j is obtained as the difference between its total deformation $\mathbf{p}^{i,M^j}$ and the inelastic strain $\boldsymbol{\pi}^i$ as

$$\boldsymbol{\eta}^{i,M^j} = \mathbf{p}^{i,M^j} - \delta_{ij}\boldsymbol{\pi}^i = (\mathbf{R}_{\pi^i}^{M^j} - \delta_{ij}\mathbf{I})\boldsymbol{\pi}^i \quad \text{(no sum)} \tag{25}$$

where $\mathbf{p}^{i,M^j}$ and $\mathbf{R}_{\pi^i}^{M^j}$ are the restriction to the mortar M^j of the fields $\mathbf{p}^i$ and $\mathbf{R}_{\pi^i}$, respectively. The elastic strain in the brick coincides with the total strain and results in

$$\boldsymbol{\eta}^{i,B} = \mathbf{p}^{i,B} = \mathbf{R}_{\pi^i}^B \boldsymbol{\pi}^i \tag{26}$$

with evident meaning of the symbols, under the condition that the overall Cauchy or Cosserat stress is zero.

When the RVE is subjected to the overall elastic strain $\mathbf{E}_e$ and to the eight inelastic strains $\boldsymbol{\pi}^i$, the superposition of the effects can be performed. In such a way, it is possible to compute: the total overall strain

$$\mathbf{E} = \mathbf{E}_e + \sum_{i=1}^{8} \mathbf{P}^i$$

where $\mathbf{P}^i$ is the macroscopic Cauchy or Cosserat strain associated to the inelastic strain $\boldsymbol{\pi}^i$, the overall stress

$$\boldsymbol{\Sigma} = \boldsymbol{\Sigma}_e + \sum_{i=1}^{8} \boldsymbol{\Sigma}_{\pi^i} = \boldsymbol{\Sigma}_e$$

the total average strains in the eight mortar joints

$$\bar{\boldsymbol{\varepsilon}}^{M^j} = \bar{\mathbf{R}}_e^{M^j} \mathbf{E}_e + \sum_{i=1}^{8} \bar{\mathbf{R}}_e^{M^j} \boldsymbol{\pi}^i$$

where $\bar{\mathbf{R}}_{\pi^i}^{M^j}$ is the average of the matrix $\mathbf{R}_{\pi^i}^{M^j}$ in the mortar joint M^j, and the average stresses in the mortar joints:

$$\bar{\boldsymbol{\sigma}}^{M^j} = \mathbf{C}^M \left(\bar{\boldsymbol{\varepsilon}}^{M^j} + \sum_{i=1}^{8} \boldsymbol{\eta}^{i,M^j} \right) = \mathbf{C}^M (\bar{\boldsymbol{\varepsilon}}^{M^j} - \boldsymbol{\pi}^i)$$

In the following it is assumed that the inelastic strain is uniform in each mortar joint and that the nonlinear behavior of the RVE depends on the average stresses and strains evaluated in each of the 8 mortar joints for all the RVEs illustrated in Figure 2.

3 Numerical Applications

In the following some numerical applications are reported. Firstly, the constitutive response of the RVE1, shown in Figure 2, obtained by applying the presented homogenization procedure is illustrated. In order to validate the proposed procedure, the results are compared with the ones obtained by a micromechanical FEA, performed by discretizing the RVE1 with 4-node quadrilateral finite elements and by adopting the constitutive laws presented above for bricks and mortar. The geometrical parameters of bricks and mortar are assumed as follows: sizes of the brick $b = 240$ mm, $h = 120$ mm; thickness of the mortar joints $s = 10$ mm. Furthermore, the mechanical parameters adopted for the computations are: for the brick $E_b = 18000$ MPa and $\nu_b = 0.15$; for the mortar $E_m = 1000$ MPa, $\nu_m = 0.15$, $\varepsilon_{N,0} = 0.0005$, $\gamma_{NT,0} = 0.001$, $G_{cl} = 0.00125$ MPa, $G_{\text{ell}} = 0.00217$ MPa and $\mu = 0.5$. From the given data and by using formulas (4)–(7), it results that $\bar{\nu}^B = 0.1043$ and $\bar{\nu}^M = 0.0058$.

In particular, a symmetric shear test is performed subjecting the RVE to a constant vertical compression $E_2 = -3.0E-4$ and to a cyclic symmetric shear history

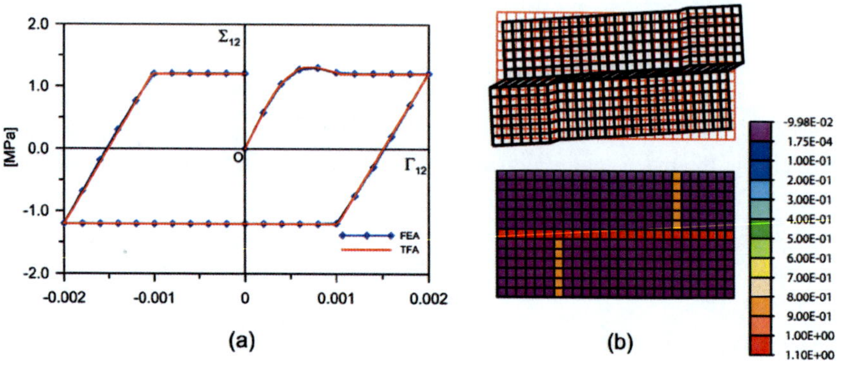

Fig. 3 Symmetric shear test on half RVE1: (a) shear stress versus shear strain; (b) deformed configuration and damage map.

with Γ_{12} varying from 0 to $2.0E-3$, then to $-2.0E-3$ and, finally, to 0. In Figure 3(a) the macroscopic shear stress component Σ_{12} versus the macroscopic shear strain Γ_{12} is shown, where the continuous curve denotes the results obtained with the proposed homogenization procedure (TFA), and the dashed curve concerns the micromechanical full FEA solution performed on half of RVE1 (336-element mesh) due to the symmetry of the geometrical scheme. A very good agreement between the two solutions can be remarked. After an initial linear elastic response, the nonlinear behavior appears due to the activation of both the damaging and plasticity mechanisms in both head and bed joints. During the subsequent unloading and reloading paths the RVE1 response is characterized by the development of the friction plasticity.

In Figure 3(b) the RVE1 deformed configuration and the damage map are reported at the final step of the loading phase ($\Gamma_{12} = 2.0E-3$). It can be seen that damage appears in head and bed joints, although it tends to localize into the bed joints.

In Figure 4(a) the response of the RVE1 under the application of the curvature varying from 0 to $4.0E-6$, to $2.0E-6$, to $10.0E-6$ and finally to 0, is depicted. In particular, the Cosserat macroscopic micro-couple M_2 versus the micro-curvature K_2 is drawn. In this case all the RVE1 is analyzed, adopting a 650-element mesh for the FEA. Again, the continuous curve refers to the results obtained with the proposed procedure and the diamond symbols to the FEA. In this application the two curves depart a little and the stiffness degrading process appears more severe in the FEA curve. This denotes a different damage evolution in the mortar joints, which is caused mainly by the simplified hypotheses on which the proposed homogenization procedure is based. After the initial linear elastic phase, the damage starts and evolves in the two right bed joints, subjected to tensile deformations, then involving also the right head joint, as it can be seen in Figure 4(b), where the deformed configuration of the RVE1 and the damage map are reported at $K_2 = 10.0E-6$.

In order to put in evidence the mechanical response of different RVEs under Cauchy and Cosserat deformation components, the three RVEs shown in Figure 2

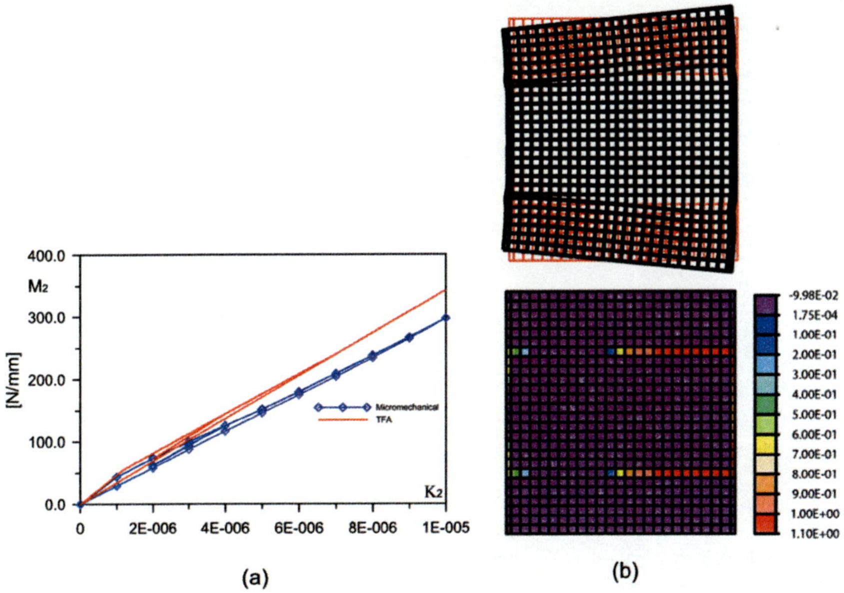

Fig. 4 Flexural test on the RVE: (a) micro-couple M_2 versus micro-curvature K_2; (b) deformed configuration and damage map.

are analyzed assigning periodicity conditions to the nodes lying on the horizontal and vertical sides.

The following sizes are considered for the brick: $b = 210$ mm, $h = 56$ mm; the thickness of the mortar joints is $s = 10$ mm. The mechanical parameters of the materials are: for the brick $E_b = 16700$ MPa and $\nu_b = 0.15$; for the mortar $E_m = 798$ MPa, $\nu_m = 0.11$, $\varepsilon_{N,0} = 0.0003$, $\gamma_{NT,0} = 0.001$, $G_{cl} = 0.0018$ MPa, $G_{\text{ell}} = 0.0126$ MPa and $\mu = 0.75$. From the given data, it results that $\bar{\nu}^B = 0.1075$ and $\bar{\nu}^M = 0.0052$.

In Figure 5 the symmetric shear test is performed subjecting the three RVEs to a constant vertical compression $E = -3.0E-4$ and to a cyclic symmetric shear history with Γ_{12} varying from 0 to $2.0E-2$, then to $-2.0E-2$ to 0. The macroscopic shear stress component Σ_{12} versus the macroscopic shear strain Γ_{12} is shown. It has to be noted that the mechanical response of the RVE2 and RVE3 are undistinguishable. This result is expected, since the two RVEs are characterized by the same texture (the periodicity conditions imposed on the RVE3 reproduce a stack bond texture), but different sizes, and are subjected to a Cauchy deformation mode. Furthermore, the damage process in the RVE2 and RVE3 starts before than in the RVE1 and is mainly localized in the head joints, where also the friction plastic mechanism is activated. On the contrary, the behavior of the RVE1 is similar to the one already observed in Figure 3.

Then, the response of the three RVEs under the two Cosserat flexural deformation components is studied. In particular, in Figure 6(a) the micro-couple is depicted

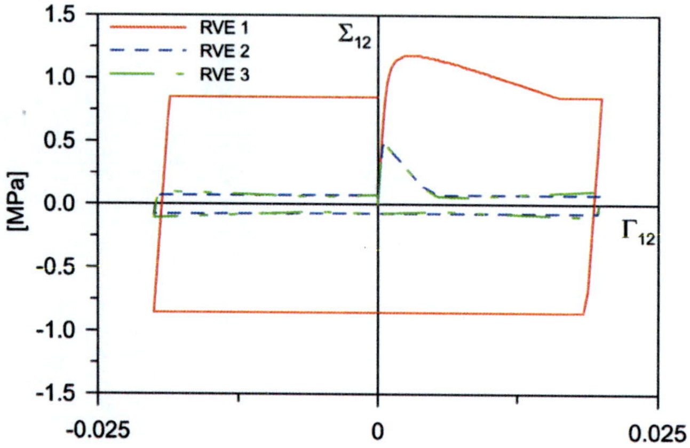

Fig. 5 Symmetric shear test on the three RVEs: shear stress Σ_{12} versus shear strain Γ_{12}.

versus the micro-curvature K_1 varying from 0 to $4.0E-5$ to 0. The behavior exhibited by the RVEs is mainly elastic and a little damage is observed for the RVE1. It has to be underlined that in this case RVE2 and RVE3 show very different responses, mainly due the different flexural stiffnesses of the two RVEs, when the K_1 component is applied. On the other hand, the RVE2 response curve appears close to the RVE1 one, although showing no damage. This fact emphasizes that the size of the RVE plays a fundamental role in the response of the masonry material, when typical Cosserat deformation modes are activated.

Instead, in Figure 6(b) the micro-couple M_2 is depicted versus the micro-curvature varying from 0 to $4.0E-6$, to $2.0E-6$, to $10.0E-6$ and finally to 0. All the response curves show a more evident nonlinear damaging behavior with respect to the previous case and, in this case, the RVE2 and RVE3 responses are undistinguishable. Moreover, the RVE1 is characterized by a higher stiffness both in the linear elastic and damaging regimes.

In order to investigate the influence of the additional Cosserat deformation components on the RVE response, a more complex loading history is applied to the three RVEs, considering the simultaneous presence of the compression strain E_2, the symmetric shear strain Γ_{12}, these two being typical Cauchy strain components, and the Cosserat micro-curvature K_2. Such loading conditions should reproduce the strain state acting on a RVE located at the bottom side of a shear pre-stressed masonry panel. In particular, a constant value of $E_2 = -3.0E-4$ is applied, together with a symmetric shear strain Γ_{12} varying from 0 to $2.0E-2$, to $-2.0E-2$ and to 0, and a micro-curvature K_2 varying from 0 to f, to $-f$, to 0, with f assuming different values. In Figures 7, 8 and 9, the symmetric shear stress Σ_{12} versus the shear strain Γ_{12} is reported for the RVE1, RVE2 and RVE3, respectively, and for the cases $K_2 = 0$, $K_2 = 5.0E-6$, $K_2 = 10.0E-6$ and $K_2 = 20.0E-6$. It clearly emerges that the presence of K_2 strongly influences the symmetric shear response of the RVE1 (Figure 7). In particular, when the RVE1 is subjected to the Cauchy strain compon-

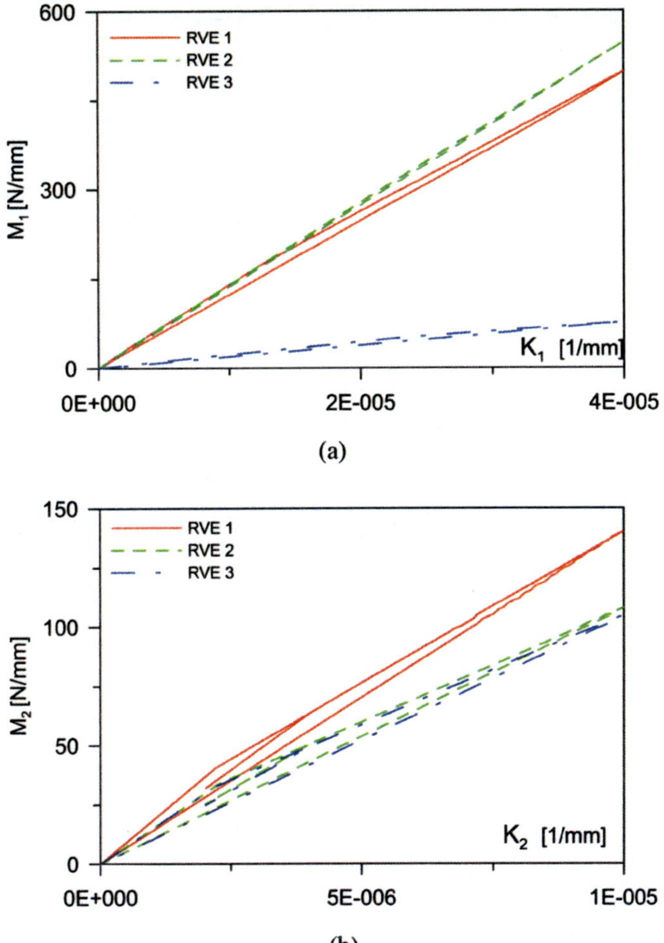

Fig. 6 Flexural tests on the three RVEs: (a) micro-couple versus micro-curvature K_1; (b) micro-couple versus micro-curvature K_2.

ents only (for $K_2 = 0$), damage appears which is located mainly in the bed joints up to the complete deterioration. The simultaneous presence of K_2 results in a more severe damage process (see dash and dash-dot curves in Figure 7), which involves and progresses also in the head joints.

On the contrary, the symmetric shear response of both the RVE2 (Figure 8) and RVE 3 (Figure 9) show a very low influence of the micro-curvature K_2, except for the highest value $K_2 = 20.0E-6$ in the case of the RVE2. These results demonstrate the very strong influence of the RVE texture on the Cosserat response, also showing that the running bond is the better texture in order to resist to loading conditions typical of shearing masonry panels.

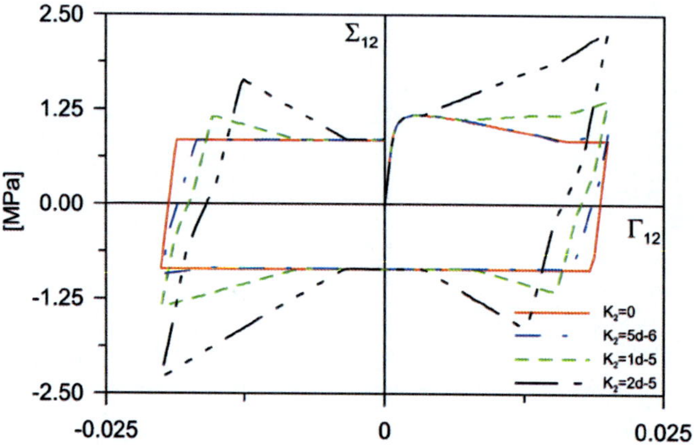

Fig. 7 Symmetric shear test on the RVE 1: comparison of the shear response subjected to different histories of K_2.

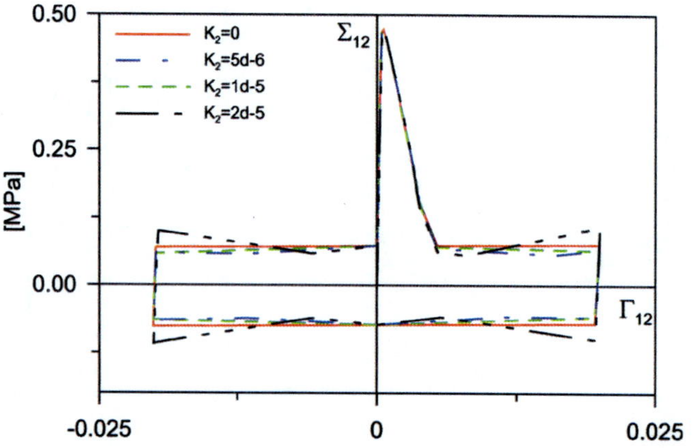

Fig. 8 Symmetric shear test on the RVE 2: comparison of the shear response subjected to different histories of K_2.

4 Conclusions

A nonlinear homogenization procedure for describing the in-plane constitutive response of regular masonry material has been presented. The equivalent medium at the macro-level has been modeled with both the Cauchy and Cosserat continuum, while the standard Cauchy model has been employed at the micro-level. Since periodic textures have been considered for masonry, the proposed homogenization technique held on the assumption of periodicity conditions imposed on the RVE, properly generalized when the Cosserat deformation modes are taken into consideration. The bridging relations between the macro and micro levels have been stated by

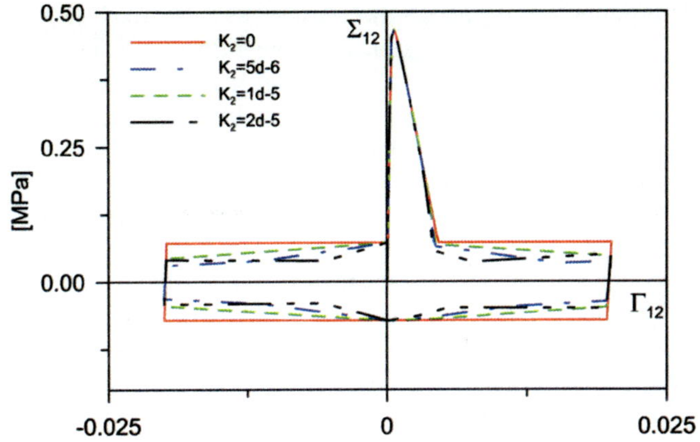

Fig. 9 Symmetric shear test on the RVE 3: comparison of the shear response subjected to different histories of K_2.

formulating a suitable kinematic map, which expresses the micromechanical displacement fields as a function of the macroscopic Cauchy and Cosserat deformation components. The higher order polynomial expansions used for the macro- micro kinematic map in the case of Cosserat macroscopic medium has allowed to analyze micromechanical deformation modes richer than in the classical first order homogenization framework, as also flexural and unsymmetric shear modes have been included. A damage-unilateral contact-friction model has been adopted for the mortar joints. Then, in the spirit of the TFA procedure, the overall elastic constitutive matrix and the localization tensors have been evaluated by linear FEAs of the RVE, on the basis of which the nonlinear damage and plasticity evolutive problems at the typical point of the macroscopic equivalent medium have been solved by a step-by-step analysis. Furthermore, the analysis of the nonlinear micromechanical response of the RVE has been carried out by means of a FEA on the basis of the coupled damage-plastic model adopted for the mortar joints. The comparison between the numerical results obtained by the proposed procedure and the ones evaluated by the nonlinear micromechanical FEA has shown a very satisfactory agreement, when macroscopic Cauchy deformation components are applied to the RVE, then validating the assumption of uniformly distributed inelastic strains along the longitudinal direction of the joints, on which TFA procedure is founded. Some discrepancies otherwise emerge, when the response of the RVE under macroscopic Cosserat typical strain components is analyzed, as a consequence of the simplified assumptions of the TFA. Also the response of three RVEs characterized by different textures has been analyzed under the application of a loading history combining Cauchy and Cosserat deformation components with the aim of reproducing loading conditions typical of structural applications on shearing masonry panels; it appeared that the nonlinear response of the RVEs is influenced by the presence of the Cosserat micro-curvatures, but it strongly affects the nonlinear behavior only in the case of the running bond

RVE, which results as the better texture to resist to shear-flexural loading conditions. Such results confirm the relevance of the use of the micro-polar Cosserat continuum for developing accurate models for masonry, since it appears clearly capable to reproduce the influence of the microstructural size, shape and texture on the overall structural response, thus overcoming the limits of the classical Cauchy formulation in describing the regular masonry behavior.

In conclusion, the homogenization technique allows one to derive the overall properties of the masonry material, in the framework of the Cauchy and Cosserat models, properly accounting for the geometrical texture and constitutive laws of the bricks and mortar. The choice of the RVE size is important for the Cosserat approach in order to correctly account for the internal length of the material. This aspect represents a realistic properties of the wall, and not just an artificial result of the modeling.

References

1. Luciano, R., Sacco, E.: Homogenization technique and damage model for old masonry material. International Journal of Solids and Structures 34(24), 3191–3208 (1997)
2. Massart, T.J., Peerlings, R.H.J., Geers, M.G.D.: An enhanced multi-scale approach for masonry wall computations with localization of damage. International Journal on Numerical Methods in Engineering 69(5), 1022–1059 (2007)
3. Brasile, S., Casciaro, R., Formica, G.: Multilevel approach for brick masonry walls. Part I: A numerical strategy for the nonlinear analysis. Computer Methods in Applied Mechanics and Engineering 196, 4934–4951 (2007)
4. Kouznetsova, V., Geers, M.G.D., Brekelmans, W.A.M.: Multi-scale constitutive modelling of heterogeneous materials with a gradient-enhanced computational homogenization scheme. International Journal on Numerical Methods in Engineering 54, 1235–1260 (2002)
5. Forest, S., Sab, K.: Cosserat overall modeling of heterogeneous materials. Mechanics Research Communications 25, 449–454 (1998)
6. Van der Sluis, O., Vosbeek, P.H.J., Schreurs, P.J.G., Meijer, H.E.H.: Homogenization of heterogeneous polymers. International Journal of Solids and Structures 36, 3193–3214 (1999)
7. Trovalusci, P., Masiani, R.: Non-linear micropolar and classical continua for anisotropic discontinuous materials. International Journal of Solids and Structures 40(5), 1281–1297 (2003)
8. Dvorak, G.J.: Transformation field analysis of inelastic composite materials. Proc. Roy. Soc. London A 437, 311–327 (1992)
9. Michel, J.C., Suquet, P.: Nonuniform transformation field analysis. International Journal of Solids and Structures 40, 6937–6955 (2003)
10. Sacco, E.: A nonlinear homogenization procedure for periodic masonry. European Journal of Mechanics A, Solids 28, 209–222 (2009)
11. Addessi, D., Sacco, E., Paolone, A.: Cosserat model for periodic masonry deduced by nonlinear homogenization. European Journal of Mechanics A, Solids 39, 724–737 (2010)
12. Anthoine, A.: Homogenization of periodic masonry: Plane stress, generalized plane strain or 3D modelling? Communications in Numerical Methods in Engineering 13, 319–326 (1997)
13. Hill, R.: Theory of mechanical properties of fibre-strengthened materials: II. Inelastic behaviour. Journal of the Mechanics and Physics of Solids 12, 213–218 (1964)
14. Mandel, J.: Plasticité Classique et Viscoplasticité. CISM Lecture Notes. Springer, Heidelberg (1971)

Coupled Friction and Roughness Surface Effects in Shallow Spherical Nanoindentation

P. Berke and T.J. Massart

Abstract When nanoindentation is used for thin film characterization, usually shallow indents are made to avoid the spurious effect of the substrate. However, surface effects stemming from surface roughness and friction can become important in shallow indentation depths, potentially resulting in the variation of nanoindentation results. A numerical study is conducted aiming for a more complete understanding of the coupled influence of friction and sample surface roughness in nanoindentation of pure nickel, using a slip rate dependent friction law. Two experimentally used post-treatment methods are applied to obtain the elastic properties from the raw numerical data. Results confirm the strong interaction between these two contributions of surface effects, and their cumulative effect leads to significant variations in both the indenter load vs. displacement curves and the evaluated elastic modulus. The resulting dispersion is somewhat higher than the one computed for a slip rate independent Coulomb friction. The velocity-weakening nature of the used friction law, is observed to induce a stick-slip behavior which has a manifestation similar to pop-ins in the load-displacement curves.

1 Introduction

Nanoindentation is in principle a nanoscale hardness test, which results in a continuously sensed indenter load vs. indenter displacement curve from which the elastic modulus of the sample is derived using post-treatment methods. It has the considerable advantage to allow the local measurement of the material properties of thin films when using sharp indenter tips in small indentation depths. The continuously sensed load-displacement curve is composed of various contributions among which: the behavior of the material (elastic-plastic [21], rate-dependent [15, 23],

P. Berke · T.J. Massart
BATir Dept. CP194/2, Université Libre de Bruxelles (ULB), Av. F.D. Roosevelt 50,
B-1050 Bruxelles, Belgium; e-mail: pberke@batir.ulb.ac.be

G. Zavarise & P. Wriggers (Eds.): Trends in Computational Contact Mechanics, LNACM 58, pp. 269–289.
springerlink.com

size-dependent [3, 57, 65], effect of residual stresses [61]), potential indenter tip deformation and misalignment [33, 48], the geometry of the contact (sample surface topography and indenter geometry [36, 37, 62]), and the effects of the contact interface behavior (adhesion, friction).

These surface effects are difficult to control experimentally. Generally only an estimation of the frictional behavior is given and its influence on the output data is unknown. The value of the coefficient of friction between two surfaces in micro and nano scale applications (potentially dependent on the actual contact area, on the relative tangential velocity and many other quantities) is generally measured by so-called scratch tests [38, 40]. Such tests however have the drawback of lacking a straightforward interpretation as the plastic behavior of the tested material is convoluted with the frictional effect, particularly when material pile-up (or plowing, depending on the sharpness of the tip) is present [7, 8, 11]. This leads to large dispersions in the value of this parameter ranging from 0.1 [16] to 1 [60] in numerical simulations.

The surface roughness of thin films can become comparable to the indenter penetration, especially when shallow indents are imposed in thin film characterization experiments in which the rule of thumb of making indents not deeper than one tenth of the thickness of the film is usually applied to avoid spurious effects of the substrate [19, 32]. In such configurations, surface effects related to the sample surface topography [58, 59] and friction [41] are the most pronounced, potentially resulting in variations of the evaluated material parameters, which may wrongly be attributed to the thin film mechanical behavior.

As a consequence nanoindentation results are not straightforward to interpret [4, 25]. This motivates numerical modeling efforts with the goal to investigate the influence of friction and its variation in indentation problems.

Friction is generally considered in numerical simulations of nanoindentation of *a perfectly smooth sample surface*, with the largest influence when using sharp indenter tips [51] in indentation depths comparable to, or larger than the radius of curvature of the indenter. This results in a variation of the local variables [2, 20, 42], as well as of the imprint geometry [12, 14, 41, 42, 56], and of the load-displacement curves, i.e., an increase in the force levels [9] and a change in the initial unloading segment of the load-displacement curves [54]. Some works conclude that the global indentation behavior is unaffected by friction [2, 20, 60], while other findings show that friction is a source of scatter in flat surface indentation [31, 41].

As a second surface effect, surface roughness is recognized to give a significant contribution to the indentation response, especially in small and moderate indentation depths, comparable to the height of the surface asperities. This indentation depth dependence is sometimes interpreted as an indentation size effect [29, 35, 66], based on the energy balance of indentation (distinguishing between the contributions of the deformation of the surface asperities and of the bulk, without taking frictional energy dissipation into account). In numerical approaches, surface roughness is generally incorporated in indentation problems with *frictionless* models, due to the already demanding computational effort of simulating multiple frictionless contacts and their (elastic-plastic) interaction [59]. The surface topography is described in two [5,58] or three-dimensions [13,59] with fractal-based and polynomial modeling assumptions [10,55]. The reader is referred to [49] for a more detailed review on rough surface contact mechanics. A qualitative agreement can be found with experimental trends [34, 50] concerning the increase of the dispersion in the results due to the presence of surface roughness with frictionless models [58].

Even though the observed dispersion is attributed to surface roughness effects only, friction is obviously naturally coupled with the effect of surface roughness in the experiments. However, numerical works analyzing both surface effects simultaneously in an indentation context (thus leading to a more realistic modeling) are rather scarce, to the best knowledge of the authors. Therefore a recent study of surface effects in nanoindentation of pure nickel was conducted, using a Coulomb friction model, with the intention to investigate how friction and surface roughness effects interact in a numerical model and to estimate their influence on the evaluated material parameters [9]. The use of pure nickel on micro and nano scales is motivated by significant advantages due to an expected better wear resistance of moving parts compared to silicon, which can potentially lead to biomedical applications (with coated nickel) on this scale.

The sample surface roughness was observed to have a twofold effect. It either increases or decreases the contact stiffness during indentation, depending on the position of the indent (on a roughness valley or on a peak, respectively). This affects the load levels necessary for reaching a given indentation depth and it also results in a variation of the evaluated elastic modulus.

The results showed a general but not uniform increase in the load levels and in the evaluated elastic modulus when adding Coulomb friction ($\mu = 0.5$) on the rough contact interface. The *topography-dependent nature of friction* is responsible for an increase in the dispersion of nanoindentation results with respect to the frictionless case, resulting in an additional dispersion giving an overall scatter in agreement with experimental measures in [50]. For valley configurations friction was found to play an important role in the load-displacement curves, while they were found to be much less important when indenting on a roughness peak. The fingerprint of friction could not be distinguished from other sources of energy dissipation (e.g. material plasticity) on nanoindentation results. This explains the higher sensitivity of work of indentation based post-treatment methods to surface effects (which is expected to be particularly high when they are used for plastic material parameters identification). These findings suggest that the coupled effect of surface roughness and friction can be partially responsible for the dispersion otherwise attributed to the tested material behavior, or other sources of nanoindentation scatter. As a general conclusion it can be shown that friction gives an important contribution to the physics of rough surface nanoindentation, therefore its significant interaction with surface roughness and its contribution to the energy balance of indentation should be accounted for in a thorough numerical analysis.

The motivation of the presented numerical work is to confirm these trends with a more complete description of the contact interface behavior, using a sliding velocity/slip rate dependent (SRD) friction approximation. Two post–treatment methods are applied to evaluate the variation of the resulting elastic modulus with respect to the reference indentation of a flat, frictionless surface, and with respect to the results obtained for the slip rate independent Coulomb friction model in [9]. The study is first revisiting the issue of nanoindentation on a smooth sample surface considering a SRD friction model in Section 3. Subsequently indentations on rough surfaces are considered in Section 4 (using an axisymmetric modeling assumption). A discussion in Section 5 focuses on the implications of the findings on the interpretation of nanoindentation data. Finally the conclusions are presented in Section 6.

2 Numerical Indentation Setups

2.1 Sliding Velocity Dependent Friction

Friction is a complex energy dissipating phenomenon, bridging different scales (from the atomic level up to the macroscale [45]), coupled to different physics at these scales, and having the common feature of hampering the relative sliding between contacting bodies. In the most general case the tangential contact behavior depends on the temperature, on the generation and presence of third bodies, on dynamic effects (vibrations, potentially induced by stick-slip), on the contact geometry, etc. [11]. At the nanoscale the frictional behavior between two materials forming a contacting surface pair can be quite different from the one witnessed on the macroscale [1] (being the average of all the potentially coupled contributions from smaller scales). This stems from the coupling to different dominant physical phenomena at different scales, and explains why the description of the frictional behavior using a phenomenological macroscopic coefficient of friction (being the ratio of the tangential load, F_{t}, to the applied normal load, F_{n}, in the contact) may become unsatisfactory.

Friction at small scales with relatively low contact pressures was shown to be highly dependent on contact adhesion, which comes into play due to the increased surface-to-volume ratio when the problem size decreases [1,39].

$$F_{\mathrm{n}}^{\mathrm{total}} = F_{\mathrm{n}}^{\mathrm{applied}} + F_{\mathrm{n}}^{\mathrm{adhesion}} \tag{1}$$

In this work the assumption is made that the applied indentation force reaches magnitudes much larger than the contact adhesion ($F_{\mathrm{n}}^{\mathrm{applied}} \gg F_{\mathrm{n}}^{\mathrm{adhesion}}$), and adhesion is not taken into account.

Experimental observations pointed out that at small scales, the friction for a single asperity elastic contact depends on the induced area of contact, A_{cont}, and a constant value of interface shear, τ (depending on the contacting material pair) [1,17,18,45]. This results in the following nanoscale friction law:

$$F_{\mathrm{t}} = \tau\, A_{\mathrm{cont}} \tag{2}$$

This friction model however induces convergence difficulties in computational approaches, related to the mapping between the global and the nodal friction behavior and to the interaction with a nonlinear material behavior.

On the other hand in macroscale contacts with high contact pressures, the state (related to the contact history), pressure (related to the normal load) and sliding velocity dependence of friction was established [28]. An increase in the slip rate or in the contact pressure were shown to result in the decrease of the coefficient of friction for very high slip velocities (up to hundreds of m/s) [6]. When considering lower slip rates (as low as some μm/s), first an increase of the apparent coefficient of friction (velocity-strengthening) was observed experimentally, followed by a decrease (velocity-weakening) with increasing slip rate [28,44]. The postulated phenomenological frictional relationship for this type of behavior takes the form

$$\mu = \mu_0 + a\, f(v_s) + b\, g(\theta) \tag{3}$$

with μ and μ_0 the nominal and the kinetic coefficients of friction, respectively. v_s is the sliding rate, a and b are constitutive parameters that measure the effects of sliding rate and of the material state respectively [28] (θ depends on time and on the cumulated displacement).

Experimental proof of such behavior in conditions comparable to nanoindentation (contact pressure in the order of several GPa, sliding velocity of some nm/s on the tip surface induced by the plastic deformation of the sample material) is apparently scarce. This may be partially due to the particular contact conditions that are difficult to reproduce in other setups on this scale, such as nanoscratching. Indeed in nanoscratching the plowing of the sample material hinders the straightforward interpretation of the resulting coefficient of friction [11, 38, 40] and its variation, as material plasticity and friction give the same effect on the nanoscratching load-displacement curves, much like in nanoindentation. In numerical simulations however, the different contributions can be easily singled out and investigated separately (e.g. the effect of surface roughness alone can be studied by deactivating friction in the model), which motivates the following numerical effort to evaluate the potential influence of SRD friction on nanoindentation results.

In a previous work a relative sliding velocity independent Coulomb friction model was used to show the important influence and interaction of surface roughness and frictional effects in shallow spherical nanoindentation [9]. This allowed to compare the obtained results to other works (in nanoindentation simulations the Coulomb friction model is used quasi exclusively [2, 12, 14, 16, 20, 22, 31, 41, 42, 51, 56, 60]). However, considering the complex physics which governs friction, the use of a single, constant coefficient of friction convoluting all friction related effects, may result in an exaggerated simplification for the representation of tip friction and potential stick-slip phenomenon in nanoindentation. Therefore as a first extension of the previous model [9], a slip rate dependent behavior of friction is included here, distinguishing between static and kinetic friction and including a simple bilinear evolution law of the friction coefficient as a function of the sliding velocity.

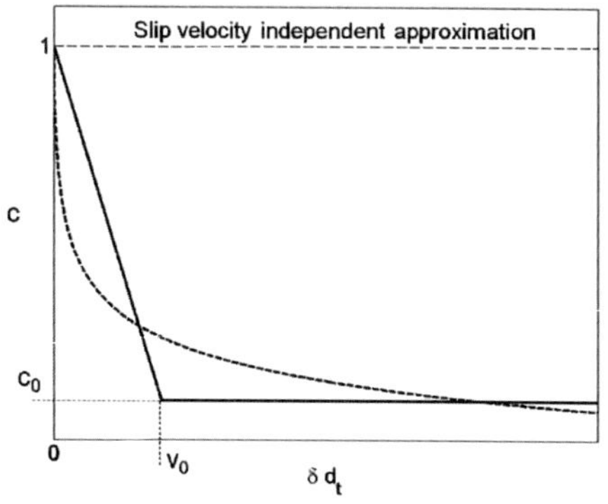

Fig. 1 Schematic evolution of the coefficient c multiplying the static coefficient of friction, as a function of the slip rate, δd_t (thick dashed line) and its simplified bilinear approximation used in the computations (thick solid line). v_0 and c_0 stand for the cutoff slip rate of the velocity-weakening portion of the bilinear law and for the constant value for sliding velocities larger than v_0, respectively.

The magnitude of sliding (or kinetic) friction is generally lower than static friction in a contact [11, 44, 45]. The transition between these two states, with $\mu_{slip} < \mu_{static}$ and the evolution of the kinetic friction as a function of slip rate $\mu(\delta d_t)$, define the tangential behavior of the contact. In the following simulations, a single coefficient of friction, equal to the static value, and its evolution as a function of the slip rate are defined.

$$\mu(\delta d_t) = c(\delta d_t) \times \mu_{static} \quad (4)$$

In order to reduce possible convergence problems due to the decreasing coefficient of friction with increasing sliding velocity in a quasi-static computation, a simplified two parameter bilinear approximation was used in the computations, composed of only one linear segment showing a velocity-weakening behavior and one with a constant, non-zero segment (Figure 1). Hence the evolution of the coefficient of friction is limited to the velocity-weakening portion of the SRD friction law.

2.2 Geometrical and Material Parameters

The indentation setup is the one studied in [9], to allow a straightforward comparison with the previously published results. In nanoindentation the applied force is specified as a function of time, generally three parts of the loading sequence are distinguished: the loading period where the applied force is increased until a peak value, the holding period where for a prescribed amount of time this peak load is maintained, and finally the unloading period where the applied force is decreased gradually to zero. A cube corner nanoindentation of pure nickel in 45 nm indentation depth is considered here, with a loading sequence of 5 s-5 s-5 s loading, holding and unloading time, respectively.

Pure polycrystalline bulk nickel is modeled as an elastic-viscoplastic material, as in [15], with the following material parameter set: the Young's modulus $E = 207$ GPa, the Poisson's ratio $\nu = 0.31$, the yield strength $\sigma_0 = 59$ MPa, the viscosity exponent $n_{vp} = 65$ and the viscosity parameter $\sigma_c = 256$ MPa $s^{1/n_{vp}}$, using an isotropic linear hardening approximation with a hardening coefficient $K = 2230$ MPa.

The rate-dependent plastic behavior of the material did not result in the rate sensitivity of the studied surface effects for a sliding velocity independent Coulomb law (verified for different loading rates corresponding to loading times between 2 and 20 s). This is due to the saturation of the viscous effects as a consequence of the high strain rates achieved during nanoindentation [15] (i.e. a further increase in the loading rate does not result in an increase of the bulk material stresses). Therefore the unique source of potential rate sensitivity of the results here is the sliding velocity dependent friction law. Potential material size effects in shallow indentation depths are not considered in the model, because the focus is set on surface effects.

The spherical part of the cube corner diamond indenter tip is modeled as a rigid spherical body of 100 nm radius, since both the elastic modulus and the yield limit of diamond largely exceed the ones of pure nickel. The numerical work is conducted using the general purpose finite element code SAMCEF [52], taking into account the material and geometrical nonlinearities due to local finite deformation and contact evolution. The frictional contact problem is solved using a Lagrange multipliers approach. For each studied case, different 2D finite element meshes of 8 noded quadratic quadrilateral elements with an axisymmetric modeling assumption were used with up to 160 contact nodes in the estimated contact area. The size of the smallest finite elements is comparable to the lattice parameters of the material, such a mesh refinement is necessary to catch accurately the contact conditions (no atomic scale related features are included). The geometrical size of the mesh in all cases is chosen sufficiently large such that a homogeneous stress distribution is obtained at the constrained boundaries of the model.

The bottom nodes of the mesh are prescribed a horizontal planarity condition and the side nodes of the model are blocked in the radial direction. The indenter has a fixed position in space and a vertical distributed load is applied on the bottom side of the deformable body (adjusted iteratively in each case) such that a 45 nm indenter penetration is reached.

2.3 Post-Treatment of the Data

The computed load-displacement curves were analyzed using two post-treatment methods applied in the experiments to assess the dispersion and the relative variation of the evaluated elastic material properties as a consequence of roughness and

SRD friction. The obtained results were compared to the reference indentation of a frictionless flat surface and to the ones obtained for the slip rate independent Coulomb friction model in [9].

The most frequently used post-treatment method of nanoindentation is the one proposed by Oliver and Pharr [47]. It is based on the assumption of the purely elastic unloading of a frictionless indenter-sample contact, and uses the unloading segment of the load-displacement curve to compute the contact stiffness for further processing. It is therefore sensitive to surface roughness. Since scanning the indent profile to measure the amount of pile-up or sink-in is rarely performed after nanoindentation experiments, to remain consistent with the standard assumptions of the experimental procedure, no particular correction is made to the numerical results to take the computed pile-up or sink-in into account.

The second post-treatment method is based on the (total and elastic) work supplied during indentation [46], using the complete load-displacement curve to obtain the elastic modulus of the sample. Such type of methodology is sometimes used to evaluate more advanced material properties as well, such as plastic flow data [22, 43, 63]. Therefore this method is more sensitive to changes in any portion of the load-displacement curve and to the variations in the load levels affecting directly the integrated work quantities.

In the following, the evaluated elastic modulus obtained for the numerical indentation of a flat, frictionless surface, satisfying the assumptions of the considered post-treatment methods $E_{\text{ref}}^{\text{Ol}-\text{Ph}} = 230.6$ GPa and $E_{\text{ref}}^{\text{Ni}} = 251.8$ GPa are considered as reference values for the post-treatment method of Oliver and Pharr and Ni et al., respectively.

3 Slip Rate Dependent Friction in Flat Surface Indentation

The nanoindentation on a perfectly flat surface with a spherical indenter of 100 nm tip radius in pure nickel is first considered with a special focus on the effects of slip rate dependent friction on the results.

The initial parameters of the SRD friction law are $\mu_{\text{static}} = 0.5$, $v_0 = 2$ nm/s and $c_0 = 0.4$ (i.e. for sliding velocities larger than 2 nm/s the apparent coefficient of friction is $0.4 \times \mu_{\text{static}} = 0.2$).

The most obvious effect of friction is an increase in the load level during the loading phase, with respect to the frictionless numerical indentation. This is however only triggered after reaching some value of the indenter penetration (around 20 nm), from which on this frictional effect exhibits a monotonic increase (which is probably related to the onset of contact slip).

Taking the slip rate dependency of the frictional behavior into account enhances the importance of the stick-slip phenomenon and results in abrupt changes in the tangent of the loading segment. These variations are first resulting in an increase of the force levels followed by a displacement burst at quasi constant force level (around 45 μN in Figure 2). Such features in the load-displacement curves are similar to

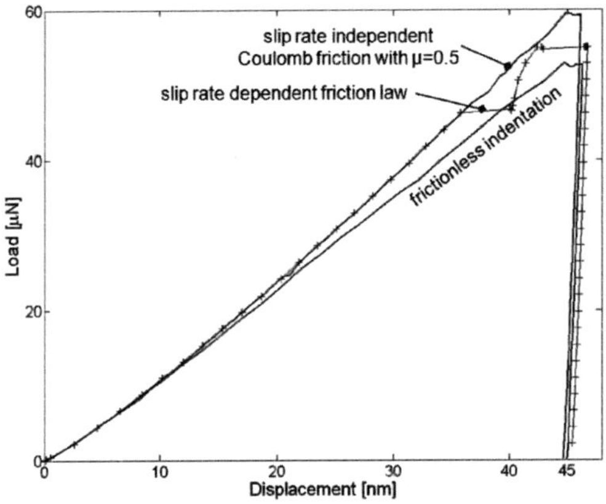

Fig. 2 Load-displacement curves of flat surface nanoindentation. The solid line with the plus marks shows the indentation response obtained for the SRD bilinear friction law.

the manifestation of pop-ins, which are associated with the instantaneous onset of material plastic yielding in the literature [27,64]. The studied surface effects are not identified as the dominant source of pop-ins, however they may constitute a potential contribution to experimentally observed nanoindentation displacement bursts. Considering the complexity of the experimental procedure (with the necessity of controlling and modifying the contact behavior) and the difficulty of decoupling different contributions such an interpretation based on experimental data was not given previously to the best knowledge of the authors.

Even though no discrete plastic events are represented in the proposed model pop-in-like displacement bursts appear on the loading curve. Here, they are caused by the build up and the relaxation of the surface traction when the contact state changes between stick and slip. Even though stick-slip is clearly present in simulations using a slip rate independent Coulomb friction law as well, its influence is more limited, which smoothens the potential displacement fluctuations on the load-displacement curves related to this transition, as opposed to SRD simulations. In this interpretation the velocity-weakening frictional behavior results in the rapid spread of the slip in the whole contact area, a general break down of the stick state at a given force level could explain the observed fluctuation. Increasing slip rate decreases the coefficient of friction (related to the slip resistance of the contact), which on one hand *enhances slip*, on the other hand results in a contact force *redistribution* leading to higher loads on the sticking nodes potentially reaching their slip limits. This may result in an autogenerated slip avalanche in an extended part of the total contact area (corresponding to the displacement bursts of the load-displacement curves) until a new global equilibrium state is reached. This seems to be supported by the fact that the SRD loading curve "oscillates" between the loading curves obtained for simulations for $\mu = 0.5$ and $\mu = 0$ with the slip rate independent Coulomb friction model.

Fine tuning the friction law parameters could potentially better reproduce the experimental observations of pop-ins, however, since other physical phenomena, such as material plasticity and its complex interaction with the frictional behavior may intervene as well, this was not attempted here.

3.1 Friction Induced Variation in the Evaluated Elastic Modulus

Although keeping the same input material parameters, friction induces variations in the elastic material properties obtained by post-treatment methods. Such variations can be defined in terms of the elastic modulus evaluated for indentations with friction and for the frictionless reference case, by

$$\gamma_{\text{fric}}^{\text{flat}} = \frac{E_{\text{fric}}^{\text{flat}}}{E_{\text{ref}}} - 1 \tag{5}$$

The work of indentation based post-treatment method performs similarly to the post-treatment method of Oliver and Pharr for the SRD friction law, showing a considerable overestimation of the reference modulus of around 15 and 16%, respectively.

The computed variation matches the value obtained for the slip rate independent friction law for the energy-based post-treatment method, showing a similarly high sensitivity to SRD friction. On the other hand, the post-treatment method of Oliver and Pharr is much more sensitive to SRD friction, compared to the results obtained for the slip rate independent friction law resulting in an overestimation of merely 9% (i.e. the variation is almost doubled).

These results confirm that friction indeed affects significantly the results of nanoindentation of pure nickel in the considered configuration with a perfectly flat sample surface, for both slip rate dependent and slip rate independent friction approximations.

3.2 Influence of the Loading Rate

The influence of friction may vary as a function of the loading rate when a rate-dependent frictional behavior is considered. In order to verify to which extent frictional effects vary, the loading time was changed to experimentally used extreme values of 2 and 20 s (keeping the same holding and unloading times of 5 s) as in [15].

For the lowest loading time the significant difference with respect to the 5 s loading results is the earlier appearance of pop-in phenomenon (at 40 μN) and the presence of an additional displacement burst in the loading phase (at around 50 μN). The length of the displacement burst increases with the indentation force (observed in a general manner in all cases with multiple displacement bursts in the loading

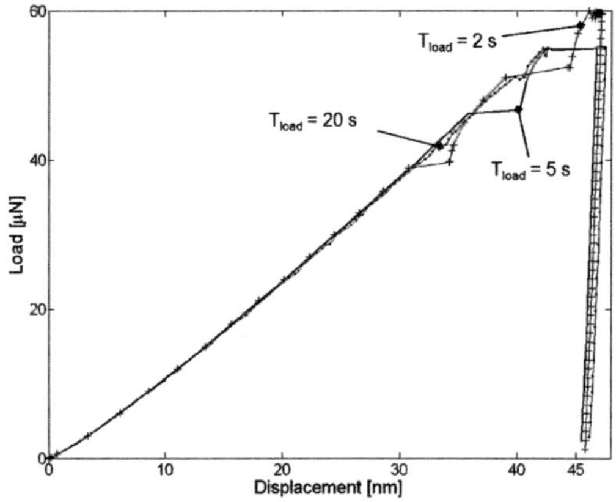

Fig. 3 Load-displacement curves of flat surface nanoindentation computed for different values of the loading time: the thick solid curve, the solid curve with plus marks and the solid curve with dot marks correspond to loading times of 5, 2 and 20 s, respectively.

phase, as later in Section 4). This seems logical considering that these fluctuations are supposed to be related to stick-slip, and at lower contact loads the surface tractions relaxed by the instantaneous slip are also of lower magnitude. This seems to show that when the loading rate is doubled stick-slip is enhanced.

The lowest loading rate results in a different, smoother indentation response, with an allure/shape of the load-displacement curve comparable to the one obtained for the slip rate independent friction assumption. This is due to the induced sliding velocity range which is much smaller during loading than in the previous cases.

Therefore it can be concluded that friction-based pop-in is mainly triggered for high slip rates. For low slip rate values, stick-slip remains a phenomenon without a pronounced influence on the nanoindentation load-displacement curve. This suggests to use the largest possible loading times in nanoindentation (however limited by the thermal drift) in order to reduce the additional dispersion due to a potential global stick-slip behavior of the contact.

3.3 Influence of the Friction Law Parameters

The friction law parameters v_0 and c_0 are varied separately in this section to study their influence on nanoindentation results.

First, results obtained by varying v_0 between 0.5 and 10 nm/s are presented, i.e. changing the slip rate at which the velocity-weakening portion of the friction law ends. Decreasing v_0 results in a more abrupt change of the coefficient of friction between the stick and the slip state, while choosing a high value results in a gradual variation extended to the range of high slip velocities.

For both considered values of v_0 the computation matches well the results obtained for the SRD friction assumption with the initial parameter set. The main difference is the onset of the displacement burst at slightly different force levels for different v_0 values. This can be understood when considering that the computations operate mainly in the velocity-weakening portion of the friction law (slip rates remain comparable to, or smaller than v_0), which explains the similar indentation response. For extremely small values of v_0 (which are not physically sound anymore), the load-displacement curve becomes smoother, comparable to the slip rate independent results. This suggests that it is indeed the velocity-weakening portion of the friction law that is responsible for the displacement burst, since for very small values of v_0 the interfacial slip rates become higher than the stabilized slip threshold and the computation operates mainly in the constant portion of the friction law.

Secondly the impact of changing c_0 to 0.8 was studied, related to the constant coefficient of friction at slip velocities larger than v_0. When increasing the value of c_0, the magnitude of velocity-weakening is decreased, i.e. the stable, high velocity coefficient of friction for slip rates larger than v_0 is nearer to the static value.

The result is a load-displacement curve with a much shorter displacement burst at a force level similar to the one obtained for the initial parameter set. Increasing c_0 thus decreases the extent of the relaxation of the contact tractions by stick-slip. Even though the resulting loading curve becomes smoother, the displacement burst is not completely canceled by decreasing the magnitude of the velocity-weakening. The c_0 friction parameter plays a more important role in this nanoindentation setup than the previously considered v_0, therefore c_0 could be considered as a dominant parameter of the bilinear SRD law, together with μ_{static}.

4 Slip Rate Dependent Friction in Rough Surface Nanoindentation

The flat nanoindentation model of Section 3 is updated here, adding a realistic roughness model on the contact interface. The goal here is to investigate how the convolution of SRD friction with surface roughness influences the global surface effects, potentially affecting the dispersion in nanoindentation results. Nanoindentation load-displacement curves were shown to be highly sensitive to the position of the indent, due to the variation of the contact stiffness related to surface roughness but also to an additional variation stemming from the topography dependence of friction [9]. In practice, the surface roughness of thin films can reach average values of 30–40 nm [24, 26, 34] which becomes comparable to the imposed indentation depth, limited by the film thickness. The nanoindentation setup in this section potentially falls in this category.

The roughness of a real surface has a multilevel nature. A roughness profile can be considered as the convolution of single profiles with various wavelengths and different amplitude to wavelength ratios. The surface roughness is chosen to have a relatively simple representation here, considering only four levels of a protuberance-

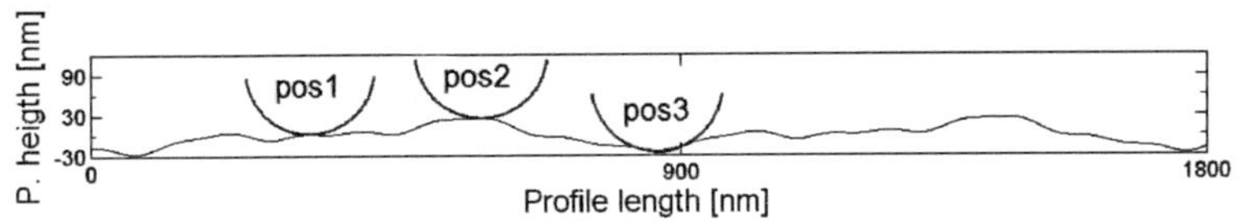

Fig. 4 Representation of the surface roughness, inspired from experimental data, used in the rough surface nanoindentation model together with the three considered indentation positions.

on-protuberance type roughness description, similar to [9]. Even though it is of great interest, a convergence study of the number of sine functions to consider for the representation of the indentation response of an experimental rough surface remains out of reach from a computational point of view. Furthermore, the assumption is made that the shape of a single roughness profile i can be well approximated by a sine function

$$y(x) = \sum_{\mathrm{i}} \frac{A_{\mathrm{i}}}{2} \sin\left(\frac{2\pi}{\lambda_{\mathrm{i}}} x + \theta_{\mathrm{i}}\right) \tag{6}$$

with A_{i} the peak to peak amplitude, λ_{i} the wavelength and θ_{i} the phase shift of the profile. The surface profile depicted in Figure 4 is described by the sum of four sine functions with different amplitudes, wavelengths and phase shifts (increasing amplitudes are associated to increasing wavelengths). Three different indentation positions are considered with an indentation depth of 45 nm after initial contact. Two main families of rough surface contact models can be distinguished in nanoindentation; the one considering the contact between an indenter tip much larger than the characteristic size of the surface asperities, resulting in a multi asperity model [59]; and single asperity models working on the scale of the surface asperities. This particular indentation scenario is part of the latter family, considering the sharpness of the indenter and the shallow indentation depth, both comparable to the size of single surface asperities.

4.1 Influence of the Frictionless Surface Roughness

Surface roughness can have a twofold effect resulting in either a higher or a lower contact stiffness, whether the indentation is performed in a roughness valley or on the tip of an asperity, respectively. As expected, the highest load level was obtained when indenting in the deepest roughness valley (position 3). Conversely, the indentation on the highest roughness peak shows the most deformable response (position 2), as depicted in Figure 5.

The variation in the evaluated elastic modulus due to the presence of surface roughness separately (frictionless indentations with the same material parameters) can be defined by

$$\gamma_{\mathrm{rough}}^{\mu=0} = \frac{E_{\mathrm{rough}}^{\mu=0}}{E_{\mathrm{ref}}} - 1 \tag{7}$$

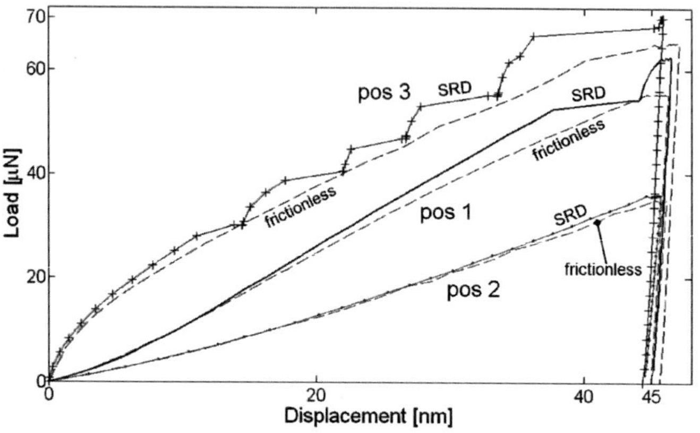

Fig. 5 Load-displacement curves for the three indentation positions of the rough surface for frictionless indentations (thin dashed lines) and considering the bilinear slip rate dependent friction law.

The variation of $\gamma^{\mu=0}_{\text{rough}}$ as a function of the indent position is depicted in Figure 6. The indentation on the highest peak (position 2) shows the least stiff indentation response, resulting in an underestimation of the evaluated elastic modulus of around 20 and 10% for the post-treatment method of Oliver and Pharr and of Ni et al., respectively. Conversely, the indentation made in the deepest valley with a shape which conforms well with the shape of the indenter tip (position 3) experiences the highest contact stiffness, resulting in an overestimation of around 10% for the Oliver and Pharr post-treatment method. The post-treatment methods of Ni et al. evaluates the elastic modulus in this valley configuration roughly 10% lower than the reference value, however the interpretation of this result is not straightforward, since it depends on other parameters related to the complete load-displacement curve, apart from the contact stiffness. Position 1 is an intermediate case (indentation of a low peak) between these two extreme configurations, with a corresponding evaluated elastic modulus between the ones obtained for positions 2 and 3 for the contact stiffness based post-treatment method, and an elastic modulus near the reference value for the energy-based post-treatment method. The post-treatment method of Ni et al. performs better with the considered frictionless indentations, since it shows a lower dispersion due to surface topography effects alone (Figure 6).

4.2 Coupled Influence of Friction and Surface Roughness

The previously discussed effects of friction on the indentation of a flat surface (Section 3) were observed for rough surfaces as well, i.e. an increase in the load level with respect to the frictionless case coupled to displacement bursts when using the

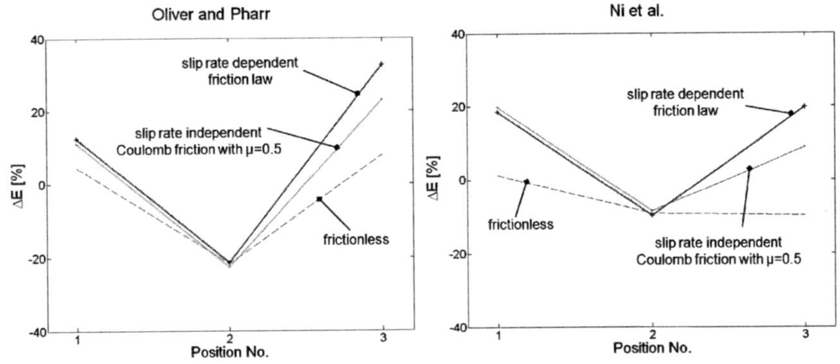

Fig. 6 Variation of the evaluated elastic modulus as a function of the indentation position for frictionless nanoindentation (dashed lines), for the slip rate independent (thin solid line with dot marks) and SRD friction approximations (thick solid line with plus marks). Results are presented for the post-treatment method of Oliver and Pharr (left) and of Ni et al. (right).

SRD friction law (Figure 5). The large influence of the surface roughness (inducing different contact angles at different indentation positions) was observed on the stick-slip activity in the load-displacement curves. When indenting the highest peak (position 2) no displacement burst appears during the loading phase, and the overall effect of friction is the smallest. Conversely, when indenting in the deepest valley (position 3) up to 5 displacement bursts (more than generally observed in experiments), increasing in length with increasing load were observed. The indentation response in position 1 is similar to the one obtained for the flat surface indentation scenario with one pop-in only.

When adding friction on the rough contact interface, the dispersion in nanoindentation results shows a general increasing trend for both friction approximations. This is due to the topography dependency of the frictional effects, illustrated in Figure 6 by the different magnitude of the variation of the evaluated elastic modulus with respect to the frictionless case, as a function of the indentation positions. This variation is defined in terms of the evaluated elastic modulus and its reference value by

$$\gamma_c = \frac{E_{\text{rough}}^{\text{fric}}}{E_{\text{ref}}} - 1 \tag{8}$$

and it shows the coupled effect of surface roughness and friction on the contact interface. The non-homogeneous and topography-dependent variation caused by friction results in a general increase in the average value of the evaluated elastic modulus for both post-treatment methods. As observed earlier, the energy-based post-treatment method shows a higher sensitivity to frictional effects, since the relative increase of the evaluated elastic modulus due to frictional effects on a rough surface is higher (reaching 33%) than for the results obtained by the Oliver and Pharr method (up to roughly 23% increase due to frictional effects only). Even though adding the slip rate dependent behavior can have a significant influence on the load-displacement curves due to the potentially induced displacement bursts, it affects to a lesser extent the evaluated elastic modulus, with respect to the values obtained for the slip rate independent friction approximation (up to roughly 10% increase is observed).

These results confirm that frictional effects significantly increase the dispersion in nanoindentation results for rough surface indentations as well. It is emphasized that the obtained results remain of qualitative nature since a number of dimensional quantities were kept fixed (material and frictional parameters, indentation depth, tip geometry).

5 Discussion

The main reason why a continuum finite element model was chosen is the ability and simplicity of decoupling the contribution of surface effects (potentially depending on contaminant and/or oxide layers) from the behavior of the bulk material, compared to a "standard" atomistic approach which could also be a viable choice on this scale [30, 53]. This obviously implies that the continuum model is incapable of reproducing pop-ins related to dislocation activity, but it allows investigating different contributions with higher computational efficiency, in a straightforward manner. No comparison was attempted between our results and atomistic models, since this is rarely possible directly [30], however keeping in mind that such low scale models give precious insight to experimentally observed complex phenomena.

When the indentation of rough surfaces is considered using an axisymmetric model, the roughness profile is described by concentric circular rings leading to a stiffer response than the real three-dimensional surface with randomly distributed surface asperities [59] (especially when indenting in the middle of a roughness ring). This affects the quantitative conclusions that can be drawn about the influence of friction. Therefore the overall dispersion in nanoindentation results cannot be estimated accurately from this set of simulations and the validity of quantitative results may be limited to this particular indentation scenario.

Other possible boundary conditions on the sides of the model were tested for the slip rate independent Coulomb friction law; by enforcing the condition of a straight vertical boundary leaving the radial displacement free and with a completely free boundary. The boundary conditions of the fine scale single asperity model could also be adjusted based on an estimation issued from computations with significantly larger geometrical size (but a coarser mesh in the contact zone), as proposed in [59]. The relative variation of the evaluated elastic modulus due to the considered surface effects remained almost identical for all boundary conditions, as shown earlier [9]. Other boundary conditions were therefore not considered here, keeping in mind however, that they can play a significant role in a direct comparison with experimental results.

The use of a general velocity-weakening frictional behavior was experienced to cause convergence problems in quasi-static simulations (hence the bilinear approximation with a cutoff using a constant coefficient of friction). A remedy to such convergence issues could be to use displacement controlled simulations (however the holding portion of the load-displacement curves could not be reproduced this way), line search techniques (driving the computation may become complex), or dynamic nonlinear computations (even though dynamic instabilities may occur [45]). The chosen time step between increments can also play a role in the solution procedure. Here relatively small time steps (down to 5 ms) were imposed to ensure a good approximation of the contact slip rate during loading, in order to capture the details of the stick-slip phenomenon. Therefore solving the presented nanoindentation problem numerically requires some fine tuning of the parameters that drive the computation and the potential necessity of small time steps results in high computational costs in all cases. Finding the right parameter set could allow computing a more realistic sliding velocity dependent behavior than the bilinear law considered here.

6 Concluding Remarks

The results show a strong interaction between roughness and frictional surface effects and allow to draw the following salient conclusions. The cumulative effect of friction and surface roughness on the dispersion in the raw nanoindentation results and in the evaluated elastic modulus was confirmed in this set of simulations for both slip rate dependent and slip rate independent friction approximations. Considering friction in a numerical model with roughness increases the scatter because of the high sensitivity of the frictional effects on the contact geometry, resulting in an additional variability of the results.

Slip rate dependent friction was observed to induce a stick-slip phenomenon with an influence that spreads on the scale of observation of nanoindentation, i.e. the load-displacement curves, resulting in an increase of the dispersion in the results with respect to slip rate independent simulations. This effect is potentially related to the velocity-weakening behavior of the postulated friction law. The friction-induced displacement bursts of the loading curve are similar to the experimentally observed pop-in phenomenon. Based on the presented numerical results, in order to decrease their impact in flat nanoindentation the use of the lowest possible loading rates is proposed. The c_0 friction parameter of the SRD friction law, related to the constant coefficient of friction at high slip velocities, was found to play the most important role in the considered nanoindentation setup (apart from μ_{static} of course).

Slip rate dependent behavior increased the evaluated elastic modulus by up to 10% with respect to the values obtained for the slip rate independent friction approximation for rough surface nanoindentation. The post-treatment method of Ni et al. was observed to perform better with frictionless rough surface indentations (since it resulted in a lower dispersion), on the other hand it is more sensitive to friction than the contact stiffness based post-treatment method. The post-treatment method of Oliver and Pharr was observed to show a much larger overestimation due to SRD friction, compared to the results obtained for the slip rate independent friction law for flat surface indentation, but this trend disappeared for rough surface indentation configurations.

Results confirm that neglecting frictional and surface roughness effects may be a debatable assumption when the indentation depth becomes comparable to the size of the surface roughness, which is an unfavorable but inevitable characteristic of the measurement of thin films material properties. If surface effects are not considered, the corresponding large dispersion and the increase in the evaluated average elastic modulus could be interpreted in terms of variations of the material behavior, or of other potential sources of nanoindentation scatter.

Taking SRD friction into account in the numerical model leads to computational difficulties (potential convergence issues and high computational cost). It was observed to play a role mainly when the stick-slip behavior of the contact is in the focus of the study, the related increase in the dispersion of nanoindentation results (especially for rough surface indentation) seems less important.

Further developments could obviously be performed using a more suitable friction law for the considered nano and micro scale (e.g. including a velocity-strengthening portion and finding a smoother description of the variation of the kinetic coefficient of friction as a function of the slip rate). A richer, more thorough description of the surface roughness using other modeling assumptions could also be investigated in a future work.

Acknowledgements The first author was sponsored by the Fonds de la Recherche Scientifique F.R.S.-FNRS of Belgium (post-doctoral research grant "Chargé de Recherches" No. 1.2.093.10.F). The authors also acknowledge the support of F.R.S.-FNRS Belgium (Grant No. 1.5.032.09.F) for the intensive computational facilities used for this work.

References

1. Achanta, S., Liskiewicz, T., Drees, D., Celis, J.-P.: Friction mechanisms at the microscale. Tribol. Int. 42, 1792–1799 (2009)
2. Antunes, J.M., Menezes, L.F., Fernandes, J.V.: Three-dimensional numerical simulation of Vickers indentation tests. Int. J. Solid Struct. 43, 784–806 (2006)
3. Abu Al-Rub, R.K.: Prediction of micro and nanoindentation size effect from conical and pyramidal indentation. Mech. Mater. 39, 787–802 (2007)
4. Baker, S.P.: Between nanoindentation and scanning force microscopy: measuring mechanical properties in the nanometer regime. Thin Solid Films 308-309, 289–296 (1997)
5. Bobji, M.S., Biswas, S.K.: Hardness of a surface containing uniformly spaced pyramidal asperities. Tribol. Lett. 7, 51–56 (1999)
6. Ben-Dor, G., Dubinsky, A., Elperin, T.: Localized interaction models with non-constant friction for rigid penetrating impactors. Int. J. Solid Struct. 44, 2593–2607 (2007)
7. Bellemare, S., Dao, M., Suresh, S.: The frictional sliding response of elasto-plastic materials in contact with a conical indenter. Int. J. Solid Struct. 44, 1970–1989 (2007)
8. Bellemare, S., Dao, M., Suresh, S.: Effects of mechanical properties and surface friction on elasto-plastic sliding contact. Mech. Mater. 40, 206–219 (2008)

9. Berke, P., El Houdaigui, F., Massart, T.J.: Coupled friction and roughness surface effects in shallow spherical nanoindentation. Wear 268, 223–232 (2010)
10. Bora, C.K., Flater, E.E., Street, M.D., Redmond, J.M., Starr, M.J., Carpick, R.W., Plesha, M.E.: Multiscale roughness and modeling of MEMS interfaces. Tribol. Lett. 19, 37–48 (2005)
11. Blau, P.J.: The significance and use of the friction coefficient. Tribol. Int. 34, 585–591 (2001)
12. Bolzon, G., Maier, G., Panico, M.: Material model calibration by indentation, imprint mapping and inverse analysis. Int. J. Solid Struct. 41, 2957–2975 (2004)
13. Bobji, M.S., Shivakumar, K., Alehossein, H., Venkateshwarlu, V., Biswas, S.K.: Influence of surface roughness on the scatter in hardness measurements – a numerical study. Int. J. Rock Mech. Min. 36, 399–404 (1999)
14. Bucaille, J.L., Stauss, S., Felder, E., Michler, J.: Determination of plastic properties of metals by instrumented indentation using different sharp indenters. Acta Mater. 51, 1663–1678 (2003)
15. Berke, P., Tam, E., Delplancke-Ogletree, M.-P., Massart, T.J.: Study of the rate-dependent behavior of pure nickel in conical nanoindentation through numerical simulation coupled to experiments. Mech. Mater. 41, 154–164 (2009)
16. Bressan, J.D., Tramontin, A., Rosa, C.: Modeling of nanoindentation of bulk and thin film by finite element method. Wear 258, 115–122 (2005)
17. Carpick, R.W., Agraït, N., Ogletree, D.F., Salmeron, M.: Measurement of interfacial shear (friction) with an ultrahigh vacuum atomic force microscope. J. Vac. Sci. Technol. B 14, 1289–1295 (1996)
18. Carpick, R.W., Agraït, N., Ogletree, D.F., Salmeron, M.: Variation of the interfacial shear strength and adhesion of a nanometer-sized contact. Langmuir 12, 3334–3340 (1996)
19. Cai, X., Bangert, H.: Hardness measurements of thin films – Determining the critical ratio of depth to thickness using fem. Thin Solid Films 264, 59–71 (1995)
20. Carlsson, S., Biwa, S., Larsson, P.-L.: On friction effects at inelastic contact between spherical bodies. Int. J. Mech. Sci. 42, 107–128 (2000)
21. Cheng, Y.-T., Cheng, C.-M.: Scaling, dimensional analysis, and indentation measurements. Mater. Sci. Eng. R 44, 91–149 (2004)
22. Cao, Y.P., Lu, J.: A new method to extract the plastic properties of metal materials from an instrumented spherical indentation loading curve. Acta Mater. 52, 4023–4032 (2004)
23. Chudoba, T., Richter, F.: Investigation of creep behaviour under load during indentation experiments and its influence on hardness and modulus results. Surf. Coat. Tech. 148, 191–198 (2001)
24. de Souza, G.B., Foerster, C.E., da Silva, S.L.R., Lepienski, C.M.: Nanomechanical properties of rough surfaces. J. Mater. Res. 9, 159–163 (2006)
25. Fischer-Cripps, A.C.: Critical review of analysis and interpretation of nanoindentation test data. Surf. Coat. Tech. 200, 4153–4165 (2006)
26. Fang, T.-H., Chang, W.-J., Lin, C.-M.: Nanoindentation characterization of ZnO thin films. Mater. Sci. Eng. A, 452–453:715–720 (2007)
27. Fujikane, M., Leszcznski, M., Nagao, S., Nakayama, T., Yamanaka, S., Niihara, K., Nowak, R.: Elastic-plastic transition during nanoindentation in bulk GaN crystal. J. Alloy. Compd. 450, 5–411 (2008)
28. Fortt, A.L., Schulson, E.M.: Velocity-dependent friction on Coulombic shear faults in ice. Acta Mater. 57, 4382–4390 (2009)

29. Gao, Y.X., Fan, H.: A micro-mechanism based analysis for size-dependent indentation hardness. J. Mater. Sci. 37, 4493–4498 (2002)
30. Gouldstone, A., Chollacoop, N., Dao, M., Li, J., Minor, A.M., Shen, Y.-L.: Indentation across size scales and disciplines: Recent developments in experimentation and modeling. Acta Mater. 55, 4015–4039 (2007)
31. Habbab, H., Mellor, B.G., Syngellakis, S.: Post-yield characterisation of metals with significant pile-up through spherical indentations. Acta Mater. 54, 1965–1973 (2006)
32. Hainsworth, S.V., Soh, W.C.: The effect of the substrate on the mechanical properties of TiN coatings. Surf. Coat. Tech. 163/164, 515–520 (2003)
33. Jeong, S.-M., Lee, H.-L.: Finite element analysis of the tip deformation effect on nanoindentation hardness. Thin Solid Films 492, 173–179 (2005)
34. Nanda Kumar, A.K., Kannan, M.D., Jayakumar, S., Rajam, K.S., Raju, V.S.: Investigations on the mechanical behaviour of rough surfaces of TiNi thin films by nanoindentation studies. Surf. Coat. Tech. 201, 3253–3259 (2006)
35. Kim, J.-Y., Kang, S.-K., Lee, J.-J., Jang, J., Lee, Y.-H., Kwon, D.: Influence of surface-roughness on indentation size effect. Acta Mater. 55, 3555–3562 (2007)
36. Kim, J.-Y., Lee, B.-W., Read, D.T., Kwon, D.: Influence of tip bluntness on the size-dependent nanoindentation hardness. Scripta Mater. 52, 353–358 (2005)
37. Lu, C.-J., Bogy, D.B.: The effect of tip radius on nano-indentation hardness tests. Int. J. Solid Struct. 32, 1759–1770 (1995)
38. Li, J., Beres, W.: Scratch test for coatings/substrate systems – A literature review. Can. Metall. Quart. 46, 155–174 (2007)
39. Lafaye, S., Gauthier, C., Schirrer, R.: A surface flow line model of a scratching tip: Apparent and true local friction coefficients. Tribol. Int. 38, 113–127 (2005)
40. Lafaye, S., Gauthier, C., Schirrer, R.: Analyzing friction and scratch tests without in situ observation. Wear 265, 664–673 (2008)
41. Mata, M., Alcalà, J.: The role of friction on sharp indentation. J. Mech. Phys. Solids 52, 145–165 (2004)
42. Mesarovic, S.D., Fleck, N.A.: Spherical indentation of elastic-plastic solids. Proc. Royal Soc. Lond. A-CONTA 455, 2707–2738 (1999)
43. Ma, D., Ong, C.W., Lu, J., He, J.: Methodology for the evaluation of yield strength and hardening behavior of metallic materials by indentation with spherical tip. J. Appl. Phys. 94, 288–294 (2003)
44. Martins, J.A.C., Oden, J.T., Simões, F.M.F.: A study of static and kinetic friction. Int. J. Eng. Sci. 28, 29–92 (1990)
45. Nosonovsky, M., Bhushan, B.: Multiscale friction mechanisms and hierarchical surfaces in nano- and bio-tribology. Mater. Sci. Eng. Rep. 58, 162–193 (2007)
46. Ni, W., Cheng, Y.-T., Cheng, C.-M., Grummon, D.S.: An energy-based method for analyzing instrumented spherical indentation experiments. J. Mater. Res. 19, 149–157 (2004)
47. Oliver, W.C., Pharr, G.M.: An improved technique for determining hardness and elastic modulus using load and displacement sensing indentation measurements. J. Mater. Res. 7, 1564–1583 (1992)
48. Pelletier, C.G.N., Dekkers, E.C.A., Govaert, L.E., den Toonder, J.M.J., Meijer, H.E.H.: The influence of indenter-surface misalignment on the results of instrumented indentation tests. Polym. Test. 26, 949–959 (2007)
49. Persson, B.N.J.: Contact mechanics for randomly rough surfaces. Surf. Sci. Rep. 61, 201–227 (2006)

50. Qasmi, M., Delobelle, P.: Influence of the average roughness r_{ms} on the precision of the Young's modulus and hardness determination using nanoindentation technique with a Berkovich indenter. Surf. Coat. Tech. 201, 1191–1199 (2006)
51. Qin, J., Huang, Y., Hwang, K.C., Song, J., Pharr, G.M.: The effect of indenter angle on the microindentation hardness. Acta Mater. 55, 6127–6132 (2007)
52. Samtech. Samcef V13.1, Samtech, Liège, Belgium, `http://www.samtech.com`
53. Szlufarska, I.: Atomistic simulations of nanoindentation, Mater. Mater. Today 9, 42–50 (2006)
54. Tsou, C., Hsu, C., Fang, W.: Interfaces friction effect of sliding contact on nanoindentation test. Sensor Actuator A 117, 309–316 (2005)
55. Tao, Q., Lee, H.P., Lim, S.P.: Contact mechanics of surfaces with various models of roughness descriptions. Wear 249, 539–545 (2001)
56. Taljat, B., Pharr, G.M.: Development of pile-up during spherical indentation of elastic-plastic solids. Int. J. Solid Struct. 41, 3891–3904 (2004)
57. Tho, K.K., Swaddiwudhipong, S., Hua, J., Liu, Z.S.: Numerical simulation of indentation with size effect. Mater. Sci. Eng. A 421, 268–275 (2006)
58. Walter, C., Antretter, T., Daniel, R., Mitterer, C.: Finite element simulation of the effect of surface roughness on nanoindentation of thin films with spherical indenters. Surf. Coat. Tech. 202, 1103–1107 (2007)
59. Walter, C., Mitterer, C.: 3D versus 2D finite element simulation of the effect of surface roughness on nanoindentation of hard coatings. Surf. Coat. Tech. 203, 3286–3290 (2009)
60. Wang, T.H., Fang, T.-H., Lin, Y.-C.: A numerical study of factors affecting the characterization of nanoindentation on silicon. Mater. Sci. Eng. A 447, 244–253 (2007)
61. Yu, N., Polycarpoua, A.A., Conry, T.F.: Tip-radius effect in finite element modeling of sub-50 nm shallow nanoindentation. Thin Solid Films 450, 295–303 (2004)
62. Warren, A.W., Guo, Y.B.: Machined surface properties determined by nanoindentation: Experimental and FEA studies on the effects of surface integrity and tip geometry. Surf. Coat. Tech. 201, 423–433 (2006)
63. Zhao, M., Ogasawara, N., Chiba, N., Chen, X.: A new approach to measure the elastic-plastic properties of bulk materials using spherical indentation. Acta Mater. 54, 23–32 (2006)
64. Zong, Z., Soboyejo, W.: Indentation size effects in face centered cubic single crystal films. Mater. Sci. Eng. A 404, 281–290 (2004)
65. Zhao, M., Slaughter, W.S., Li, M., Mao, S.X.: Material-length-scale-controlled nanoindentation size effects due to strain-gradient plasticity. Acta Mater. 51, 4461–4469 (2003)
66. Zhang, T.-Y., Xu, W.-H., Zhao, M.-H.: The role of plastic deformation on rough surfaces in the size-dependent hardness. Acta Mater. 52, 57–68 (2004)

Application of the Strain Rate Intensity Factor to Modeling Material Behavior in the Vicinity of Frictional Interfaces

Elena Lyamina and Sergei Alexandrov

Abstract This paper concerns with a novel approach to predicting the formation of a narrow layer of intensive plastic deformation in the vicinity of frictional interfaces. Theoretical solutions based on several conventional rigid plastic models are singular near maximum friction surfaces. In particular, the equivalent strain rate approaches infinity near such surfaces. This is in qualitative agreement with experimental observations that material properties in the vicinity of frictional interfaces are often quite different from the properties in the bulk. The new theory relates the strain rate intensity factor, which is the coefficient of the main singular term in a series expansion of the equivalent strain rate in the vicinity of maximum friction surfaces, and the thickness of the layer of intensive plastic deformation. Moreover, new constitutive equations involving the strain rate intensity factor are proposed for some parameters which characterize the structure of material. The process of plane strain extrusion is considered in some detail as an illustrative example. The strain rate intensity factor is determined from an approximate solution. Using this solution and experimental data the thickness of the layer of intensive plastic deformation is found. A non-local ductile fracture criterion is adopted to predict the initiation of ductile fracture in the layer. A numerical method for determining the strain rate intensity factor in the case of plane strain flow of rigid perfectly plastic material is discussed.

1 Introduction

Material properties of thin surface layers in the vicinity of frictional interfaces are affected by manufacturing processes much greater than in the bulk material. As a result, the distribution of material properties is very non-uniform. Such distributions

Elena Lyamina · Sergei Alexandrov
A. Ishlinsky Institute for Problems in Mechanics RAS, 119526, pr. Vernadskogo 101-1, 119526 Moscow, Russia; e-mail: {lyamina@ipmnet.ru, sergei_alexandrov@yahoo.com

G. Zavarise & P. Wriggers (Eds.): Trends in Computational Contact Mechanics, LNACM 58, pp. 291–320.
springerlink.com

have been found experimentally for different types of material and under different process conditions [7, 23, 24, 29, 36]. The evolution equations for internal variables characterizing this or that material property usually involve the equivalent strain rate in such a manner that a higher gradient of the equivalent strain rate results in a higher gradient of the internal variables (see, for example, [26, 28]). Based on this observation, a general method to predict the evolution of material properties in a thin layer in the vicinity of frictional interfaces in manufacturing processes has been proposed in [6]. The theoretical background of the method is the strain rate intensity factor which has been defined in [20] as the coefficient of the principal singular term in a series expansion of the equivalent strain rate in the vicinity of maximum friction surfaces. The model of rigid perfectly plastic material obeying an arbitrary pressure-independent yield criterion and its associated flow rule has been adopted in [20]. The maximum friction law for this model is defined by the condition that the friction stress at sliding is equal to the shear yield stress of the material. An alternative formulation of the maximum friction law has been proposed for the double shearing model and the double slip and rotation model in [9] and [8], respectively. The former model has been proposed in [45] and the latter in [31]. Both account for pressure-dependency of the yield criterion but include the equation of incompressibility. Such a combination of properties of the models is advantageous for some metallic materials [35, 42]. Particular solutions for plastically anisotropic materials [25] and viscoplastic materials [5, 17, 18] show that the aforementioned singular behaviour of the equivalent strain rate may occur in the case of such models as well, although it depends on the specific constitutive equations adopted.

The present paper summarizes results related to singular behaviour of the equivalent strain rate in the vicinity of maximum friction surfaces and methods developed to use the strain rate intensity factor in applications. The specific derivation is restricted to plane strain deformation.

2 Strain Rate Intensity Factor

The original formulation of the maximum friction law at sliding for rigid perfectly plastic solids is

$$\tau_f = \tau_s, \tag{1}$$

where τ_f is the friction stress and τ_s is the shear yield stress, a material constant for rigid perfectly plastic materials. An external boundary where the condition (1) is valid will be referred to as a maximum friction surface. The mathematical meaning of the maximum friction law has led to its alternative formulation. The best starting point here is to consider plane strain deformation of rigid perfectly plastic material. In this case the shear stress is equal to the shear yield stress in characteristic coordinates [33]. Therefore, the alternative formulation of the maximum friction law is: a characteristic direction is tangential to a maximum friction surface. It will be shown later that of special interest are solutions where an envelope of characteristics

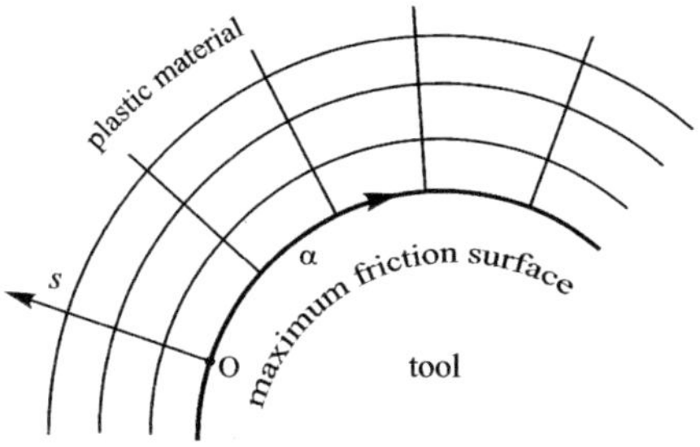

Fig. 1 Local coordinate system for plane-strain deformation.

coincides with a maximum friction surface and, consequently, no analytic solution can be extended beyond this surface. The alternative formulation meets no difficulty in the case of hyperbolic equations. Examples are plane strain deformation of rigid perfectly plastic material, any mode of deformation for the double-shearing model and the double slip and rotation model. However, the complete system of equations may be not hyperbolic in the case of non-plane deformation of rigid perfectly plastic solids (there are exceptions such as rigid perfectly plastic material obeying Tresca's yield criterion and its associated flow rule) and, consequently, no real characteristics exist. The original formulation (1) should be used for such models, though isolated characteristics may exist even in the case of non-plane deformation of rigid perfectly plastic solids [32]. It is believed that maximum friction surfaces coincide with such isolated characteristics.

The maximum friction law in the form of (1) has been adopted in many classical problems of plasticity theory [33, 44]. The alternative formulation has been used in [4, 8–11, 14, 38, 41].

2.1 Basic Relations for Plane Strain Deformation

In order to perform asymptotic analysis of equations in the vicinity of the maximum friction surface, it is convenient to introduce a local orthogonal coordinate system $s\alpha$ shown in Figure 1. The coordinate line $s = 0$ coincides with the maximum friction surface (it is actually a curve in the plane of flow), and $s-$coordinate lines are straight, orthogonal to this curve and directed from tool towards plastic material. Therefore, the solution is sound in the domain $s \geq 0$. The equilibrium equations in this coordinate system take the following form

$$\frac{\partial \sigma_{\alpha\alpha}}{\partial \alpha} + H\frac{\partial \sigma_{\alpha s}}{\partial s} + \frac{2\sigma_{\alpha s}}{R} = 0$$

$$\frac{\partial \sigma_{\alpha s}}{\partial \alpha} + H\frac{\partial \sigma_{ss}}{\partial s} + \frac{\sigma_{ss} - \sigma_{\alpha\alpha}}{R} = 0 \qquad (2)$$

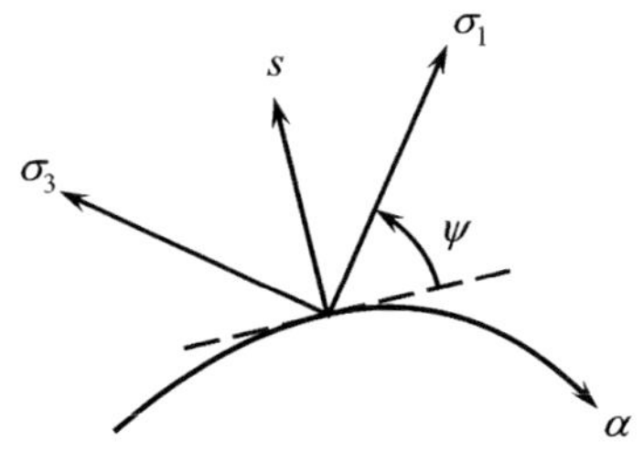

Fig. 2 Orientation of principal stress axes.

where $\sigma_{\alpha\alpha}$, σ_{ss}, and $\sigma_{\alpha s}$ are the components of the stress tensor, $R(\alpha)$ is the radius of curvature of the curve $s = 0$, and $H = 1 + s/R(\alpha)$. The notation for $R(\alpha)$ emphasizes that this quantity may depend on α. The non-zero strain rate components are expressed in terms of the velocity components u_α and u_s as

$$\xi_{\alpha\alpha} = \frac{1}{H}\left(\frac{\partial u_\alpha}{\partial \alpha} + \frac{u_s}{R}\right), \quad \xi_{ss} = \frac{\partial u_s}{\partial s}, \quad \xi_{\alpha s} = \frac{1}{2}\left[\frac{1}{H}\left(\frac{\partial u_s}{\partial \alpha} - \frac{u_\alpha}{R}\right) + \frac{\partial u_\alpha}{\partial s}\right] \quad (3)$$

The only non-zero component of the spin involved in the double shearing model is given by

$$\omega_{\alpha s} = \frac{1}{2}\left[\frac{\partial u_\alpha}{\partial s} - \frac{1}{H}\left(\frac{\partial u_s}{\partial \alpha} - \frac{u_\alpha}{R}\right)\right] \quad (4)$$

The stress components can be represented in the following form:

$$\sigma_{\alpha\alpha} = \sigma + q\cos 2\psi, \quad \sigma_{ss} = \sigma - q\cos 2\psi, \quad \sigma_{\alpha s} = q\sin 2\psi \quad (5)$$

where ψ is the angle which the principal axis corresponding to the major principal stress σ_1 (the other principal stresses will be denoted by σ_2 and σ_3 such that $\sigma_1 \geq \sigma_2 \geq \sigma_3$ and thus the principal stress σ_2 is orthogonal to the plane of flow) makes with the α-direction, measured anti-clockwise from the positive α-direction (Figure 2). Also,

$$\sigma = \frac{\sigma_{\alpha\alpha} + \sigma_{ss}}{2}, \quad q = \frac{1}{2}\sqrt{(\sigma_{\alpha\alpha} - \sigma_{ss})^2 + 4\sigma_{\alpha s}^2} \quad (6)$$

The equivalent strain rate under plane strain conditions is defined as

$$\xi_{\text{eq}} = \sqrt{\frac{2}{3}\left(\xi_{\alpha\alpha}^2 + \xi_{ss}^2 + 2\xi_{\alpha s}^2\right)} \quad (7)$$

2.2 Rigid Perfectly Plastic Material

The constitutive equations of this model include the yield criterion in the form

$$\sqrt{(\sigma_{\alpha\alpha}-\sigma_{ss})^2+4\sigma_{\alpha s}^2}=2\tau_s \tag{8}$$

and its associated flow rule

$$\xi_{\alpha\alpha}=\lambda\tau_{\alpha\alpha},\quad \xi_{ss}=\lambda\tau_{ss},\quad \xi_{\alpha s}=\lambda\tau_{\alpha s},\quad \lambda\geq 0 \tag{9}$$

where $\tau_{\alpha\alpha}=\sigma_{\alpha\alpha}-\sigma$, $\tau_{ss}=\sigma_{ss}-\sigma$, and $\tau_{\alpha s}=\sigma_{\alpha s}$. It follows from (6) and (8) that $q=\tau_s$. Then, substituting (5) into (2) gives

$$\begin{aligned}&\frac{\partial p}{\partial\alpha}-\sin 2\psi\frac{\partial\psi}{\partial\alpha}+H\cos 2\psi\frac{\partial\psi}{\partial s}+\frac{\sin 2\psi}{R}=0\\&H\frac{\partial p}{\partial s}+\cos 2\psi\frac{\partial\psi}{\partial\alpha}+H\sin 2\psi\frac{\partial\psi}{\partial s}-\frac{\cos 2\psi}{R}=0\end{aligned} \tag{10}$$

where $p=\sigma/2\tau_s$. Excluding λ in (9) and using (3) and (5) result in

$$\begin{aligned}&\frac{1}{H}\left(\frac{\partial u_\alpha}{\partial\alpha}+\frac{u_s}{R}\right)+\frac{\partial u_s}{\partial s}=0\\&\left[\frac{1}{H}\left(\frac{\partial u_s}{\partial\alpha}-\frac{u_\alpha}{R}\right)+\frac{\partial u_\alpha}{\partial s}\right]\cot 2\psi=\frac{1}{H}\left(\frac{\partial u_\alpha}{\partial\alpha}+\frac{u_s}{R}\right)-\frac{\partial u_s}{\partial s}\end{aligned} \tag{11}$$

Equations (10) and (11) constitute the system for four unknowns, p, ψ, u_α, and u_s. This system will be investigated in the vicinity of the curve $s=0$ (Figure 1) assuming that the boundary condition (1) is satisfied over a finite length of this line. With no loss of generality, tool can be regarded as motionless and the velocity component u_α can be assumed to be positive at the friction surface. Therefore,

$$u_s=0 \tag{12}$$

and

$$\tau_{\alpha s}>0 \tag{13}$$

at $s=0$. It follows from (1) and (13) that $\tau_{\alpha s}=\tau_s$ at the friction surface. However, it is more convenient to use the maximum friction law in the alternative form. In particular, it is known that the characteristic directions make angles $\pi/4$ with the direction of the principal stress σ_1 [33]. Therefore, it follows from (13) and Figure 2 that

$$\psi=\pi/4 \tag{14}$$

at $s=0$. Equation (14) represents the maximum friction law in the case under consideration. Substituting (14) into (11) shows that

$$\frac{\partial u_s}{\partial s}=0,\quad \frac{1}{H}\left(\frac{\partial u_\alpha}{\partial\alpha}+\frac{u_s}{R}\right)=0 \tag{15}$$

at the friction surface, unless

$$\left[\frac{1}{H}\left(\frac{\partial u_s}{\partial \alpha}-\frac{u_\alpha}{R}\right)+\frac{\partial u_\alpha}{\partial s}\right]\to\infty \tag{16}$$

as $s\to 0$. As follows from (3), equations (15) can be rewritten in the form $\xi_{\alpha\alpha}=\xi_{ss}=0$. The latter are the characteristic relations [33]. Equations (15) place severe restrictions on the velocity field. In particular, it is seen from (12) and $(15)^2$ that the velocity component u_α must be constant over the entire friction surface. In many cases other boundary conditions would result in $u_\alpha=0$ at $s=0$, which contradicts the assumption that the regime of sliding occurs. Therefore, it is natural to expect that in most cases the condition (16) is satisfied at maximum friction surfaces. This is equivalent to the statement that an envelope of characteristics coincides with the maximum friction surface.

Reasonable assumptions concerning the velocity field are

$$u_\alpha<\infty,\qquad \left|\frac{\partial u_s}{\partial \alpha}\right|<\infty \tag{17}$$

at $s=0$. Then it follows from (16) and (17) that

$$\frac{\partial u_\alpha}{\partial s}\to\infty \tag{18}$$

as $s\to 0$. It has been taken into account here that the derivative $\partial u_\alpha/\partial s$ is positive, as follows from (9) and (13). The final assumption concerning the velocity field is that its components are represented in the form of power series in the vicinity of the maximum friction surface. Then

$$u_\alpha=U_0+U_1 s^\beta+o(s^\beta),\quad s\to 0 \tag{19}$$

It is obvious that $\beta>0$, otherwise the representation (19) would contradict the assumption $(17)^1$. On the other hand, $\beta<1$, otherwise it is not possible to obtain (18) from (19). Thus

$$1>\beta>0 \tag{20}$$

Taking into account (12) the velocity component u_s can be represented in the form

$$u_s=V_1 s^\gamma+o\left(s^\gamma\right),\quad s\to 0 \tag{21}$$

where $\gamma>0$. Substituting (19) and (21) into $(11)^1$ shows that the terms of order $O(1)$ as $s\to 0$ are compatible if and only if $\gamma=1$. Therefore, the representation (21) transforms to

$$u_s=V_1 s+o(s),\quad s\to 0 \tag{22}$$

Substituting (19) and (22) into $(11)^2$ gives

$$\cot 2\psi=O(s^{1-\beta}),\quad s\to 0 \tag{23}$$

Taking into account (14) and expanding $\cot 2\psi$ in a power series in the vicinity of $\psi = \pi/4$ equation (23) can be rewritten in the form

$$\frac{\pi}{4} - \psi = As^{1-\beta} + o(s^{1-\beta}), \quad s \to 0 \tag{24}$$

Then

$$\sin 2\psi = 1 + O(s^{2-2\beta}), \quad \cos 2\psi = 2As^{1-\beta} + O(s^{3-3\beta}), \quad s \to 0 \tag{25}$$

Assuming that

$$p = p_0 + p_1 s^{\delta} + o(s^{\delta}), \quad \delta > 0, \quad s \to 0 \tag{26}$$

and substituting (24), (25), and (26) into $(10)^1$ lead to

$$\frac{dp_0}{d\alpha} + 1 + O(s^{1-2\beta}) = 0, \quad s \to 0 \tag{27}$$

The last term is compatible with the remainder of equation (27) if and only if

$$\beta = 1/2 \tag{28}$$

Using (28) and substituting (24), (25), and (26) into $(10)^2$ show that this equation is satisfied in the vicinity of the friction surface for $\delta = 1/2$. Thus, the asymptotic representation of the solution in the vicinity of maximum friction surfaces is determined from (19), (22), (24), (27) and (28) in the form

$$\begin{aligned} u_\alpha &= U_0 + U_1 s^{1/2} + o(s^{1/2}), \quad u_s = V_1 s + o(s), \\ \psi &= \frac{\pi}{4} - As^{1/2} + o(s^{1/2}), \quad p = p_0 + p_1 s^{1/2} + o(s^{1/2}) \end{aligned} \tag{29}$$

as $s \to 0$. Of special interest is behaviour of the equivalent strain rate in the vicinity of maximum friction surfaces. Substituting (29) into (3) and, then, the result into (7) give

$$\xi_{\text{eq}} = \frac{D}{\sqrt{s}} + o\left(\frac{1}{\sqrt{s}}\right), \quad s \to 0 \tag{30}$$

where D is the strain rate intensity factor.

The equivalent strain rate is also represented in the form of (30) in the case of three-dimensional flow of rigid perfectly plastic material obeying an arbitrary smooth yield criterion [20] and axisymmetric flow of rigid perfectly plastic material obeying Tresca's yield criterion [19].

2.3 Double Shearing Model

The constitutive equations of this model [45] include the Coulomb–Mohr yield criterion in the form

$$(\sigma_{\alpha\alpha}+\sigma_{ss})\sin\varphi+\sqrt{(\sigma_{\alpha\alpha}-\sigma_{ss})^2+4\sigma_{\alpha s}^2}=2k\cos\varphi, \tag{31}$$

the incompressibility equation

$$\xi_{\alpha\alpha}+\xi_{ss}=0, \tag{32}$$

and the following equation that connects stresses and velocities:

$$2\cos 2\psi\xi_{\alpha s}-\sin 2\psi\,(\xi_{\alpha\alpha}-\xi_{ss})+2\sin\varphi\,(\omega_{\alpha s}+\dot{\psi})=0 \tag{33}$$

where k is the cohesion, φ is the angle of internal friction, and $\dot{\psi}$ is the convected derivative of ψ, so that

$$\dot{\psi}=\frac{\partial\psi}{\partial t}+\frac{u_\alpha}{H}\frac{\partial\psi}{\partial\alpha}+u_s\frac{\partial\psi}{\partial s} \tag{34}$$

where $\partial\psi/\partial t$ is the derivative at a point which is fixed relative to the αs-coordinates. The physical meaning of equation (33) is that the deformation consists of two simultaneous superimposed shearing deformations on the characteristic curves of the stress equations including (31) and the equations of equilibrium. By definition, a motion consists of a shearing deformation along one family of characteristic curves if, in a coordinate system in which the characteristic direction in the neighborhood of a generic particle M is fixed, the relative velocity of two adjacent particles near M is directed in the characteristic direction if the particles do not lie on the same characteristic line, and is zero if they do lie on the same characteristic line [45]. The definition for the shearing deformation along the other family of characteristics can be introduced in a similar manner. It follows from (6) and (31) that

$$\sigma=k\cot\varphi-q\operatorname{cosec}\varphi \tag{35}$$

Then, excluding σ in (5) by means of (35) and substituting the result into (2) give

$$\begin{gathered}(\cos 2\psi-\operatorname{cosec}\varphi)\frac{\partial\ln q}{\partial\alpha}+H\sin 2\psi\frac{\partial\ln q}{\partial s}-2\sin 2\psi\frac{\partial\psi}{\partial\alpha}+\\+2H\cos 2\psi\frac{\partial\psi}{\partial s}+\frac{2\sin 2\psi}{R}=0,\\ \sin 2\psi\frac{\partial\ln q}{\partial\alpha}-H(\cos 2\psi+\operatorname{cosec}\varphi)\frac{\partial\ln q}{\partial s}+2\cos 2\psi\frac{\partial\psi}{\partial\alpha}+\\+2H\sin 2\psi\frac{\partial\psi}{\partial s}-\frac{2\cos 2\psi}{R}=0\end{gathered} \tag{36}$$

Using (3), (4) and (34), equations (32) and (33) can be transformed to

$$\frac{1}{H}\left(\frac{\partial u_\alpha}{\partial\alpha}+\frac{u_s}{R}\right)+\frac{\partial u_s}{\partial s}=0,\quad -\frac{\sin 2\psi}{H}\frac{\partial u_\alpha}{\partial\alpha}+(\cos 2\psi+\sin\varphi)\frac{\partial u_\alpha}{\partial s}$$

$$+\frac{(\cos 2\psi-\sin\varphi)}{H}\frac{\partial u_s}{\partial\alpha}+\sin 2\psi\frac{\partial u_s}{\partial s}+\frac{u_\alpha}{HR}(\sin\varphi-\cos 2\psi)$$

$$-\frac{u_s}{HR}\sin 2\psi + 2\sin\varphi\left(\frac{\partial\psi}{\partial t} + \frac{u_\alpha}{H}\frac{\partial\psi}{\partial\alpha} + u_s\frac{\partial\psi}{\partial s}\right) = 0 \tag{37}$$

Equations (36) and (37) constitute the system for four unknowns, q, ψ, u_α, and u_s. This system will be investigated in the vicinity of the maximum friction surface, $s = 0$ in Figure 1. As in the previous section, it is assumed that the conditions (12) and (13) are valid as well as the other assumptions concerning the velocity field. It is known that the characteristic directions make angles $\pm(\pi/4 + \varphi/2)$ with the direction of the principal stress σ_1 [45]. Therefore, it follows from (13) and Figure 2 that

$$\psi = \psi_s = \frac{\pi}{4} + \frac{\varphi}{2} \tag{38}$$

at $s = 0$. It is seen from this equation that ψ is constant at points of the maximum friction surface. Therefore, $\partial\psi/\partial t = 0$ and $\partial\psi/\partial\alpha = 0$ at $s = 0$. Combining these relations and (12) shows that

$$\frac{\partial\psi}{\partial t} + \frac{u_\alpha}{H}\frac{\partial\psi}{\partial\alpha} + u_s\frac{\partial\psi}{\partial s} = 0 \tag{39}$$

at $s = 0$. Substituting (38) and (39) into equation $(37)^2$ shows that the maximum friction surface coincides with a characteristic of the system of equations, unless

$$\frac{\partial u_\alpha}{\partial s} \to \infty \tag{40}$$

as $s \to 0$. As before, it is reasonable to assume that equation (40) is satisfied in most cases (i.e. the maximum friction surface coincides with an envelope of characteristics). Then, the velocity component u_α is represented by (19) where the value of β belongs to the range (20). Equation $(11)^1$ coincides with $(37)^1$. Therefore, the velocity component u_s is represented by (22). Substituting (19), (22), and (39) into $(37)^2$ gives

$$\cos 2\psi + \sin\varphi = O\left(s^{1-\beta}\right), \quad s \to 0 \tag{41}$$

Taking into account (38) and expanding $\cos 2\psi$ in a power series in the vicinity of $\psi = \psi_s$ equation (41) can be rewritten in the form

$$\psi_s - \psi = As^{1-\beta} + o\left(s^{1-\beta}\right), \quad s \to 0 \tag{42}$$

Substituting (22) and (42) into (39) shows that the latter is satisfied even though $|\partial\psi/\partial s| \to \infty$ as $s \to 0$. Excluding the derivative $\partial\ln q/\partial s$ between equations (36) leads to

$$2H\left(1 + \cos 2\psi \operatorname{cosec}\varphi\right)\frac{\partial\psi}{\partial s} - 2\sin 2\psi \operatorname{cosec}\varphi\frac{\partial\psi}{\partial\alpha}$$
$$-\cot^2\varphi\frac{\partial\ln q}{\partial\alpha} + \frac{2}{R}\sin 2\psi \operatorname{cosec}\varphi = 0 \tag{43}$$

It follows from (42) that

$$\cos 2\psi + \operatorname{cosec}\varphi = \frac{\cos^2\varphi}{\sin\varphi} + O(s^{1-\beta}),$$

$$\sin 2\psi = \cos\varphi + O(s^{1-\beta}), \quad \cos 2\psi = -\sin\varphi + O(s^{1-\beta}),$$

$$\cos 2\psi - \operatorname{cosec}\varphi = -(\operatorname{cosec}\varphi + \sin\varphi) + O(s^{1-\beta}),$$

$$1 + \cos 2\psi \operatorname{cosec}\varphi = 2A\cot\varphi s^{1-\beta} + o(s^{1-\beta}) \tag{44}$$

as $s \to 0$. Substituting (42) and (44) into (43) shows that the first term in (43) is compatible with other terms of this equation if and only if $1 - 2\beta = 0$. Therefore, equation (28) is obtained and the asymptotic representation of the velocity component u_α is given by equation (29). Assuming that

$$\ln q = Q_0 + Q_1 s^\delta + o(s^\delta), \quad \delta > 0, \quad s \to 0 \tag{45}$$

and substituting (42), (44), and (45) into any of equations (36) give, with the use of (28), $\delta = 1/2$. Therefore, (45) becomes

$$q = q_0 + q_1 s^{1/2} + o(s^{1/2}), \quad s \to 0 \tag{46}$$

Taking into account (35) and (46) it is possible to conclude that the asymptotic representation for σ is given by (29) where p should be understood as $p = \sigma/k$.

Thus the representation for the equivalent strain rate in the vicinity of maximum friction surfaces is given by (30) and the strain rate intensity factor can be introduced for the double-shearing model. In particular, the representation for the equivalent strain rate in the form of (30) for axisymmetric flow has been found in [3]. On the other hand, other models of pressure-dependent plasticity may or may not lead to the asymptotic representation in the form of (30), for example [2, 8, 13].

3 Thickness of the Layer of Intensive Plastic Deformation

Plane strain and axisymmetric extrusion/drawing processes through long dies with no lubricant are ideal for the development of special constitutive equations for material property evolution in the vicinity of frictional interfaces. For, the distance travelled by an infinitesimal volume of material along the friction surface is quite large and a layer of intensive plastic deformation is seen very clearly in the final product. To illustrate this, a micrograph of a small piece of material cut in the vicinity of the friction surface after plane strain extrusion through a wedge-shape die is shown in Figure 3 [7]. The important geometric parameters of the process are presented in Figure 4. The material tested was brass. The narrow layer of material where its structure was sharply changed is clearly seen in Figure 3. The thickness of this layer is about 100 μm. A layer of about 400 μm in the vicinity of the friction

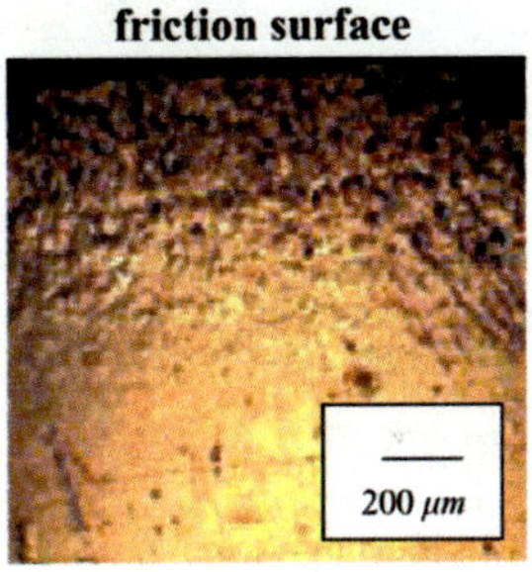

Fig. 3 Structure of sub-surface layer.

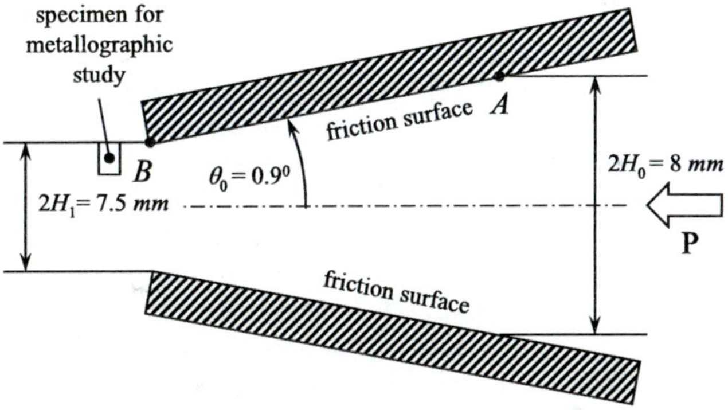

Fig. 4 Schematic of extrusion process.

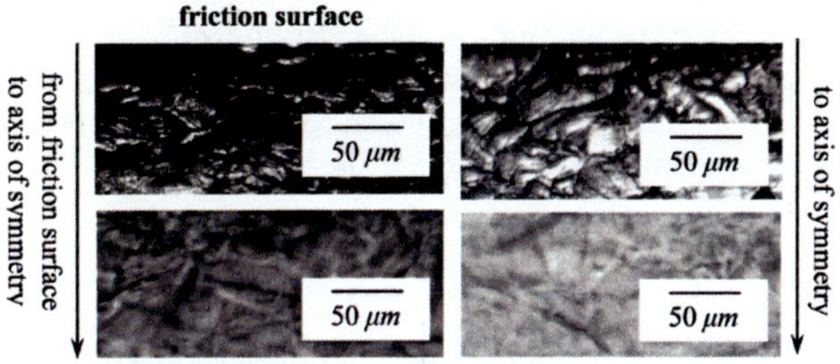

Fig. 5 Distribution of grain size and shape in the vicinity of friction surface.

surface divided into four sub-layers is shown in Figure 5 at a finer resolution than that used in Figure 3. A difference in the grain size and shape between the different sub-layers is seen in Figure 5. The main goal of the theory under development is to predict the thickness of such layers as well as the evolution of material properties inside the layers.

3.1 General Approach

The approach [7] for predicting the thickness of a layer of intensive plastic deformation in the vicinity of frictional interfaces where the friction stress is high enough is based on the main assumption that the evolution of material properties in this layer is solely controlled by the strain rate intensity factor which includes an effect of all process conditions. As follows from (30) the unit for the strain rate intensity factor is $\sqrt{m}/s$. The simplest model is obtained when it is supposed that the time derivative of the thickness of the layer of intensive plastic deformation depends on its current thickness and the strain rate intensity factor. Then

$$\frac{dh}{dt} = \Phi(h, D) \tag{47}$$

where h is the thickness of the layer of intensive plastic deformation and $\Phi(h, D)$ is an arbitrary function of its arguments. However, because the unit for s is m and the unit for t is s, it follows from the π-theorem that equation (47) must have the following form

$$\frac{dh}{dt} = \alpha D\sqrt{h} \tag{48}$$

where α is a dimensionless constant. In general, it is sufficient to combine (48) and one experiment to determine α. Using the experimental result illustrated in Figure 3, it has been obtained in [7] that $\alpha \approx 6.17$. However, it has also been mentioned in [7] that equation (48) predicts the increase in h with no limit as the deformation proceeds. It seems that it is not realistic, though equation (48) can be used for predicting the increase in h as long as its value is less or equal to the maximum value of h reached in the experiment from which the value of α has been determined.

A more complicated and realistic model is obtained by assuming that the value of h cannot exceed a given value h_s, a material constant. Since the unit for h_s is m, equation (48) does not follows from (47) and the latter can be rewritten in the form

$$\frac{dh}{dt} = D\sqrt{h_s} f\left(\frac{h}{h_s}\right) \tag{49}$$

where $f(h/h_s)$ is an arbitrary function of its argument. Obviously, several different experiments should be carried out to approximate the function $f(h/h_s)$. The simplest function $f(h/h_s)$ satisfying all the necessary conditions is a linear function of its argument. Then, equation (49) becomes

$$\frac{dh}{dt} = \alpha D\sqrt{h_s}\left(1 - \frac{h}{h_s}\right) \tag{50}$$

The initial condition to this equation, as well as to equation (49), is

$$h = 0 \tag{51}$$

for $t = 0$.

3.2 Plane Strain Drawing/Extrusion

In order to apply (50) to the plane-strain extrusion process illustrated in Figure 4, it is necessary to find the distribution of the strain rate intensity factor along the friction surface. To the best of authors' knowledge, no numerical solution is available for strain rate intensity factors. However, because the length of the die is rather long, the solution for plastic flow through an infinite wedge-shape channel given, for example, in [33] can be used to predict an approximate value of D [7] assuming that the friction stress is equal to the shear yield stress. It follows from the solution [33] that the equivalent strain rate in the plane polar coordinate system $r\theta$ in which the equation for the friction surface AB (Figure 4) is $\theta = \theta_0$ and the axis of symmetry of the process coincides with the axis $\theta = 0$ is given by

$$\xi_{eq} = \frac{2}{\sqrt{3}} \frac{U}{r^2 (c - \cos 2\psi) \cos 2\psi} \tag{52}$$

where c and U are constants and ψ has the same meaning as in (5) and is determined as a function of θ from the following equation

$$\frac{d\psi}{d\theta} = \frac{c - \cos 2\psi}{\cos 2\psi} \tag{53}$$

The maximum friction law at the surface AB, equation (14), requires

$$\psi = \frac{\pi}{4} \tag{54}$$

at $\theta = \theta_0$. Also, $\psi = 0$ at $\theta = 0$ (symmetry condition). Substituting this condition and (54) into the general solution to equation (53) gives the equation for c in the form

$$\frac{c}{\sqrt{c^2 - 1}} \arctan \sqrt{\frac{c+1}{c-1}} = \frac{\pi}{4} + \theta_0 \tag{55}$$

Expanding the right-hand side in (52) in a series in the vicinity of $\psi = \pi/4$ leads to

$$\xi_{eq} = \frac{U}{\sqrt{3} c r^2} \left(\frac{\pi}{4} - \psi \right)^{-1} + o \left[\left(\frac{\pi}{4} - \psi \right)^{-1} \right], \quad \psi \to \frac{\pi}{4} \tag{56}$$

Expanding the numerator and denominator in (53) in a series in the vicinity of $\psi = \pi/4$ and integrating give, with the use of (54),

$$\frac{\pi}{4} - \psi = c^{1/2} (\theta_0 - \theta)^{1/2} + o \left[(\theta_0 - \theta)^{1/2} \right], \quad \theta \to \theta_0 \tag{57}$$

Substituting (57) into (56) results in

$$\xi_{\text{eq}} = \frac{U\left(\theta_0 - \theta\right)^{-1/2}}{\sqrt{3}c^{3/2}r^2} + o\left[\left(\theta_0 - \theta\right)^{-1/2}\right], \quad \theta \to \theta_0 \tag{58}$$

Comparing (30) and (58) leads to

$$D = \frac{U}{\sqrt{3}c^{3/2}r^{3/2}} \tag{59}$$

In the case of the steady process under consideration, equation (50) becomes

$$u\frac{\partial h}{\partial r} = \alpha D\sqrt{h_s}\left(1 - \frac{h}{h_s}\right) \tag{60}$$

where u is the radial velocity at $\theta = \theta_0$. It follows from the solution [33] that this velocity is

$$u = -\frac{U}{rc} \tag{61}$$

Substituting (61) into (60) gives

$$\frac{\partial h}{\partial r} = -\alpha\sqrt{\frac{h_s}{3rc}}\left(1 - \frac{h}{h_s}\right) \tag{62}$$

Also, the initial condition (51) becomes

$$h = 0 \tag{63}$$

for $r = R_A$ where R_A is the radial coordinate of point A (Figure 4). The solution to equation (62) satisfying the boundary condition (63) has the form

$$\frac{h}{h_s} = 1 - \exp\left[-\frac{2\alpha}{\sqrt{3ch_s}}\left(\sqrt{R_A} - \sqrt{r}\right)\right] \tag{64}$$

The thickness of the layer of intensive plastic deformation in the final product is determined from (64) as

$$\frac{h_f}{h_s} = 1 - \exp\left[-\frac{2\alpha}{\sqrt{3ch_s}}\left(\sqrt{R_A} - \sqrt{R_B}\right)\right] \tag{65}$$

where R_B is the radial coordinate of point B (Figure 4). The parameters c, R_A and R_B are expressed in terms of H_0, H_1 and θ_0. In particular,

$$R_A = \frac{H_0}{\sin\theta_0}, \quad R_B = \frac{H_1}{\sin\theta_0} \tag{66}$$

From (65) and (66)

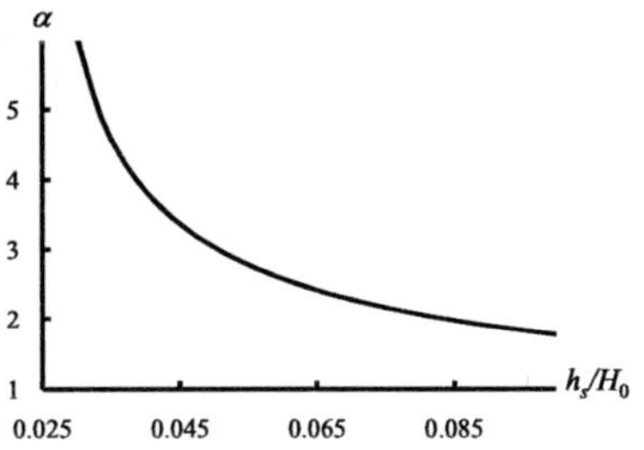

Fig. 6 Relation between α and h_s/H_0 for extrusion process illustrated in Figure 4.

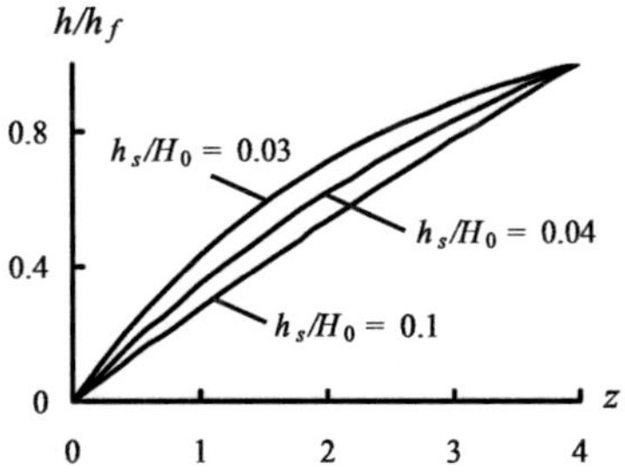

Fig. 7 Effect of model parameters on development of layer of intensive plastic deformation for extrusion process illustrated in Figure 4.

$$\frac{h_f}{H_0} = \frac{h_s}{H_0}\left\{1 - \exp\left[-\frac{2\alpha}{\sqrt{3c\sin\theta_0}}\sqrt{\frac{H_0}{h_s}}\left(1-\sqrt{\frac{H_1}{H_0}}\right)\right]\right\} \tag{67}$$

The value of c is determined from (55) as a function of θ_0 numerically.

Experimental data are only available for one set of geometric parameters shown in Figure 4. Therefore, it is impossible to determine both h_s and α. The relation between these parameters can be found from (67) assuming that $h_f = 100\ \mu\text{m}$, as follows from the experiment. This relation is illustrated in Figure 6 in the range $0.03 \le h_s/H_0 \le 0.1$. Using (64) it is now possible to find an effect of the model parameters on the development of the layer of intensive plastic deformation. To this end, it is convenient to introduce the dimensionless distance from point A along the friction surface in the following form:

$$z = \frac{R_A - r}{H_0} \tag{68}$$

Using (66), (68) and the relation between h_s and α illustrated in Figure 6 the dependence of h/h_f on z has been found from (64) and is illustrated for several values of h_s/H_0 in Figure 7. It is seen from this figure that the smaller value of h_s/H_0, the greater its effect is.

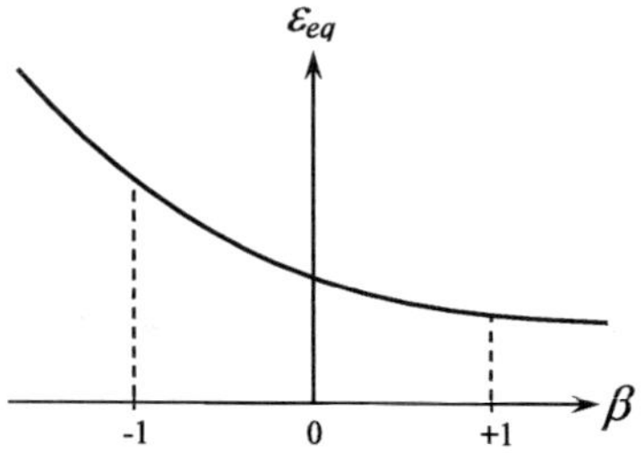

Fig. 8 Original workability diagram.

4 Ductile Fracture near Frictional Interfaces

Empirical ductile fracture criteria are often adopted to predict fracture initiation in metal forming processes [22, 30]. A difficulty with application of these fracture criteria in the vicinity of maximum friction surfaces where (30) is valid is somehow analogous to that in the mechanics of cracks [1]. In particular, most ductile fracture criteria in metal forming involve the equivalent strain rate in such a manner that they would immediately predict fracture initiation at the maximum friction surface since the behavior of a fracture parameter is singular and its value approaches infinity at the friction surface. To overcome a similar difficulty in linear elastic fracture mechanics, the stress intensity factor and its critical value are used to predict crack propagation. The stress intensity factor appears in asymptotic analyses performed in the vicinity of a sharp crack-tip and is the coefficient of the singular term. In spite of the fact that the assumptions under which the stress intensity factor is determined are not satisfied in real materials (the crack-tip is not sharp and a region of inelastic deformation exists in its vicinity), this approach is effective in structural design (among numerous textbooks and monographs on the subject, see [34, 39]). Therefore, by analogy to this approach, it is natural to assume that the strain rate intensity factor can be used to predict fracture in the vicinity of maximum friction surfaces in metal forming processes.

One of widely used ductile fracture criteria is based on the workability diagram [43] schematically shown in Figure 8 where $\beta = 3\sigma/\sigma_{\text{eq}}$ is the triaxiality factor, σ is the hydrostatic stress, σ_{eq} is the equivalent stress, and ε_{eq} is the equivalent strain. σ, σ_{eq} and ε_{eq} are defined by

$$\sigma = \frac{\sigma_{ij}\delta_{ij}}{3}, \quad \sigma_{\text{eq}} = \sqrt{\frac{3}{2}\left(\sigma_{ij} - \sigma\delta_{ij}\right)\left(\sigma_{ij} - \sigma\delta_{ij}\right)}, \quad \varepsilon_{\text{eq}} = \int_0^t \xi_{\text{eq}} dt \tag{69}$$

where integration should be carried out over strain path and δ_{ij} is Kroneker's delta. Here and in what follows the summation convection, according to which a recurring letter suffix indicates that the sum must be formed of all terms obtainable by assigning to the suffix the values 1, 2, and 3, is adopted. Similarly, in a quantity containing two repeated suffixes, say i and j, the summation must be carried out for all values 1, 2, 3 of both i and j. The equivalent strain rate involved in (69) is defined by

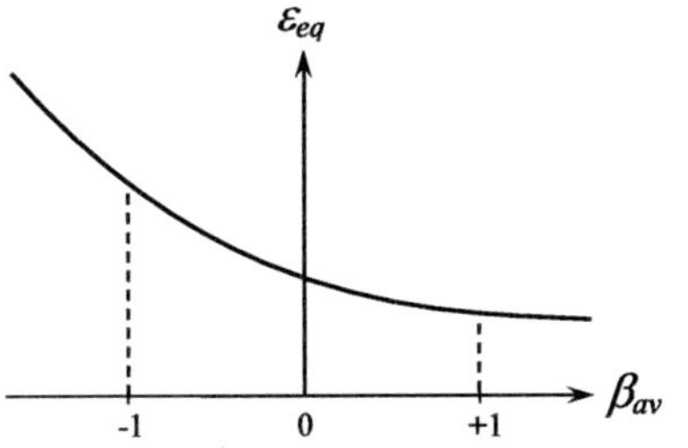

Fig. 9 Workability diagram applicable to any process.

$$\xi_{eq} = \sqrt{\frac{2}{3}\xi_{ij}\xi_{ij}} \tag{70}$$

This definition reduces to (7) in the case of plane strain deformation. By definition, the workability diagram predicts the equivalent strain to fracture, ε_f, when β = constant at a given material particle throughout the process. The latter condition is approximately satisfied in some standard tests (for example, $\beta = 1$ in uniaxial tension, $\beta = 0$ in torsion, and $\beta = -1$ in uniaxial compression) but is not realistic in metal forming processes. In order to account for the variation of β, an average value of the triaxiality factor is usually introduced by

$$\beta_{av} = \frac{1}{\varepsilon_{eq}}\int_0^t \beta \xi_{eq} dt \tag{71}$$

Then, β is replaced with β_{av} in the workability diagram. The new diagram (Figure 9) is, by assumption, applicable to any process. Using a solution to the boundary value problem describing a metal forming process ε_{eq} and β_{av} can be calculated from (69) and (71) as functions of t. Then, excluding t, the relation between ε_{eq} and β_{av} is obtained which determines a curve in the space used in Figure 9. Fracture initiates when this curve intersects the diagram shown in Figure 9. The ductile fracture criterion can be written as

$$\varepsilon_{eq} = \varepsilon_f = \Phi(\beta_{av}) \tag{72}$$

where $\Phi(\beta_{av})$ is the analytical representation of the curve shown in Figure 9. The function $\Phi(\beta_{av})$ should be determined experimentally as the curve shown in Figure 8 follows from experiment.

This ductile fracture criterion has been modified to account for the singular distribution of the equivalent strain rate (30) and then applied to several metal forming processes in [12, 15, 16].

4.1 General Approach

A non-local fracture criterion in the vicinity of crack tips accounting for the singular behaviour of the stress field from the linear elastic solution has been proposed in [40]. An analogous approach can be used to introduce a non-local fracture criterion generalizing the ductile fracture criterion described in the previous section.

Introduce an average value of the equivalent strain rate in the vicinity of maximum friction surfaces by

$$\Xi_{\mathrm{eq}} = \frac{1}{h}\int_0^h \xi_{\mathrm{eq}} ds \tag{73}$$

where h is the thickness of the layer of intensive plastic deformation introduced in Section 3. Substituting (30) into (73) gives

$$\Xi_{\mathrm{eq}} = \frac{2D}{\sqrt{h}} \tag{74}$$

to leading order. The quantities $\varepsilon_{\mathrm{eq}}$ and β_{av} introduced in (69) and (71) can be generalized as

$$\mathrm{E}_{\mathrm{eq}} = \int_0^t \Xi_{\mathrm{eq}} dt, \quad \chi_{\mathrm{av}} = \frac{1}{\mathrm{E}_{\mathrm{eq}}}\int_0^t \beta \Xi_{\mathrm{eq}} dt \tag{75}$$

Substituting (74) into (75) leads to

$$\mathrm{E}_{\mathrm{eq}} = 2\int_0^t \frac{D}{\sqrt{h}} dt, \quad \chi_{\mathrm{av}} = \left(\int_0^t \frac{\beta D}{\sqrt{h}} dt\right)\left(\int_0^t \frac{D}{\sqrt{h}} dt\right)^{-1} \tag{76}$$

Then, the fracture criterion (72) can be modified as

$$\mathrm{E}_{\mathrm{eq}} = \mathrm{E}_f = \Psi(\chi_{\mathrm{av}}) \tag{77}$$

where E_f is the value of E_{eq} at fracture. The function $\Psi(\chi_{\mathrm{av}})$ should be determined from special experiment designed to study the process of ductile fracture in the vicinity of frictional interfaces. However, qualitative behaviour of this function should be somehow similar to that of the function $\Phi(\beta_{\mathrm{av}})$ shown in Figure 9. In particular, it follows from (69), (71) and (76) that $\chi_{\mathrm{av}} = \beta_{\mathrm{av}} = \beta$ for $\beta = \text{constant}$.

4.2 Plane Strain Drawing/Extrusion

The example given in this section should be understood as illustrative because there are no experimental data for determining the function $\Psi(\chi_{\mathrm{av}})$ involved in (77), though some data from the real experiment described in Section 3 are used in this section.

Using the theory developed in Section 3 the thickness of the layer of intensive plastic deformation in the vicinity of the friction surface in plane strain extrusion

(Figure 4) is given by (64). Moreover, equations (76) become

$$\mathrm{E}_{\mathrm{eq}} = 2\int_{R_A}^{r} \frac{D}{\sqrt{hu}} dr, \quad \chi_{\mathrm{av}} = \left(\int_{R_A}^{r} \frac{\beta D}{\sqrt{hu}} dr\right)\left(\int_{R_A}^{r} \frac{D}{\sqrt{hu}} dr\right)^{-1} \tag{78}$$

The shear stress in the polar coordinate system $r\theta$ introduced in Section 3.2 is equal to the shear yield stress at the friction surface by the definition for the maximum friction surface. Then, it follows from (8) that $\sigma_{rr} = \sigma_{\theta\theta} = \sigma$. Thus, using the definition for β in the case of Mises yield criterion when $\sigma_{\mathrm{eq}} = \sqrt{3}\tau_s$ it is possible to find that

$$\beta = \frac{\sqrt{3}\sigma}{\tau_s} \tag{79}$$

The stress distribution in plastic flow through an infinite wedge-shaped channel found in [33] can be represented in the form

$$\frac{\sigma_{rr}}{\tau_s} = -2c\ln\left(\frac{r}{R_A}\right) + \cos 2\psi - c\ln(c - \cos 2\psi) + G,$$
$$\frac{\sigma_{\theta\theta}}{\tau_s} = -2c\ln\left(\frac{r}{R_A}\right) - \cos 2\psi - c\ln(c - \cos 2\psi) + G,$$
$$\frac{\sigma_{r\theta}}{\tau_s} = \sin 2\psi \tag{80}$$

Here c is given by (55), ψ by (53) and R_A is the radial coordinate of point A (Figure 4). Also, G is a new constant of integration not involved in the velocity solution given in Section 3.2. Its value can be found from the condition that no force is applied to the exit end in the process of extrusion (Figure 4). This condition is

$$\int_{0}^{\theta_0} \left(\sigma_{rr}\cos\theta - \sigma_{r\theta}\sin\theta\right) d\theta = 0 \tag{81}$$

where σ_{rr} should be taken at $r = R_B$ (Figure 4). Substituting (80) into (81) and using (53) and (66) give

$$G\sin\theta_0 = 2c\sin\theta_0 \ln\left(\frac{H_1}{H_0}\right)$$
$$+ \int_{0}^{\pi/4} \frac{\cos\psi}{(c - \cos 2\psi)} \left[c\cos\theta\ln(c - \cos 2\psi) - \cos(2\psi + \theta)\right] d\psi \tag{82}$$

In the case of plane strain deformation $\sigma = \left(\sigma_{rr} + \sigma_{\theta\theta}\right)/2$. Therefore, it follows from (80) and (79) that at the friction surface where $\psi = \pi/4$ the distribution of β is given by

$$\beta = -2\sqrt{3}c\ln\left(\frac{r}{R_A}\right) - \sqrt{3}c\ln c + \sqrt{3}G \tag{83}$$

The integrals involved in (78) are improper because $h = 0$ at $r = R_A$. However, it is easy to show convergence. Equation (64) in the vicinity of $r = R_A$ is represented as

$$\frac{h}{h_s} = \frac{\alpha}{\sqrt{3c}}\sqrt{\frac{R_A}{h_s}}\left(1-\frac{r}{R_A}\right) + o\left(1-\frac{r}{R_A}\right), \quad r \to R_A \tag{84}$$

Then

$$\int_{R_A}^{r} \frac{D}{\sqrt{hu}}dr = -\frac{2D_A(3c)^{1/4}R_A^{3/4}}{u_A\sqrt{\alpha}h_s^{1/4}}\sqrt{1-\frac{r}{R_A}}$$

$$\int_{R_A}^{r} \frac{\beta D}{\sqrt{hu}}dr = -\frac{2\beta_A D_A(3c)^{1/4}R_A^{3/4}}{u_A\sqrt{\alpha}h_s^{1/4}}\sqrt{1-\frac{r}{R_A}} \tag{85}$$

to leading order in the range $R_A \geq r \geq R_A(1-\delta)$ where $\delta \ll 1$. Here the subscript A indicates that the corresponding quantity should be taken at point A (Figure 4). In particular, it follows from (59), (61), and (83) that

$$D_A = \frac{U}{\sqrt{3}c^{3/2}R_A^{3/2}}, \quad u_A = -\frac{U}{R_A c}, \quad \beta_A = -\sqrt{3}c\ln c + \sqrt{3}G \tag{86}$$

Substituting (85) into (78) with the use of (66), (68), and (86) results in

$$\mathrm{E}_{\mathrm{eq}} = \frac{4}{\sqrt{\alpha}}\left(\frac{H_0\sin\theta_0}{3ch_s}\right)^{1/4}\sqrt{z}, \quad \chi_{\mathrm{av}} = \beta_A \tag{87}$$

in the range $0 \leq z \leq \delta/\sin\theta_0$.

Using the value of c found from (55) and the relation between ψ and θ found from (53) equation (82) can be solved for G numerically with no difficulty. Using (83), (85), (86), and (87) the variation of E_{eq} and χ_{av} with r can be found numerically. Excluding r in these relations gives the dependence of E_{eq} on χ_{av} along the friction surface involved in the fracture criterion. Since the specific form of the function $\Psi(\chi_{\mathrm{av}})$ involved in (77) and specific values of the parameters involved in the evolution equation (50) are still not available from experiment, a parametric analysis is provided in the remainder of this section. It has been assumed in all calculations that $\delta = 10^{-6}$. Even though the function $\Psi(\chi_{\mathrm{av}})$ is unknown, it is reasonable to expect that Ψ is a decreasing and convex down function of χ_{av}, by analogy to the function $\Phi(\beta_{\mathrm{av}})$ involved in (72). These assumptions can be expressed as

$$\frac{d\Psi}{d\chi_{\mathrm{av}}} < 0, \quad \frac{d^2\Psi}{d\chi_{\mathrm{av}}^2} > 0 \tag{88}$$

for all χ_{av}.

In order to show an effect of α and h_s involved in (50), calculation has been performed for $H_1/H_0 = 0.8$ and $\theta_0 = 10°$. The relation between α and h_s/H_0 shown in Figure 6 has been taken into account. The variation of E_{eq} with χ_{av} for three values

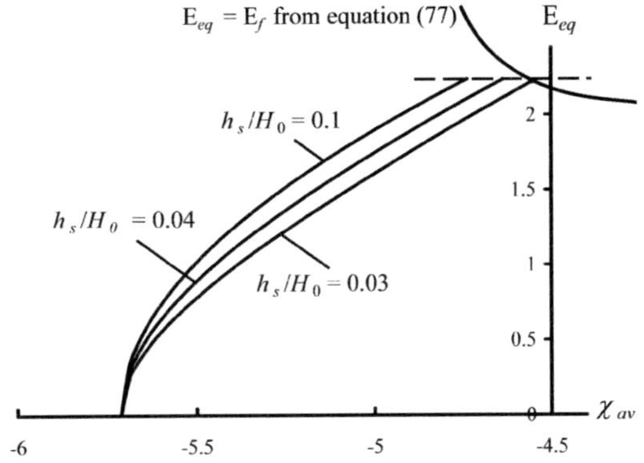

Fig. 10 Dependence of E_{eq} on χ_{av} along friction surface in plane strain extrusion for several values of h_s/H_0.

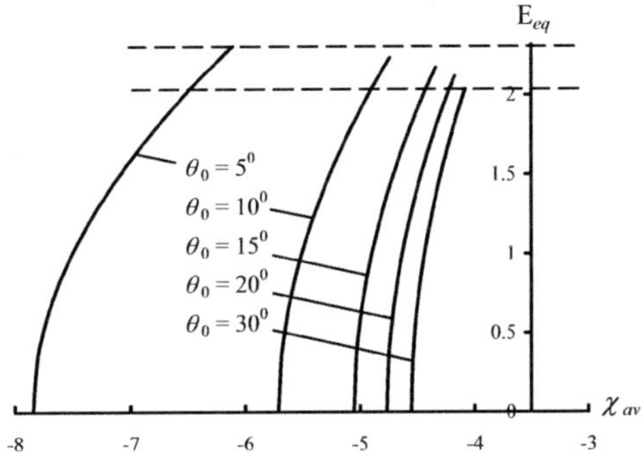

Fig. 11 Dependence of E_{eq} on χ_{av} along friction surface in plane strain extrusion for several values of θ_0.

of h_s/H_0 is depicted in Figure 10. When a material particle enters the plastic zone (point A in Figure 4), $E_{eq} = 0$ by definition and χ_{av} is independent of h_s/H_0, as is seen from (87). Therefore, all the three curves start at the same point (Figure 10). The other end point of these curves corresponds to point B (Figure 4). It is seen from Figure 10 that the value of E_{eq} at this point slightly increases with h_s/H_0. However, a decrease in the corresponding value of χ_{av} is more pronounced. Therefore, it is expected that materials with lower values of h_s/H_0 will fracture faster. In particular, a typical dependence of E_f on χ_{av} satisfying the conditions (88) is schematically shown in Figure 10 for the case when the fracture initiation occurs at point B (Figure 4) if $h_s/H_0 = 0.03$. It is seen that the curves $h_s/H_0 = 0.04$ and $h_s/H_0 = 0.1$ do not reach the curve $E_{eq} = E_f$ and, therefore, no fracture occurs in the process with the geometric parameters considered if $h_s/H_0 > 0.03$. Figure 11 illustrates an effect of the die angle on the behaviour of curves $E_{eq}(\chi_{av})$ for $H_1/H_0 = 0.8$ and $h_s/H_0 = 0.1$. The tendency here is not so obvious as in the previous case. The value of E_{eq} at point B increases as the value of θ_0 decreases. However, the corresponding decrease in the value of χ_{av} may compensate this increase in E_{eq}. Therefore, an exact shape of the curve $E_{eq} = E_f$ is required to find the most dangerous angle. The existence of such an angle is confirmed by results of calculation shown in Figure 12 where the relation between E_{eq} and χ_{av} at point B (Figure 4) is depicted in the range

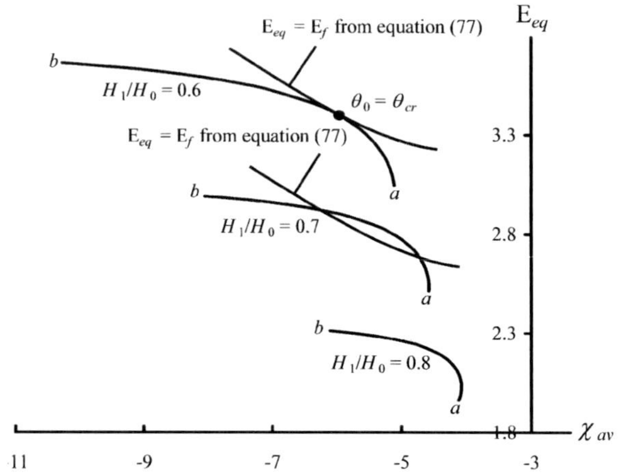

Fig. 12 Relation between E_{eq} and χ_{av} at point B (Figure 4) in plane strain extrusion when θ_0 varies from 5° to 45° for several values of h_s/H_0.

$5° \leq \theta_0 \leq 45°$ for three values of H_1/H_0 at $h_s/H_0 = 0.1$. Point b in Figure 12 denotes the end of each curve corresponding to $\theta_0 = 5°$. As the value of θ_0 increases, the point representing E_{eq} and χ_{av} moves from b to a along the curve and reaches point a when $\theta_0 = 45°$. Assuming that the conditions (88) are satisfied, it is clear from Figure 12 that for each curve shown there is just one angle $\theta_0 = \theta_{cr}$ such that the corresponding curve has just one common point with the curve $E_{eq} = E_f$, and $5° < \theta_{cr} < 45°$. This case is illustrated for the curve corresponding to $H_1/H_0 = 0.6$. In all other cases when fracture occurs the corresponding curve shown in Figure 12 has two common points with the curve $E_{eq} = E_f$. This case is illustrated for the curve corresponding to $H_1/H_0 = 0.7$. Fracture occurs for the range of angles θ_0 corresponding to the arc of the curve between the points of intersection. The angle θ_{cr} always belongs to this range. It is of course possible that no fracture occurs. In this case the corresponding curve shown in Figure 12 has no common point with the curve $E_{eq} = E_f$. This case is illustrated for the curve corresponding to $H_1/H_0 = 0.8$ when the equation (77) is represented by any of the two curves shown in Figure 12.

The process of plane strain drawing can be treated in a similar manner. To this end, it is necessary to find G by means of the condition (81) at $r = R_A$ (no force is applied to the entrance end of the sheet). Therefore, equation (82) becomes

$$G \sin\theta_0 = \int_0^{\pi/4} \frac{\cos\psi}{(c - \cos 2\psi)} \left[c \cos\theta \ln(c - \cos 2\psi) - \cos(2\psi + \theta)\right] d\psi \qquad (89)$$

Replacing G found by means of (82) with G found by means of (89) results in shifting the curves representing relations between E_{eq} and χ_{av} (Figures 10 to 12) along the χ_{av}-axis in its positive direction (Figures 13 to 15). Therefore, the general conclusions drawn for extrusion are also applicable to the process of drawing. It

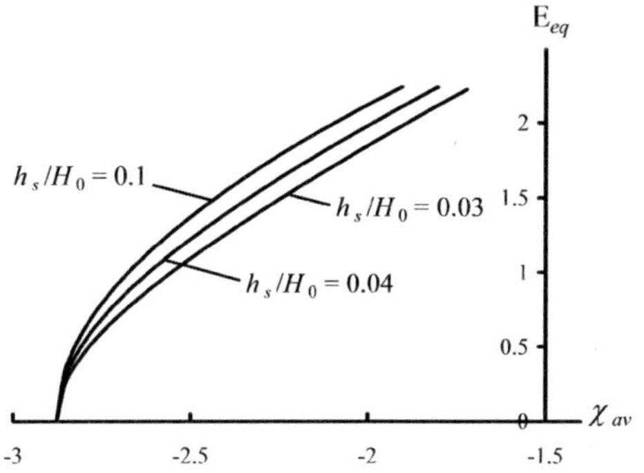

Fig. 13 Dependence of E_{eq} on χ_{av} along friction surface in plane strain drawing for several values of h_s/H_0.

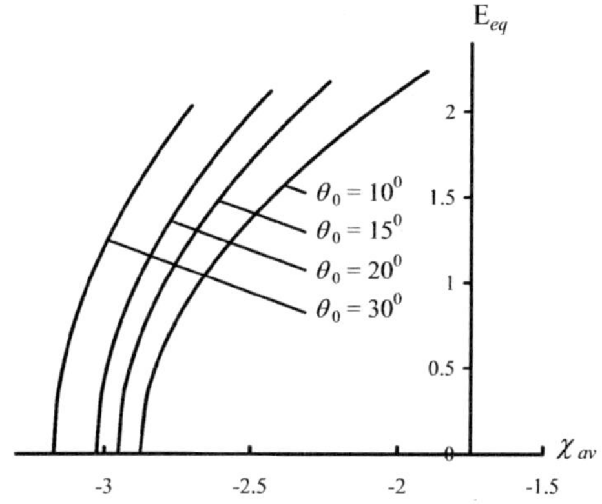

Fig. 14 Dependence of E_{eq} on χ_{av} along friction surface in plane strain drawing for several values of θ_0.

is however necessary to take into account that equation (77) is independent of the process. Therefore, assuming that the inequalities (88) are valid a trivial conclusion which follows from Figures 10, 11, 13, and 14 is that the process of ductile fracture is more intensive in drawing than in extrusion. Moreover, the average tensile stress applied to the exit end in the case of drawing cannot exceed $2\tau_s$. It follows from the yield condition (8). The average stress in question can be determined from

$$f_B = \frac{1}{H_1}\int_0^{\theta_0} \left(\sigma_{rr}\cos\theta - \sigma_{r\theta}\sin\theta\right) R_B d\theta = \frac{1}{\sin\theta_0}\int_0^{\theta_0} \left(\sigma_{rr}\cos\theta - \sigma_{r\theta}\sin\theta\right) d\theta \tag{90}$$

where (66) has been used and the stress components should be taken at $r = R_B$. Substituting (80) into (90) and using (53) and (66) give

$$\frac{f_B}{\tau_s} = G - 2c\ln\left(\frac{H_1}{H_0}\right)$$
$$-\frac{1}{\sin\theta_0}\int_0^{\pi/4} \frac{\cos\psi}{(c-\cos 2\psi)}\left[c\cos\theta\ln(c-\cos 2\psi) - \cos(2\psi+\theta)\right] d\psi \tag{91}$$

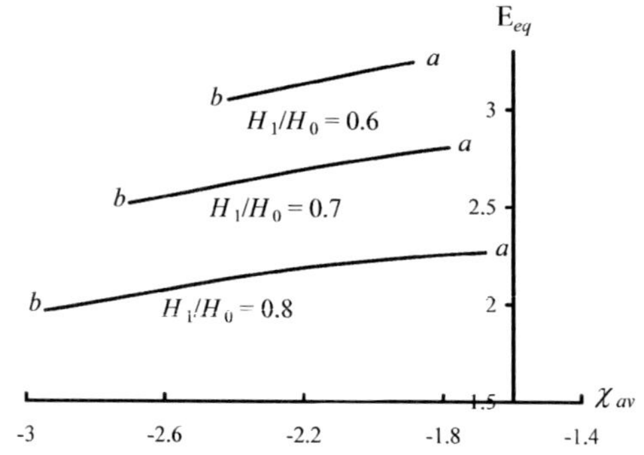

Fig. 15 Relation between E_{eq} and χ_{av} at point B (Figure 4) in plane strain drawing when θ_0 varies from 5° to 45° for several values of h_s/H_0.

Excluding the integral in (91) by means of (89) and assuming that $f_B \leq 2\tau_s$ lead to

$$c \ln\left(\frac{H_1}{H_0}\right) + 1 \geq 0 \tag{92}$$

This condition has been verified in course of calculation. In particular, the condition (92) is not satisfied for $\theta_0 = 5°$ and $H_1/H_0 = 0.8$. Therefore, the corresponding curve is not depicted in Figure 14. In general, in the range of parameters considered the condition (92) is not satisfied if $\theta_0 < \theta_m$ where $\theta_m \approx 8°$, $\theta_m \approx 15°$ and $\theta_m \approx 25°$ for $H_1/H_0 = 0.8$, $H_1/H_0 = 0.7$ and $H_1/H_0 = 0.6$, respectively. Points a in Figure 15 correspond to $\theta_0 = \theta_m$, but not $\theta_0 = 5°$ as in Figure 12.

5 Computational Aspects

A lack of special purpose numerical codes is the main difficulty with application of the theory under consideration to engineering problems. Commercial FE packages are not useful because of the singularity in the velocity field. Moreover, traditional numerical methods even do not predict high velocity gradients near friction surfaces occurring in experiment [21, 29]. Numerical codes based on the method of characteristics may be very efficient for calculating the strain rate intensity factor [37]. Derivation in [37] and in this section is restricted to rigid perfectly plastic solids under plane strain deformation. However, it is believed that similar results can be obtained for other material models characterized by hyperbolic systems of equations because the mathematical structure of the equations rather than their physical meaning is important for the numerical method. Plane strain flow of rigid perfectly plastic material is discussed in detail in many monographs and textbooks on plasticity theory, for example [33]. The system of equations can be referred to slip-lines α and β considered as right-handed curvilinear orthogonal coordinates. All derivations will be restricted to domains where both α and β are curved. Let ϕ be the angle between the α-direction at the origin O of a Cartesian coordinate system xy and the

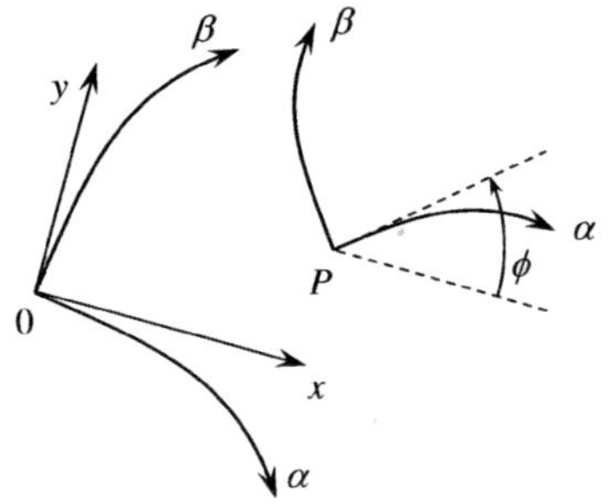

Fig. 16 Illustration of definition for ϕ angle.

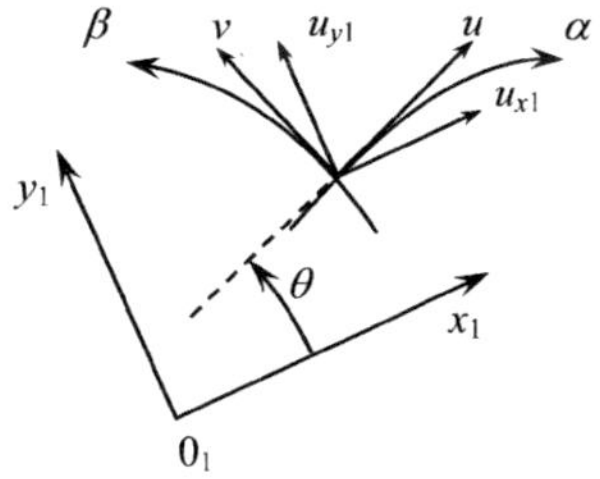

Fig. 17 Illustration of definition for θ angle.

α-direction at any point P, measured anti-clockwise (Figure 16). The Cartesian coordinate system is chosen such that the x-axis is tangent to the α-line and the y-axis is tangent to the β-line at O. The angle ϕ can be expressed in terms of α and β as

$$\phi = \alpha + \beta \tag{93}$$

The radii of curvature of the α- and β-lines, R and S, are defined by

$$\frac{1}{R} = \frac{\partial \phi}{\partial s_\alpha}, \qquad \frac{1}{S} = -\frac{\partial \phi}{\partial s_\beta} \tag{94}$$

where $\partial/\partial s_\alpha$ and $\partial/\partial s_\beta$ are space derivatives along the α- and β-lines respectively, their relative sense being such that they form a right-handed pair. Introduce a new arbitrary Cartesian coordinate system x_1y_1 whose orientation is determined by angle θ, measured anti-clockwise from the x_1-axis to the tangent to the α-line. It follows from geometric considerations (Figure 17) that

$$\frac{\partial}{\partial x_1} = \cos\theta \frac{\partial}{\partial s_\alpha} - \sin\theta \frac{\partial}{\partial s_\beta}, \quad \frac{\partial}{\partial y_1} = \sin\theta \frac{\partial}{\partial s_\alpha} + \cos\theta \frac{\partial}{\partial s_\beta} \tag{95}$$

Using (93) and (94) and rotating the x_1y_1-coordinate system such that its x_1-axis coincides with the tangent to the α-line (i.e. putting $\theta = 0$) equations (95) can be reduced to

$$\frac{\partial}{\partial x_1} = \frac{\partial}{R\partial \alpha}, \quad \frac{\partial}{\partial y_1} = -\frac{\partial}{S\partial \beta} \tag{96}$$

These equations are relations between the derivatives along the slip-lines and the derivatives along the Cartesian coordinates tangent to the slip-lines at a given point.

It is convenient to introduce velocity components u and v referred to the α- and β-lines. The velocity components in the Cartesian coordinates x_1 and y_1, u_{x1} and u_{y1}, are given by (Figure 17)

$$u_{x1} = u\cos\theta - v\sin\theta, \quad u_{y1} = u\sin\theta + v\cos\theta \tag{97}$$

The shear strain rate in the Cartesian coordinates is defined by

$$\xi_{x1y1} = \frac{1}{2}\left(\frac{\partial u_{x1}}{\partial y_1} + \frac{\partial u_{y1}}{\partial x_1}\right) \tag{98}$$

If $\theta = 0$, then $\xi_{x1y1} = \xi_{\alpha\beta}$ where $\xi_{\alpha\beta}$ is the shear strain rate in the $\alpha\beta$-coordinates. Substituting (97) into (98), putting $\theta = 0$ and using (96) it is possible to arrive at

$$2\xi_{\alpha\beta} = \frac{1}{R}\left(\frac{\partial v}{\partial \alpha} + u\right) - \frac{1}{S}\left(\frac{\partial u}{\partial \beta} - v\right) \tag{99}$$

The velocity field is singular at maximum friction surfaces if the surface coincides with an envelope of characteristics. With no loss of generality, it is possible to assume that an α-line is tangent to the friction surface. Then,

$$S = 0 \tag{100}$$

on the friction surface where $s = 0$ (see equation (30) and Figure 1). It is known that the normal strain rates in the $\alpha\beta$-coordinates vanish [33]. Therefore, equation (7) transforms to

$$\xi_{eq} = \frac{2}{\sqrt{3}}\left|\xi_{\alpha\beta}\right| \tag{101}$$

It follows from (100) that the first term on the right-hand side of (99) is negligible near the maximum friction surfaces. Therefore, equation (99) can be represented in the form

$$2\xi_{\alpha\beta} = -\frac{1}{S}\left(\frac{\partial u}{\partial \beta} - v\right) \tag{102}$$

as $s \to 0$. Substituting (102) into (101) and comparing the resulting expression and (30) it is possible to conclude that

$$S = S_0\sqrt{s} + o\left(\sqrt{s}\right) \quad \text{as} \quad s \to 0 \tag{103}$$

Since the y_1-direction in perpendicular to the friction surface, y_1 can be replaced with s in (96) and, then, the second equation of this system gives

$$\frac{\partial S}{\partial s} = -\frac{\partial S}{S\partial \beta} \tag{104}$$

With the use of (100) and (103) equation (104) leads to

$$S = -\frac{S_0^2}{2}(\beta - \beta_0) + o(\beta - \beta_0) \quad \text{as} \quad \beta \to \beta_0 \tag{105}$$

where β_0 is the value of β at the friction surface. It follows from (101)–(103) and (30) that

$$D = \frac{1}{\sqrt{3}Q}\left|\left(\frac{\partial u}{\partial \beta} - v\right)_{s=0}\right|$$
$$Q = \sqrt{-2(\partial S/\partial \beta)_{s=0}} \tag{106}$$

Thus the strain rate intensity factor can be calculated with the use of the distribution of S, u and v along β-lines in the vicinity of friction surfaces. All these quantities can be found from numerical solutions based on the method of characteristics with no conceptual difficulty.

A numerical technique based on the method of characteristics for axisymmetric flow of material obeying Tresca's yield criterion has been developed in [27].

6 Conclusions

An overview of a novel approach to predict the evolution of material properties in a narrow layer in the vicinity of frictional interfaces affected by manufacturing processes has been presented in this chapter. The key point of the approach is the strain rate intensity factor which naturally appears in solutions for several rigid plastic material models when the maximum friction law is one of the boundary conditions. The general concept of the approach is somehow analogous to that in the linear-elastic fracture mechanics where the stress intensity factor is widely adopted to predict crack propagation. First, the strain rate intensity factor is used to find the thickness of the layer of intensive plastic deformation near frictional interfaces. Then, it is supposed that the magnitude of the strain rate intensity factor controls the intensity of physical processes within this layer. In particular, the evolution of plastic damage has been studied in some detail by means of a non-local ductile fracture criterion. An illustrative example has been given with the use of some experimental data on plane strain extrusion of a brass specimen. An efficient numerical technique for calculating the strain rate intensity factor in plane strain problems for rigid perfectly plastic material has been proposed.

Acknowledgements The research described has been supported by the Russian Foundation for Basic Research (project 10-08-00083).

References

1. Alexandrov, S.: Interrelation between constitutive laws and fracture criteria in the vicinity of friction surfaces. In: Bouchaud, E., Jeulin, D., Prioul, C., Roux, S. (eds.) Physical Aspects of Fracture. Kluwer, Dordrecht (2001)
2. Alexandrov, S.: Comparison of double-shearing and coaxial models of pressure-dependent plastic flow at frictional boundaries. Trans. ASME J. Appl. Mech. 70, 212–219 (2003)
3. Alexandrov, S.: Singular solutions in an axisymmetric flow of a medium obeying the double shear model. J. Appl. Mech. Techn. Physics 46, 766–771 (2005)
4. Alexandrov, S.: Steady penetration of a rigid cone into pressure-dependent plastic material. Int. J. Solids Struct. 43, 193–205 (2006)
5. Alexandrov, S.E., Baranova, I.D., Mishuris, G.: Compression of a viscoplastic layer between rough parallel plates. Mech. Solids 43, 863–869 (2008)
6. Alexandrov, S., Grabco, D., Lyamina, E., Shikimaka, O.: An approach to prediction of evolution of material properties in the vicinity of frictional interfaces in metal forming. Key Engng. Mater. 345/346, 741–744 (2007)
7. Alexandrov, S., Grabko, D., Shikimaka, O.: The determination of the thickness of a layer of intensive deformations in the vicinity of the friction surface in metal forming processes. J. Mach. Manuf. Reliab. 38, 277–282 (2009)
8. Alexandrov, S., Harris, D.: Comparison of solution behaviour for three models of pressure-dependent plasticity: A simple analytical example. Int. J. Mech. Sci. 48, 750–762 (2006)
9. Alexandrov, S., Lyamina, E.: Singular solutions for plane plastic flow of pressure-dependent materials. Dokl. Phys. 47, 308–311 (2002)
10. Alexandrov, S., Lyamina, E.: Compression of a mean-stress sensitive plastic material by rotating plates. Mech. Solids 38, 40–48 (2003)
11. Alexandrov, S., Lyamina, E.: Plane-strain compression of material obeying the double-shearing model between rotating plates. Int. J. Mech. Sci. 45, 1505–1517 (2003)
12. Alexandrov, S., Lyamina, E.: Application of a non-local criterion to fracture prediction in plane-strain drawing. J. Technol. Plasticity 30, 53–59 (2005)
13. Alexandrov, S., Lyamina, E.: Qualitative distinctions in the solutions based on the plasticity theories with Mohr–Coulomb yield criterion. J. Appl. Mech. Techn. Phys. 46, 883–890 (2005)
14. Alexandrov, S., Lyamina, E.: Flow of pressure-dependent plastic material between two rough conical walls. Acta Mech. 187, 37–53 (2006)
15. Alexandrov, S., Lyamina, E.: Prediction of fracture in the vicinity of friction surfaces in metal forming processes. J. Appl. Mech. Techn. Phys. 47, 757–761 (2006)
16. Alexandrov, S., Lyamina, E.: A Nonlocal criterion of fracture near a friction surface and its application to analysis of drawing and extrusion processes. J. Mach. Manufact. Reliabil. 36, 262–267 (2007)
17. Alexandrov, S., Mishuris, G.: Viscoplasticity with a saturation stress: distinguished features of the model. Arch. Appl. Mech. 77, 35–47 (2007)
18. Alexandrov, S., Mishuris, G.: Qualitative behaviour of viscoplastic solutions in the vicinity of maximum-friction surfaces. J. Engng. Math. 65, 143–156 (2009)
19. Alexandrov, S., Richmond, O.: Asymptotic behavior of the velocity field in the case of axially symmetric flow of a material obeying the Treska condition. Dokl. Phys. 43, 362–364 (1998)

20. Alexandrov, S., Richmond, O.: Singular plastic flow fields near surfaces of maximum friction stress. Int. J. Non-Linear Mech. 36, 1–11 (2001)
21. Appleby, E.J., Devenpeck, M.K., Lu, C.Y., Rao, R.S., Richmond, O., Wright, P.K.: Strip drawing: A theoretical-experimental comparison. Int. J. Mech. Sci. 26, 351–362 (1984)
22. Atkins, A.G.: Fracture in forming. J. Mater. Process. Technol. 56, 609–618 (1996)
23. Aukrust, T., LaZghab, S.: Thin shear boundary layers in flow of hot aluminium. Int. J. Plast. 16, 59–71 (2000)
24. Chandrasekar, S., Farris, T.N., Kompella, S., Moylan, S.P.: A new approach for studying mechanical properties of thin surface layers affected by manufacturing processes. Trans. ASME J. Manuf. Sci. Engng. 125, 310–315 (2003)
25. Collins, I.F., Meguid, S.A.: On the influence of hardening and anisotropy on the plane-strain compression of thin metal strip. Trans. ASME J. Appl. Mech. 44, 271–278 (1977)
26. Dean, T.A., Lin, J.: Modelling of microstructure evolution in hot forming using unified constitutive equations. J. Mater. Process. Technol. 167, 354–362 (2005)
27. Druyanov, B.A., Nepershin, R.I.: Problems of Technological Plasticity. Elsevier, Amsterdam (1994)
28. Duan, X., Sheppard, T.: Simulation of substructural strengthening in hot flat rolling. J. Mater. Process. Technol. 125/126, 179–187 (2002)
29. Dutton, R.E., Goetz, R.L., Shamasundar, S., Semiatin, S.L.: The ring test for P/M materials. Trans. ASME J. Manuf. Sci. Engng. 120, 764–769 (1998)
30. El-Domiaty, A.A., Kandil, M.A., Shabara, M.A.: Validity assessment of ductile fracture criteria in cold forming. J. Mater. Engng. Perform. 5, 478–488 (1996)
31. Grekova, E.F., Harris, D.: A hyperbolic well-posed model for the flow of granular materials. J. Engng. Math. 52, 107–135 (2005)
32. Graggs, J.W.: Characteristic surfaces in ideal plasticity in three dimensions. Quart. J. Mech. Appl. Math. 7, 35–39 (1954)
33. Hill, R.: The Mathematical Theory of Plasticity. Clarendon Press, Oxford (1950)
34. Kanninen, M.F., Popelar, C.H.: Advanced Fracture Mechanics. Oxford University Press, New York (1985)
35. Kao, A.S., Kuhn, H.A., Richmond, O., Spitzig, W.A.: Influence of superimposed hydrostatic pressure on bending fracture and formability of a low carbon steel containing globular sulfides. Trans. ASME J. Engng. Mater. Technol. 112, 26–30 (1990)
36. Kokovkhin, E.A., Trunina, T.A.: Formation of a finely dispersed structure in steel surface layers under combined processing using hydraulic pressing. J. Mach. Manuf. Reliab. 37, 160–162 (2008)
37. Lyamina, E.A.: Application of the method of characteristics to finding the strain rate intensity factor. In: Onate, E., Owen, R., Suarez, B. (eds.) Computational Plasticity Fundamentals and Applications. CIMNE, Barcelona (2007)
38. Marshall, E.A.: The compression of a slab of ideal soil between rough plates. Acta Mech. 3, 82–92 (1967)
39. Meguid, S.A.: Engineering Fracture Mechanics. Elsevier Applied Science, London (1989)
40. Moran, B., Norris, D.M., Quinones, D.F., Reaugh, J.E.: A plastic-strain, mean-stress criterion for ductile fracture. Trans. ASME J. Engng. Mater. Technol. 100, 279–286 (1978)
41. Pemberton, C.S.: Flow of imponderable granular materials in wedge-shaped channels. J. Mech. Phys. Solids 13, 351–360 (1965)

42. Richmond, O., Spitzig, W.A., Sober, R.J.: The effect of hydrostatic pressure on the deformation behavior of maraging and HY-80 steels and its implications for plasticity theory. Metall. Trans. 7A, 1703–1710 (1976)
43. Shabaik, A., Vujovic, V.: Workability criteria for ductile fracture. Trans. ASME J. Engng. Mater. Technol. 108, 245–249 (1986)
44. Shield, R.T.: Plastic flow in a converging conical channel. J. Mech. Phys. Solids 3, 246–258 (1955)
45. Spencer, A.J.M.: A theory of the kinematics of ideal soils under plane strain conditions. J. Mech. Phys. Solids 12, 337–351 (1964)

Unilateral Problems for Laminates: A Variational Formulation with Constraints in Dual Spaces

Franco Maceri and Giuseppe Vairo

Abstract In this paper, models for laminated composite plates accounting for elastic bimodular constitutive behavior and frictionless unilateral contact conditions are established and rationally deduced from the three-dimensional elasticity by means of a variational constrained approach. Consistent internal constraints on both stress and strain dual fields are enforced through a modified Hu–Washizu-type functional, defined on the convex set of the compatible displacements. A bimodular strain energy density is adopted and for the first-order shear-deformable (Reissner–Mindlin type) laminate model a variational formulation of Signorini's problem is recovered. The rational deduction of a Lo–Christensen–Wu-type model for bimodular laminate on unilateral support is also outlined and briefly discussed.

1 Introduction

Technological development of composite laminates, used in a variety of complex structures, e.g. in space, automotive and civil applications, may be clearly related to their increasingly better performance-to-weight ratios in comparison with the homogeneous case. Such better performance is deeply related to the constitutive anisotropic response of laminated plates. As an example, a complex behavior involving extension-bending coupling appears. Furthermore, composite materials usually employed to obtain laminate's layers are characterized by small values of shear moduli along the thickness direction in comparison with the longitudinal in-plane ones and, as a consequence, non-negligible shear deformations in the thickness are often induced. Accordingly, suitable and consistent modelling of such effects is necessary.

Because of their specific geometry (thickness dimension fairly smaller than the others) the analysis is generally carried out by means of approximate two-

Franco Maceri · Giuseppe Vairo
Department of Civil Engineering, University of Rome "Tor Vergata", via del Politecnico 1, 00133 Rome, Italy; e-mail: franco.maceri@lagrange.it, vairo@ing.uniroma2.it

G. Zavarise & P. Wriggers (Eds.): Trends in Computational Contact Mechanics, LNACM 58, pp. 321–338.
springerlink.com

dimensional models. Many different theoretical and numerical formulations, based on classical plate theories, can be found in the specialized literature, accounting for different refinements of shear effects [1–6]. Nevertheless, the rational deduction of these theories from three-dimensional elasticity (in particular for anisotropic materials as well as for non-conventional cases such as unilateral material behavior and contact problems) can be truly considered as an open task yet, to achieve a safer and more guaranteed technical use of such theories.

Rational deduction of structural theories is performed mainly through two strategies: the asymptotic method and the constrained approach. The first one is based on the main idea that the three-dimensional solution of the elasticity equations can be approximated through successive terms of a series where, for plates, the slenderness ratio between plate thickness and its characteristic in-plane dimension is taken as a small parameter. Accordingly, under hypotheses ensuring series convergence and varying series truncation order, different structural theories can be rationally deduced as approximate solutions of an exactly-stated problem [7–10]. On the contrary, constrained methods aim to find exact solutions of simplified constrained problems, i.e. problems formulated enforcing approximate representations of the unknown functions. Accordingly, the three-dimensional elastostatic problem is reduced to a consistent simplified one (two-dimensional in the case of plates and shells) by adopting suitable assumptions, regarded as internal constraints, on strain and/or stress fields. That approach was successfully employed for deducing classical plate and shell theories [11–13], layer-wise laminate theories [14], as well as theories of beams with solid [15, 16] or thin-walled [17] cross-sections. Podio-Guidugli and co-workers [12, 13, 15] proposed a constrained approach based on strain assumptions and on the concept of constrained material, by introducing an ad hoc constitutive law. On the contrary, the constrained approach proposed in [11] leaves unchanged the constitutive material response, involving contemporaneously consistent assumptions on both strain and stress dual fields. Nevertheless, while strain assumptions can be easily identified when the problem is characterized by special geometries, effective and consistent stress assumptions can be sometimes not obvious. In order to overcome this difficulty, Maceri and co-workers [16, 17] showed how plate, shell and beam theories can be justified enforcing in a consistent way the same constraints on both stress and strain dual fields.

In the context of the classical theory of elasticity many theoretical and numerical formulations relevant to plate and beam models accounting for continuous contact with rigid or elastic foundation as well as for infinite or finite beam/plate dimensions have been proposed in the last century. Many references and arguments can be found for example in [18–26]. Nevertheless, such classical approaches enforce a-priori kinematic representations which in turn are straightly coupled with contact assumptions, often without a consistent and rational deduction within a general three-dimensional framework. As confirmed in [27] focusing beam models, the dual-constraint approach can be successfully employed for developing by a 3D variational formulation general structural theories which naturally include unilateral material behavior as well as contact constraints.

In this paper, this dual-constraint approach is proved to be effective for deducing in a consistent way equivalent single-layer classical laminate plate theories, and for generalizing them to the case of anisotropic elastic bimodular materials and frictionless unilateral contact problems. In detail, we refer to the case of elastic Bert-type bimodular materials [28–30], roughly described by a linear relationship between stress and strain both in tension and in compression but with different elastic moduli. Bimodular behavior usually characterizes the constitutive response of many composite materials employed for structural and biomechanical applications [31–34], and belong to the class of conewise constitutive responses addressed in [35], based on continuous (not necessarily differentiable) and convex elastic potentials.

This paper is organized as follows. In Section 2 the equilibrium problem of the laminated plate, thought as a three-dimensional body and unilaterally supported by a frictionless rigid foundation, is formulated and recast by a Hu–Washizu-type variational formulation, obtained by including a bimodular strain energy density and introducing the contact constraint in the convex set of the admissible displacements. In Section 3, the three-dimensional problem is reduced to a bi-dimensional one by introducing simultaneous constraints on dual spaces (stress and strain) by a non-standard application of the Lagrange multipliers theory [36]. Accordingly, field and boundary equations for equivalent single-layer laminate models are rationally deduced as stationary conditions of a constrained Lagrangian functional, recovering a variational formulation of Signorini's problem [37, 38]. In detail, in Section 3.1 this approach is successfully applied to the first-order shear-deformable laminate plate theory, whereas Section 3.2 presents the rational deduction of a more refined Lo–Christensen–Wu-type model for bimodular laminate on unilateral support, highlighting effectiveness and generality of the employed constrained approach.

2 Unilateral Problems for a Laminated Plate

Let a plate-like cylindrical body $\Omega = \mathscr{P} \times]-h,h[$ be considered, whose uniform thickness is $2h$ and whose undeformed middle cross section is $\mathscr{P}$. A Cartesian frame (O,x_1,x_2,x_3) is introduced with x_1 and x_2-axes parallel to $\mathscr{P}$. The body Ω is assumed to be a laminate, made of n perfectly bonded layers such that

$$\Omega = \cup_{k=1}^{n} \Omega^{(k)} = \cup_{k=1}^{n} \mathscr{P} \times]x_3^{(k)}, x_3^{(k+1)}[\tag{1}$$

with $x_3^{(1)} = -h$ and $x_3^{(n+1)} = h$.

As a notation rule, where necessary, Cartesian components will be denoted by subscripts: Latin indices imply values $\{1,2,3\}$, while Greek indices imply values $\{1,2\}$. Einstein's summation convention will be adopted and the partial derivative of a function f with respect to x_i will be denoted as $f_{/i}$. Moreover, quantities relevant to the kth layer will be indicated by the superscript (k).

Top and bottom surfaces of Ω will be denoted as $\mathscr{P}^+ = \mathscr{P} \times \{h\}$ and $\tilde{\mathscr{P}}^- = \mathscr{P} \times \{-h\}$, respectively. The boundary $\partial\mathscr{P}$ of $\mathscr{P}$ is assumed to be subdivided in

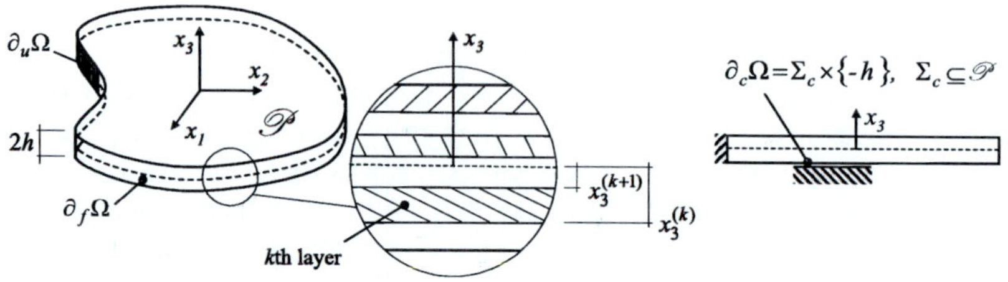

Fig. 1 Laminated plate on unilateral support: notation.

two complementary parts, $\partial_u\mathscr{P}$ and $\partial_f\mathscr{P}$. Such a subdivision subordinates a partition of lateral boundary of the laminate Ω in $\partial_u\Omega = \partial_u\mathscr{P}\times]-h,h[$, meas$(\partial_u\Omega) > 0$, and $\partial_f\Omega = \partial_f\mathscr{P}\times]-h,h[$, where the displacement $\mathbf{u}_o$ and the surface tractions $\hat{\mathbf{p}}$ are given, respectively.

Moreover, the laminate is unilaterally supported by a frictionless rigid foundation with a zero initial gap and we assume that the contact surface is a part $\partial_c\Omega$ of the bottom surface $\tilde{\mathscr{P}}^-$, such that $\partial_c\Omega = \Sigma_c\times\{-h\}\subseteq\tilde{\mathscr{P}}^-$, with $\Sigma_c\subseteq\mathscr{P}$ (see Figure 1). In the following we assume that the body undergoes small displacements from its reference configuration. As a consequence, the unit normal vector to the actual configuration of the plate supported boundary can be identified as the unit vector $\mathbf{n}$ normal to $\partial_c\Omega$.

The laminated plate is assumed to be in equilibrium when body forces $\mathbf{b}$ act upon Ω, surface tractions $\hat{\mathbf{p}}$ act upon the part $\partial_f\Omega$ of the lateral boundary, and surface forces $\mathbf{p}^+$ and $\mathbf{p}^-$ are given at top and bottom surfaces $\mathscr{P}^+$ and $\mathscr{P}^-$, respectively, with $\mathscr{P}^- = \tilde{\mathscr{P}}^-\setminus\partial_c\Omega$.

Each layer (lamina) k is assumed to be homogeneous and made of an elastic (in general, bimodular) material, characterized by a continuous and convex strain energy density $\Phi^{(k)}(\boldsymbol{\varepsilon})$, $\boldsymbol{\varepsilon}$ being the strain field.

The equilibrium problem of the laminated plate Ω belongs to the class of contact problems studied by Signorini [38] and, in the framework of the infinitesimal deformation theory and regarding Ω as a three-dimensional body, it can be recast by adopting the following constrained Hu–Washizu-type variational formulation:

Find *the displacement* $\mathbf{u}$, *the strain field* $\boldsymbol{\varepsilon}$ *and the stress field* $\boldsymbol{\sigma}$ *that make stationary the functional*

$$\begin{aligned}\mathscr{W}(\mathbf{u},\boldsymbol{\varepsilon},\boldsymbol{\sigma}) &= \int_\Omega \Phi(\boldsymbol{\varepsilon})\,dv + \int_\Omega \boldsymbol{\sigma}\cdot(\hat{\nabla}\mathbf{u}-\boldsymbol{\varepsilon})\,dv - \int_\Omega \mathbf{b}\cdot\mathbf{u}\,dv \\ &\quad - \int_{\partial_f\Omega}\hat{\mathbf{p}}\cdot\mathbf{u}\,da - \int_{\mathscr{P}}\mathbf{p}^+\cdot\mathbf{u}|_{x_3^{(n+1)}}\,da - \int_{\mathscr{P}\setminus\Sigma_c}\mathbf{p}^-\cdot\mathbf{u}|_{x_3^{(1)}}\,da \qquad (2)\end{aligned}$$

under the constraint $\mathbf{u}\in\mathbf{K}$, *where the convex set* $\mathbf{K}$ *is defined as*

$$\mathbf{K} = \{\mathbf{v} \in \mathbf{V} \mid \mathbf{v} \cdot \mathbf{n} \leq 0 \ on \ \partial_c \Omega\} \tag{3}$$

In Eq. (3) the space $\mathbf{V}$ of the admissible displacements is a normed linear space of real, vector-valued, adequately regular measurable functions defined on Ω, such that $\mathbf{v} \in \mathbf{V}$ implies $\mathbf{v} = \mathbf{u}_o$ on $\partial_u \Omega$. The space $\mathbf{V}$ is partially ordered, and the statement '$\mathbf{w} \cdot \mathbf{n} \leq \mathbf{v} \cdot \mathbf{n}$ on $\partial_c \Omega$' for $\mathbf{v}, \mathbf{w} \in \mathbf{V}$ is meaningful [37].

In Eq. (2) the symbol $\hat{\nabla}$ denotes the symmetrical part of the gradient operator, '$\cdot$' the inner product, $\mathbf{n}$ the outward normal unit vector to $\partial \Omega$, and $\Phi(\boldsymbol{\varepsilon})$ the material strain energy density, such that $\Phi = \Phi^{(k)}$ on $\Omega^{(k)}$. Stationary conditions of $\mathscr{W}$ with respect to $\mathbf{u}$, $\boldsymbol{\sigma}$ and $\boldsymbol{\varepsilon}$ yield equilibrium, compatibility and constitutive equations governing the three-dimensional elastostatic problem for the unilaterally-supported laminate Ω.

In the following we assume that each layer comprises an elastic bimodular material, having at least a monoclinic symmetry, with symmetry plane parallel to $\mathscr{P}$. Accordingly, the elastic response is identified by a linear relationship between stress and strain both in tension and in compression, but with different elastic moduli. Referring to the case of fiber-reinforced composite layers, we consider a Bert-type constitutive behavior where the bimodularity depends on the sign of the unit elongation in the fiber-direction. As proved by Bisegna and co-workers [29], for such a class of materials it is possible to deduce a consistent constitutive law by assuming the existence of an elastic potential, the strain energy density Φ.

Let $\mathbf{f}$ be the unit vector along the fiber direction and $\varepsilon_f = \boldsymbol{\varepsilon}\mathbf{f} \cdot \mathbf{f}$ the extension along $\mathbf{f}$. Moreover, let the following definitions be introduced:

$$E^+ = \{\boldsymbol{\varepsilon} : \varepsilon_f > 0\} \qquad E^o = \{\boldsymbol{\varepsilon} : \varepsilon_f = 0\} \qquad E^- = \{\boldsymbol{\varepsilon} : \varepsilon_f < 0\} \tag{4}$$

Accordingly, restrictions of Φ to E^+ and E^- are the potentials for the mappings:

$$\boldsymbol{\varepsilon} \in E^+ \mapsto \mathscr{C}^+ \boldsymbol{\varepsilon} \qquad \boldsymbol{\varepsilon} \in E^- \mapsto \mathscr{C}^- \boldsymbol{\varepsilon} \tag{5}$$

and the fourth order constitutive tensors $\mathscr{C}^+$ and $\mathscr{C}^-$ (relevant to tension and compression response, respectively) satisfy major and minor symmetries. As a direct consequence of this conservative (or hyperelastic) behavior, the potential $\Phi(\boldsymbol{\varepsilon})$ is continuous [29]. Therefore, restrictions of Φ to E^+ and E^- can be extended by continuity to E^o, provided that the following equality is satisfied:

$$\mathscr{C}^+ \boldsymbol{\varepsilon} \cdot \boldsymbol{\varepsilon} = \mathscr{C}^- \boldsymbol{\varepsilon} \cdot \boldsymbol{\varepsilon}, \quad \forall\ \boldsymbol{\varepsilon} \in E^o \tag{6}$$

Equations (5) and (6) imply that the material strain energy density Φ is convex and it can be written (omitting constant contributions) in the form

$$\Phi(\boldsymbol{\varepsilon}) = \frac{1}{2} \mathscr{C}(\varepsilon_f) \boldsymbol{\varepsilon} \cdot \boldsymbol{\varepsilon} \tag{7}$$

where

$$\mathscr{C}(\varepsilon_f) = [\mathscr{H} \mathscr{C}^+ + (1 - \mathscr{H}) \mathscr{C}^-] \tag{8}$$

$\mathscr{H}(\varepsilon_f)$ being the Heaviside function, such that $\mathscr{H}=1$ when ε_f is positive, and $\mathscr{H}=0$ otherwise.

It is worth pointing out that different fourth order constitutive tensors $\mathscr{C}^+$ and $\mathscr{C}^-$ can be in general introduced for each layer, so that $\mathscr{C}=\mathscr{C}^{(k)}$ on $\Omega^{(k)}$ and the first integral in functional (2) can be arranged as

$$\begin{aligned}\int_\Omega \Phi(\boldsymbol{\varepsilon})\,dv &= \sum_{k=1}^n \int_{\Omega^{(k)}} \Phi^{(k)}(\boldsymbol{\varepsilon})\,dv \\ &= \sum_{k=1}^n \int_{\Omega^{(k)}} \frac{1}{2}[\mathscr{H}^{(k)}\mathscr{C}^{(k)+} + (1-\mathscr{H}^{(k)})\mathscr{C}^{(k)-}]\,\boldsymbol{\varepsilon}\cdot\boldsymbol{\varepsilon}\,dv \\ &= \sum_{k=1}^n \left[\int_{\Omega^{(k)}|_{\mathscr{C}^+}} \frac{1}{2}\mathscr{C}^{(k)+}\,\boldsymbol{\varepsilon}\cdot\boldsymbol{\varepsilon}\,dv + \int_{\Omega^{(k)}|_{\mathscr{C}^-}} \frac{1}{2}\mathscr{C}^{(k)-}\,\boldsymbol{\varepsilon}\cdot\boldsymbol{\varepsilon}\,dv\right] \qquad (9)\end{aligned}$$

wherein $\mathscr{H}^{(k)}=\mathscr{H}(\varepsilon_f^{(k)})$ and $\Omega^{(k)}|_{\mathscr{C}^\pm}$ denotes the region of the kth layer where $\mathscr{C}^\pm$ applies. Since the previously-stated material symmetry, tension/compression constitutive fourth order tensors satisfy in each layer the condition $\mathscr{C}^\pm_{\alpha\beta\gamma 3}=\mathscr{C}^\pm_{\alpha 333}=0$.

3 Laminated Models by Using Constraints in Dual Spaces: A Single-Layer Approach

Assuming $(2h)\ll \mathrm{diam}(\mathscr{P})$, the three-dimensional unilateral equilibrium problem for Ω can be approximated using suitable assumptions on strain and/or stress fields. If these assumptions are regarded as internal frictionless constraints, reactive fields arise and the original three-dimensional elastic problem can be replaced by a constrained problem which can be often solved more easily.

In order to enforce constraints on both strain and stress dual fields, the constrained equilibrium problem can be suitably formulated employing Lagrange multipliers, representing reactive actions belonging to the dual space of the one where the constrained variable lives. As proved in previous works [11, 16], the consistent representation of such reactive fields arises as a consequence of the enforced constraints and it is not postulated a priori.

Let the following definitions be introduced: the total strain field is the symmetrical part of the gradient of the displacement field; the total stress field satisfies the equilibrium equations; elastic stresses and strains are related to each other by the elastic constitutive law; the total stress (or strain) is sum of its elastic and reactive parts.

In order to build up structural theories, a special representation law for the displacement field is adopted (that is equivalent to impose constraints on the total strains), and some assumptions on the stress field are made at the constitutive law level (i.e., on the elastic stress field).

Let the total strain and the elastic stress fields be constrained to belong to the kernel of the linear (possibly differential) operators **G** and **H**, respectively. These constraints act on dual spaces and can be enforced by introducing the following Lagrangian functional [11]:

$$\mathscr{L}(\mathbf{u},\boldsymbol{\varepsilon},\boldsymbol{\sigma},\boldsymbol{\chi},\boldsymbol{\omega}) = \mathscr{W}(\mathbf{u},\boldsymbol{\varepsilon},\boldsymbol{\sigma}) - \int_\Omega \boldsymbol{\chi}\cdot\mathbf{G}\boldsymbol{\varepsilon}\,dv - \int_\Omega \boldsymbol{\omega}\cdot\mathbf{H}\boldsymbol{\sigma}\,dv - \int_\Omega \mathbf{G}^*\boldsymbol{\chi}\cdot\mathbf{H}^*\boldsymbol{\omega}\,dv \tag{10}$$

under the constraint $\mathbf{u}\in\mathbf{K}$ and where vectors $\boldsymbol{\chi}$, $\boldsymbol{\omega}$ are Lagrange multipliers, $\mathbf{G}^*$ and $\mathbf{H}^*$ denoting the adjoint operators of **G** and **H**, respectively.

Following Maceri and co-workers [16, 17] we assume that dual constraints on the elastic stress field are the same as those imposed on the total strain field, that is operators **G** and **H** are chosen such that

$$\mathbf{HA} = \mathbf{GA} \tag{11}$$

for every symmetrical second order tensor **A**, resulting in Lagrange multipliers $\boldsymbol{\chi}$ and $\boldsymbol{\omega}$ which belong to dual vector subspaces characterized by the same dimensions. In this way, once kinematic constraints are chosen, consistent stress assumptions directly arise.

Employing the rule (11) and Eqs. (7) and (8), the stationary condition of $\mathscr{L}$ with respect to $\boldsymbol{\sigma}$ yields the compatibility equation

$$\boldsymbol{\varepsilon} + \mathbf{G}^*\boldsymbol{\omega} = \hat{\nabla}\mathbf{u} \quad \text{in } \Omega \tag{12}$$

the one with respect to $\boldsymbol{\varepsilon}$ yields the constitutive equation

$$\boldsymbol{\sigma} + \mathbf{G}^*\boldsymbol{\chi} = \mathscr{C}\boldsymbol{\varepsilon} \quad \text{in } \Omega \tag{13}$$

the one with respect to **u**, constrained by $\mathbf{u}\in\mathbf{K}$, yields the equilibrium equations

$$\begin{aligned} &\operatorname{div}\boldsymbol{\sigma} + \mathbf{b} = \mathbf{0} && \text{in } \Omega \\ &\boldsymbol{\sigma}\mathbf{n} = \hat{\mathbf{p}} && \text{on } \partial_f\Omega \\ &\boldsymbol{\sigma}\mathbf{n} = \mathbf{p}^{\pm} && \text{on } \mathscr{P}^{\pm} \end{aligned} \tag{14}$$

and Signorini's conditions

$$\begin{aligned} &\sigma_n \le 0 \\ &\boldsymbol{\sigma}_T = \mathbf{0} \qquad \text{on } \partial_c\Omega \\ &\sigma_n u_n = 0 \end{aligned} \tag{15}$$

where $\sigma_n = (\boldsymbol{\sigma}\mathbf{n})\cdot\mathbf{n}$, $\boldsymbol{\sigma}_T = \boldsymbol{\sigma}\mathbf{n} - \sigma_n\mathbf{n}$ and $u_n = \mathbf{u}\cdot\mathbf{n}$.

Finally, the stationary conditions of $\mathscr{L}$ with respect to the Lagrange multipliers $\boldsymbol{\chi}$ and $\boldsymbol{\omega}$ yield, respectively, the constraint equations

$$\begin{aligned} \mathbf{G}(\boldsymbol{\varepsilon}+\mathbf{G}^*\boldsymbol{\omega}) &= \mathbf{0} \\ \mathbf{G}(\boldsymbol{\sigma}+\mathbf{G}^*\boldsymbol{\chi}) &= \mathbf{0} \end{aligned} \quad \text{in } \Omega \tag{16}$$

Accordingly, by Eqs. (12), (13), and (14), $\boldsymbol{\sigma}$ and $\boldsymbol{\varepsilon}+\mathbf{H}^*\boldsymbol{\omega}$ turn out to be total stress and strain fields, respectively, and $\boldsymbol{\sigma}+\mathbf{G}^*\boldsymbol{\chi}$ and $\boldsymbol{\varepsilon}$ elastic stress and strain fields, respectively. As a consequence, reactive stress and strain fields are $-\mathbf{G}^*\boldsymbol{\chi}$ and $\mathbf{G}^*\boldsymbol{\omega}$, respectively. It is worth noting that the reactive stress field is orthogonal to every admissible total strain field and likewise the reactive strain field is orthogonal to every admissible elastic stress field, as it appears from Eq. (16).

It is pointed out that, as a consequence of the continuity condition (6), making $\mathscr{L}$ in (10) stationary with respect to $\boldsymbol{\varepsilon}$ yields Eq. (13) without jumping terms (depending on the difference between $\mathscr{C}^+$ and $\mathscr{C}^-$). Moreover, strain along the fiber-direction $\varepsilon_f^{(k)}$ for the kth layer, whose sign discriminates the unilateral constitutive behavior, has to be considered as deduced from the elastic strain field.

For what follows, functional $\mathscr{L}$ in (10) is conveniently transformed in a potential energy functional, by enforcing a priori satisfied stationary conditions of $\mathscr{L}$ with respect to $\boldsymbol{\sigma}$ and $\boldsymbol{\varepsilon}$, that is Eqs. (12) and (13). Accordingly, functional $\mathscr{L}$ becomes

$$\begin{aligned} \mathscr{E}(\mathbf{u},\boldsymbol{\chi},\boldsymbol{\omega}) = & \int_\Omega \Phi(\hat{\nabla}\mathbf{u}-\mathbf{G}^*\boldsymbol{\omega})dv - \int_\Omega \boldsymbol{\chi}\cdot\mathbf{G}\hat{\nabla}\mathbf{u}\,dv - \int_\Omega \mathbf{b}\cdot\mathbf{u}\,dv \\ & - \int_{\partial_f\Omega} \hat{\mathbf{p}}\cdot\mathbf{u}\,da - \int_{\mathscr{P}} \mathbf{p}^+\cdot\mathbf{u}\big|_{x_3^{(n+1)}}da - \int_{\mathscr{P}\setminus\Sigma_c} \mathbf{p}^-\cdot\mathbf{u}\big|_{x_3^{(1)}}da \end{aligned} \tag{17}$$

under the constraint $\mathbf{u}\in\mathbf{K}$ and where Φ is defined by Eq. (7). It can be verified that total strain and elastic strain are $\boldsymbol{\Lambda}=\hat{\nabla}\mathbf{u}$ and $\boldsymbol{\varepsilon}=\hat{\nabla}\mathbf{u}-\mathbf{G}^*\boldsymbol{\omega}$, respectively, and total stress and elastic stress are $\boldsymbol{\sigma}=\mathscr{C}(\hat{\nabla}\mathbf{u}-\mathbf{G}^*\boldsymbol{\omega})-\mathbf{G}^*\boldsymbol{\chi}$ and $\boldsymbol{\sigma}^{(\mathrm{el})}=\mathscr{C}(\hat{\nabla}\mathbf{u}-\mathbf{G}^*\boldsymbol{\omega})$, respectively.

It should be noted that $\mathscr{E}$ depends on the reactive fields. In order to obtain a potential-energy functional that does not depend on the Lagrange multipliers it is sufficient to make stationary conditions of $\mathscr{E}$ with respect to $\boldsymbol{\chi}$ and $\boldsymbol{\omega}$ a priori satisfied.

3.1 First-Order Shear-Deformable Laminated-Plate Model

In order to characterize an equivalent single-layer Reissner–Mindlin-type model, the following hypotheses on the total strain field are considered:

(i) The total strain component ε_{33} in the thickness direction is assumed to vanish everywhere.
(ii) The shear total strain between the x_3-axis and $\mathscr{P}$ is assumed to be constant in x_3.

Accordingly, operator $\mathbf{G}$ is such that

$$\mathbf{G}\boldsymbol{\varepsilon} = \{\varepsilon_{13/3}\ \varepsilon_{23/3}\ \varepsilon_{33}\}^T \tag{18}$$

Therefore (see Eq. (11)) dual constraints on the elastic stress field turn out to be directly expressed as:

(iii) The elastic out-of-plane stress σ_{33} is assumed vanish everywhere;
(iv) The shear elastic stress between the x_3-axis and $\mathscr{P}$ is assumed to be constant in x_3 at every position in $\Omega^{(k)}|_{\mathscr{C}\pm}$.

Accordingly, the functional (17) can be written as

$$\begin{aligned}\mathscr{E}(\mathbf{u},\boldsymbol{\chi},\boldsymbol{\omega}) = & \frac{1}{2}\int_\Omega \{\mathscr{C}_{\alpha\beta\gamma\delta}\Lambda_{\alpha\beta}\Lambda_{\gamma\delta} + 2\mathscr{C}_{\alpha\beta33}\Lambda_{\alpha\beta}(u_{3/3}-\omega_{33}) \\ & + \mathscr{C}_{3333}(u_{3/3}-\omega_{33})^2 + 4\mathscr{C}_{\alpha3\beta3}(\Lambda_{\alpha3}+\omega_{\alpha3/3})(\Lambda_{\beta3}+\omega_{\beta3/3})\}\,dv \\ & - \int_\Omega (2\chi_{\alpha3}\Lambda_{\alpha3/3} + \chi_{33}u_{3/3})\,dv - \Pi_{\text{ext}}\end{aligned} \tag{19}$$

defined on the manifolds $\chi_{\alpha3}|_{\pm h} = 0$, $\omega_{\alpha3} = 0$ on $\partial\Omega^{(k)}|_{\mathscr{C}\pm}$, and under the constraint $\mathbf{u} \in \mathbf{K} = \{\mathbf{v} \in \mathbf{V} \mid v_3 n_3 \leq 0 \text{ on } \partial_c\Omega\}$. Moreover, $\boldsymbol{\Lambda} = \hat{\nabla}\mathbf{u}$, and Π_{ext} accounts for external loads:

$$\begin{aligned}\Pi_{\text{ext}} = & \int_\Omega (b_\alpha u_\alpha + b_3 u_3)\,dv + \int_{\partial_f\Omega} (\hat{p}_\alpha u_\alpha + \hat{p}_3 u_3)\,d\rho\, dx_3 \\ & + \int_{\mathscr{P}} (\overset{+}{p}_\alpha u_\alpha|_{x_3^{(n+1)}} + \overset{+}{p}_3 u_3|_{x_3^{(n+1)}})\,da + \int_{\mathscr{P}\setminus\Sigma_c} (\overset{-}{p}_\alpha u_\alpha|_{x_3^{(1)}} + \overset{-}{p}_3 u_3|_{x_3^{(1)}})\,da\end{aligned} \tag{20}$$

$d\rho$ being the arc element along $\partial\mathscr{P}$.

Stationary conditions of the functional (19) with respect to $\chi_{\alpha3}$ and χ_{33} give, respectively, the following constraints on the displacement field, i.e. on the total strain field:

$$\begin{aligned}(u_{\alpha/3} + u_{3/\alpha})_{/3} &= 0 \\ u_{3/3} &= 0\end{aligned} \tag{21}$$

from which, by integration, the displacement field is obtained:

$$\begin{aligned}u_\alpha(x_1,x_2,x_3) &= s_\alpha(x_1,x_2) + x_3\,\varphi_\alpha(x_1,x_2) \\ u_3(x_1,x_2,x_3) &= w(x_1,x_2)\end{aligned} \tag{22}$$

constrained by the condition

$$w(x_1,x_2) \in \{f(x_1,x_2) \mid f(x_1,x_2)n_3 \leq 0 \text{ on } \Sigma_c\} \tag{23}$$

where $w(x_1,x_2)$ describes the deflection of $\mathscr{P}$, $s_\alpha(x_1,x_2)$ are the displacement components parallel to $\mathscr{P}$, and $\varphi_\alpha(x_1,x_2)$ describe the rotation of fibers parallel to the x_3-axis.

The stationary conditions of functional $\mathscr{E}$ with respect to $\omega_{\alpha 3}$ and ω_{33} give, respectively:

$$\left\{\mathscr{C}_{\alpha 3\beta 3}\left[(u_{\beta/3}+u_{3/\beta})/2+\omega_{\beta 3/3}\right]\right\}_{/3}=0$$

$$\mathscr{C}_{3333}(u_{3/3}-\omega_{33})+\mathscr{C}_{\alpha\beta 33}(u_{\alpha/\beta}+u_{\beta/\alpha})/2=0 \tag{24}$$

Solving Eqs. (24) with respect to $\omega_{\alpha 3}$ and ω_{33} and substituting into the functional (19) together with (22), the potential energy functional $\mathscr{E}$ is obtained in terms of pure displacement unknowns $s_\alpha(x_1,x_2)$, $\varphi_\alpha(x_1,x_2)$, $w(x_1,x_2)$. In detail, performing the integration over the thickness, it can be expressed as

$$\hat{\mathscr{E}}(s_\alpha,\varphi_\alpha,w)=\frac{1}{2}\int_{\mathscr{P}}[\mathbf{A}\mathbf{e}\cdot\mathbf{e}+2\mathbf{B}\boldsymbol{\kappa}\cdot\mathbf{e}+\mathbf{D}\boldsymbol{\kappa}\cdot\boldsymbol{\kappa}+\mathbf{H}\boldsymbol{\gamma}\cdot\boldsymbol{\gamma}]\,da$$

$$-\int_{\mathscr{P}}\mathbf{q}\cdot\mathbf{d}\,da-\int_{\partial_f\mathscr{P}}\hat{\mathbf{q}}\cdot\mathbf{d}\,d\rho \tag{25}$$

under the constraint (23), and where

- $\mathbf{e}$, $\boldsymbol{\kappa}$ and $\boldsymbol{\gamma}$ denote the membranal, curvature and shear generalized strain tensors, respectively

$$e_{\alpha\beta}=(s_{\alpha/\beta}+s_{\beta/\alpha})/2$$

$$\kappa_{\alpha\beta}=(\varphi_{\alpha/\beta}+\varphi_{\beta/\alpha})/2$$

$$\gamma_\alpha=w_{/\alpha}+\varphi_\alpha \tag{26}$$

- $\mathbf{A}$, $\mathbf{B}$, $\mathbf{D}$ (fourth-order) and $\mathbf{H}$ (second-order) are the laminate constitutive tensors (membrane, membrane-bending coupling, bending and shear, respectively), defined as

$$A_{\alpha\beta\gamma\delta}=\sum_{k=1}^{n}\int_{x_3^{(k)}}^{x_3^{(k+1)}}\hat{\mathscr{C}}^{(k)}_{\alpha\beta\gamma\delta}\,dx_3$$

$$B_{\alpha\beta\gamma\delta}=\sum_{k=1}^{n}\int_{x_3^{(k)}}^{x_3^{(k+1)}}x_3\hat{\mathscr{C}}^{(k)}_{\alpha\beta\gamma\delta}\,dx_3$$

$$D_{\alpha\beta\gamma\delta}=\sum_{k=1}^{n}\int_{x_3^{(k)}}^{x_3^{(k+1)}}x_3^2\hat{\mathscr{C}}^{(k)}_{\alpha\beta\gamma\delta}\,dx_3$$

$$H_{\alpha\beta}=\sum_{k=1}^{n}\int_{x_3^{(k)}}^{x_3^{(k+1)}}\mathscr{C}^{(k)}_{\alpha 3\beta 3}\,dx_3 \tag{27}$$

- **d** is the vector collecting the generalized displacement components

$$\mathbf{d}(x_1,x_2) = \{s_1 \quad s_2 \quad w \quad \varphi_1 \quad \varphi_2\}^T \tag{28}$$

- **q** and $\hat{\mathbf{q}}$ are the vectors of the generalized external forces

$$\mathbf{q}(x_1,x_2) = \{r_1 \quad r_2 \quad r_3 \quad m_1 \quad m_2\}^T$$
$$\hat{\mathbf{q}}(x_1,x_2) = \{\hat{r}_1 \quad \hat{r}_2 \quad \hat{r}_3 \quad \hat{m}_1 \quad \hat{m}_2\}^T \tag{29}$$

with

$$r_i = \int_{x_3^{(1)}}^{x_3^{(n+1)}} b_i\, dx_3 + p_i^+ + p_i^- \qquad \hat{r}_i = \int_{x_3^{(1)}}^{x_3^{(n+1)}} \hat{p}_i\, dx_3$$

$$m_\alpha = \int_{x_3^{(1)}}^{x_3^{(n+1)}} x_3\, b_\alpha\, dx_3 + 2h(p_\alpha^+ - p_\alpha^-) \qquad \hat{m}_\alpha = \int_{x_3^{(1)}}^{x_3^{(n+1)}} x_3\, \hat{p}_\alpha\, dx_3 \tag{30}$$

In Eqs. (27), the so-called reduced elastic law has been employed, by introducing for each layer and in agreement with the position (8) the relationship

$$\hat{\mathscr{C}}^{(k)}_{\alpha\beta\gamma\delta} = \mathscr{C}^{(k)}_{\alpha\beta\gamma\delta} - \mathscr{C}^{(k)}_{\alpha\beta 33}\mathscr{C}^{(k)}_{\gamma\delta 33}/\mathscr{C}^{(k)}_{3333} \tag{31}$$

As it is customary in plate theories, the generalized stress (that is the stress resultants over the thickness) can be introduced as

$$\mathbf{S} = \begin{bmatrix} \mathbf{N} \\ \mathbf{M} \\ \mathbf{Q} \end{bmatrix} = \begin{bmatrix} \mathbf{A} & \mathbf{B} & \mathbf{0} \\ \mathbf{B} & \mathbf{D} & \mathbf{0} \\ \mathbf{0} & \mathbf{0} & \mathbf{H} \end{bmatrix} \begin{bmatrix} \mathbf{e} \\ \boldsymbol{\kappa} \\ \boldsymbol{\gamma} \end{bmatrix} \tag{32}$$

where

$$N_{\alpha\beta} = \sum_{k=1}^{n} \int_{x_3^{(k)}}^{x_3^{(k+1)}} \sigma^{(k)}_{\alpha\beta}\, dx_3$$

$$M_{\alpha\beta} = \sum_{k=1}^{n} \int_{x_3^{(k)}}^{x_3^{(k+1)}} x_3\, \sigma^{(k)}_{\alpha\beta}\, dx_3$$

$$Q_\alpha = \sum_{k=1}^{n} \int_{x_3^{(k)}}^{x_3^{(k+1)}} \sigma^{(k)}_{\alpha 3}\, dx_3 \tag{33}$$

The stationary conditions of $\hat{\mathscr{E}}$ with respect to the unknown displacement functions, under the constraint (23), give the governing equilibrium equations in $\mathscr{P}$

$$
\begin{aligned}
&\left[A_{\alpha\beta\gamma\delta}(s_{\gamma/\delta}+s_{\delta/\gamma})+B_{\alpha\beta\gamma\delta}(\varphi_{\gamma/\delta}+\varphi_{\delta/\gamma})\right]_{/\beta}+2r_\alpha=0\\
&\left[B_{\alpha\beta\gamma\delta}(s_{\gamma/\delta}+s_{\delta/\gamma})+D_{\alpha\beta\gamma\delta}(\varphi_{\gamma/\delta}+\varphi_{\delta/\gamma})\right]_{/\beta}-2H_{\alpha\beta}(\varphi_\beta+w_{/\beta})+2m_\alpha=0\\
&H_{\alpha\beta}(\varphi_\beta+w_{/\beta})_{/\alpha}+r_3+r_c=0
\end{aligned}
\tag{34}
$$

and the boundary conditions on $\partial_f\mathscr{P}$

$$
\begin{aligned}
&\left[A_{\alpha\beta\gamma\delta}(s_{\gamma/\delta}+s_{\delta/\gamma})+B_{\alpha\beta\gamma\delta}(\varphi_{\gamma/\delta}+\varphi_{\delta/\gamma})\right]n_\beta=2\hat{r}_\alpha\\
&\left[B_{\alpha\beta\gamma\delta}(s_{\gamma/\delta}+s_{\delta/\gamma})+D_{\alpha\beta\gamma\delta}(\varphi_{\gamma/\delta}+\varphi_{\delta/\gamma})\right]n_\beta=2\hat{m}_\alpha\\
&H_{\alpha\beta}(\varphi_\beta+w_{/\beta})n_\alpha=\hat{r}_3
\end{aligned}
\tag{35}
$$

where the surface reaction density

$$
r_c=(\boldsymbol{\sigma}\mathbf{n})_3 \tag{36}
$$

corresponding to the unilateral constraint has to satisfy $r_c \le 0$ on Σ_c and $r_c = 0$ otherwise.

Solution of (34) under conditions (23) and (35) gives the unknown functions s_α, w and φ_α, from which total, reactive and elastic strain fields can be computed. Therefore, the elastic stress field $\boldsymbol{\sigma}^{(\mathrm{el})}=\mathscr{C}(\hat{\nabla}\mathbf{u}-\mathbf{G}^*\boldsymbol{\omega})$ in the kth layer turns out to be

$$
\begin{aligned}
&\sigma^{(\mathrm{el},k)}_{\alpha\beta}=\hat{\mathscr{C}}^{(k)}_{\alpha\beta\gamma\delta}\left[(s_{\gamma/\delta}+s_{\delta/\gamma})/2+x_3(\varphi_{\gamma/\delta}+\varphi_{\delta/\gamma})/2\right]\\
&\sigma^{(\mathrm{el},k)}_{\alpha3}=\mathscr{C}^{(k)}_{\alpha3\beta3}(\varphi_\beta+w_{/\beta})\\
&\sigma^{(\mathrm{el},k)}_{33}=0
\end{aligned}
\tag{37}
$$

It is worth pointing out that resultant normal forces **N** and moments **M** obtained by means of the elastic stress field satisfy global equilibrium conditions. On the other hand, the correct evaluation through the elastic stress field of the equilibrated resultant shear forces **Q** requires the use of shear correction factors. They can be introduced starting from the exact profiles of the shear stresses $\sigma_{\alpha3}$, and are known a priori only for homogeneous plate or for simple problems [39]. This aspect represents a clear limitation of first-order shear-deformation theories, although several approaches can be found in the specialized literature to overcome this drawback (e.g., [40–42]).

The total stress field (locally equilibrated) coincides with the elastic one only for the component $\sigma_{\alpha\beta}$, whereas total stress components $\sigma_{\alpha3}$ and σ_{33} can be recovered, identifying the corresponding reactive parts, by involving the local equilibrium (14) and Signorini's conditions (15). In detail, they result in

$$\sigma_{\alpha 3}^{(k)} = \sigma_{\alpha 3(0)}^{(k)} - \int_{x_3^{(k)}}^{x_3} \hat{\mathscr{C}}_{\alpha\beta\gamma\delta}(e_{\gamma\delta/\beta} + z\,\kappa_{\gamma\delta/\beta})\,dz - \int_{x_3^{(k)}}^{x_3} b_\alpha\,dz$$

$$\sigma_{33}^{(k)} = \sigma_{33(0)}^{(k)} - \sigma_{\alpha 3(0)/\alpha}(x_3 - x_3^{(k)}) - \int_{x_3^{(k)}}^{x_3} (b_3 - \int_{x_3^{(k)}}^{z} b_{\alpha/\alpha}\,dt)\,dz$$
$$+ \int_{x_3^{(k)}}^{x_3} \left[\int_{x_3^{(k)}}^{z} \hat{\mathscr{C}}_{\alpha\beta\gamma\delta}(e_{\gamma\delta/\alpha\beta} + t\,\kappa_{\gamma\delta/\alpha\beta})\,dt \right] dz \tag{38}$$

with $x_3 \in]x_3^{(k)}, x_3^{(k+1)}[$ and where $\sigma_{\alpha 3(0)}^{(k)}$ and $\sigma_{33(0)}^{(k)}$ denote the stress components at $x_3 = x_3^{(k)}$, that is $\sigma_{\alpha 3(0)}^{(k)} = \sigma_{\alpha 3}^{(k)}|_{x_3^{(k)}}$ and $\sigma_{33(0)}^{(k)} = \sigma_{33}^{(k)}|_{x_3^{(k)}}$ with

$$\sigma_{\alpha 3(0)}^{(1)} = \begin{cases} p_\alpha^- & \text{on } \mathscr{P}^- \\ 0 & \text{on } \partial_c \Omega \end{cases} \qquad \sigma_{33(0)}^{(1)} = \begin{cases} p_3^- & \text{on } \mathscr{P}^- \\ r_c & \text{on } \partial_c \Omega \end{cases} \tag{39}$$

It is worth observing that, in agreement with the kinematic assumptions, in-plane strains are linear functions of the thickness coordinate x_3 and their integrals (employed to compute the total shear stress components $\sigma_{\alpha 3}$) are quadratic functions in each layer.

We emphasize that present derivation clearly shows as the reduced constitutive law (31) in $\hat{\mathscr{E}}$ is a straightforward and rational consequence of constraints on dual fields, without contradiction because they act on fields (total strain and elastic stress) which are not related by constitutive law. In other words, the reduced constitutive law comes out from the procedure adopted and is not a priori enforced by means of a constrained constitutive law.

It is worth pointing out that, independently of the lamination stack sequence, bimodular constitutive behavior always induces in general the coupling between flexural and extensional problems. Moreover, the laminated constitutive tensors previously introduced in Eqs. (27) have to be evaluated taking for each layer $\Omega^{(k)} = \Omega^{(k)}|_{\mathscr{C}^+} \cup \Omega^{(k)}_{\mathscr{C}^-}$. Clearly, this partition of $\Omega^{(k)}$ needs the preliminary knowledge at every point $P \in \mathscr{P}$ of the x_3-coordinate value where $\varepsilon_f^{(k)} = 0$. Therefore, a free-boundary problem underlies this formulation and an iterative procedure has to be employed in order to evaluate components of **A**, **B**, **D** and **H**. To this aim, finite elements can be used. Nevertheless, due to the hypothesis $(2h) \ll \text{diam}(\mathscr{P})$, in the case that $(x_3^{(k+1)} - x_3^{(k)}) \ll 2h\ \forall k \in \{1, 2, \ldots, n\}$, it is possible to consider

$$\Omega^{(k)}|_{\mathscr{C}\pm} = \mathscr{P}^{(k)}|_{\mathscr{C}\pm} \times]x_3^{(k)}, x_3^{(k+1)}[\tag{40}$$

$\mathscr{P}^{(k)}|_{\mathscr{C}\pm}$ denoting the subregion of the midsurface $\mathscr{P}$ in which $\mathscr{C}^\pm$ applies for the layer k. Furthermore, to perform the iterative procedure on each layer, the average along-the-thickness measure $\bar{\varepsilon}_f^{(k)}$ based on the elastic strain component $\varepsilon_f^{(k)}$ and relevant to the local fiber direction $\mathbf{f}^{(k)}$, that is

$$\bar{\varepsilon}_f^{(k)} = \frac{1}{\left(x_3^{(k+1)} - x_3^{(k)}\right)} \int_{x_3^{(k)}}^{x_3^{(k+1)}} \boldsymbol{\varepsilon}\,\mathbf{f}^{(k)} \cdot \mathbf{f}^{(k)}\, dx_3 \tag{41}$$

can be considered as a control value of the along-the-fiber strain $\varepsilon_f^{(k)}$. Relationships (40) and (41) allow to obtain an approximate solution for the bimodular laminate problem up to terms of order $|x_3^{(k+1)} - x_3^{(k)}| \approx 2h/n$.

3.2 Lo–Christensen–Wu-Type Laminate Model: An Outline

In order to account for a more refined kinematics and following the classical Lo–Christensen–Wu plate model, the total strain field can be constrained by imposing that

(i) The total strain component ε_{33} is assumed to be linear in the thickness coordinate x_3.
(ii) The shear total strain between the x_3-axis and $\mathscr{P}$ is assumed to be quadratic in x_3.

Accordingly, operator $\mathbf{G}$ is such that

$$\mathbf{G}\boldsymbol{\varepsilon} = \{\varepsilon_{13/333}\ \ \varepsilon_{23/333}\ \ \varepsilon_{33/33}\}^T \tag{42}$$

Therefore (see Eq. (11)) dual constraints on the elastic stress field turn out to be directly expressed as

(iii) The elastic out-of-plane stress σ_{33} is assumed to be linear with x_3 at every position in $\Omega^{(k)}|_{\mathscr{C}\pm}$.
(iv) The shear elastic stress between the x_3-axis and $\mathscr{P}$ is assumed to be quadratic in x_3 at every position in $\Omega^{(k)}|_{\mathscr{C}\pm}$.

Accordingly, the functional (17) can be arranged as

$$\begin{aligned}\mathscr{E}(\mathbf{u},\boldsymbol{\chi},\boldsymbol{\omega}) = \frac{1}{2}\int_\Omega \{&\mathscr{C}_{\alpha\beta\gamma\delta}\Lambda_{\alpha\beta}\Lambda_{\gamma\delta} + 2\mathscr{C}_{\alpha\beta 33}\Lambda_{\alpha\beta}(u_{3/3} - \omega_{33/33}) \\ &+ \mathscr{C}_{3333}(u_{3/3} - \omega_{33/33})^2 + 4\mathscr{C}_{\alpha 3\beta 3}(\Lambda_{\alpha 3} + \omega_{\alpha 3/333})(\Lambda_{\beta 3} + \omega_{\beta 3/333})\}\, dv \\ &- \int_\Omega (2\chi_{\alpha 3}\Lambda_{\alpha 3/333} + \chi_{33}u_{3/333})\, dv - \Pi_{\text{ext}}\end{aligned} \tag{43}$$

defined on the manifolds $\chi_{\alpha 3}|_{\pm h} = \chi_{\alpha 3/3}|_{\pm h} = \chi_{\alpha 3/33}|_{\pm h} = \chi_{33}|_{\pm h} = \chi_{33/3}|_{\pm h} = 0$, and $\omega_{\alpha 3} = \omega_{\alpha 3/3} = \omega_{\alpha 3/33} = \omega_{33} = \omega_{33/3} = 0$ on $\partial\Omega^{(k)}|_{\mathscr{C}\pm}$, as well as under the constraint $\mathbf{u} \in \mathbf{K} = \{\mathbf{v} \in \mathbf{V} \mid v_3 n_3 \leq 0 \text{ on } \partial_c\Omega\}$, and where Π_{ext} is expressed by Eq. (20).

Stationary conditions of the functional (43) with respect to $\chi_{\alpha 3}$ and χ_{33} give the constraints on the displacement field:

$$
\begin{aligned}
&(u_{\alpha/3}+u_{3/\alpha})_{/333}=0\\
&u_{3/333}=0
\end{aligned}
\tag{44}
$$

from which, by integration, the admissible displacement representation is obtained:

$$
\begin{aligned}
u_\alpha(x_1,x_2,x_3) &= s_\alpha(x_1,x_2)+x_3\,\varphi_\alpha(x_1,x_2)+g_\alpha(x_1,x_2)[3(x_3/h)^2-1]/2\\
&\quad+\psi_\alpha(x_1,x_2)x_3\,[5(x_3/h)^2-3]/2\\
u_3(x_1,x_2,x_3) &= w(x_1,x_2)+x_3\,\lambda(x_1,x_2)+a(x_1,x_2)\,[3(x_3/h)^2-1]/2
\end{aligned}
\tag{45}
$$

constrained by the condition

$$
u_3|_{-h}=[w(x_1,x_2)-h\lambda(x_1,x_2)+a(x_1,x_2)]\in\tilde{\mathbf{K}}
\tag{46}
$$

with $\tilde{\mathbf{K}}=\{f(x_1,x_2)\mid f(x_1,x_2)n_3\le 0 \text{ on } \Sigma_c\}$. In Eqs. (45), functions $s_\alpha(x_1,x_2)$ identify the over-the-thickness average membranal displacements in the plane (x_1,x_2); the term containing $g_\alpha(x_1,x_2)$ defines a membranal correction contribution with null mean value over the thickness; functions $\varphi_\alpha(x_1,x_2)$ and $\psi_\alpha(x_1,x_2)$ describe the average rotation of fibers parallel to the x_3-axis; $w(x_1,x_2)$ describes the mean deflection of $\mathscr{P}$; $\lambda(x_1,x_2)$ is the over-the-thickness average total strain component along the direction x_3; the term containing function $a(x_1,x_2)$ defines a bending correction contribution with a null mean value over the thickness.

The stationary conditions of functional $\mathscr{E}$ with respect to $\omega_{\alpha 3}$ and ω_{33} give, respectively,

$$
\begin{aligned}
&\left\{\mathscr{C}_{\alpha 3\beta 3}\left[(u_{\beta/3}+u_{3/\beta})/2+\omega_{\beta 3/333}\right]\right\}_{/333}=0\\
&\left[\mathscr{C}_{3333}(u_{3/3}-\omega_{33/33})+\mathscr{C}_{\alpha\beta 33}(u_{\alpha/\beta}+u_{\beta/\alpha})/2\right]_{/33}=0
\end{aligned}
\tag{47}
$$

from which, taking into account Eqs. (44) and (45), Lagrange multipliers $\omega_{\alpha 3}$ and ω_{33} can be uniquely determined. Therefore, following the same machinery previously highlighted, the dependency of $\mathscr{E}$ on the Lagrangian multipliers $\boldsymbol{\chi}$ and $\boldsymbol{\omega}$ can be eliminated and a pure energy-type functional $\hat{\mathscr{E}}$ is obtained, whose stationary conditions with respect to the displacement unknowns s_α, φ_α, g_α, ψ_α, w, λ, and a lead to the governing equations defined on $\mathscr{P}$ of the unilateral laminated plate, and the corresponding boundary conditions on $\partial_f\mathscr{P}$, under the constraint (46). For the sake of compactness, such a deduction is herein omitted.

Owing to the kinematic assumptions and the corresponding admissible displacement field (45), such a model allows to obtain a quadratic representation along the thickness of the elastic shear stresses $\sigma_{\alpha 3}$. Accordingly, it does not require the use of shear correction factors, as in the case of first-order shear-deformation models.

4 Concluding Remarks

This paper presents a consistent deduction of unilateral laminate plate models from three-dimensional elasticity. Following an approach involving constraints on both stress and strain dual fields, a variational formulation able to account for contact and unilateral material problems has been discussed. This approach, based on a modified constrained Hu–Washizu functional, has been specialized to the cases of a unilateral fiber-governed (Bert-type) material constitutive model and of Signorini's contact problem. A non-standard Lagrange multipliers technique is involved to rationally deduce equivalent single-layer laminated-plate models, accounting for different shear refinements. In the case of the classical first-order shear-deformation theory, field and boundary equations governing the equilibrium of the unilateral laminate are deduced, resulting in a strong coupling between flexural and extensional problems and involving a free-boundary problem related to the determination of the neutral surface (where the along-the-fiber strain component vanishes). Moreover, in order to account for a more refined kinematics and to avoid the use of shear correction factors, the rational deduction of a Lo–Christensen–Wu-based unilateral laminated-plate model is briefly presented.

It is worth pointing out that present dual-constraint framework allows to incorporate frictionless contact in plate problems without significant complications, by constraining the three-dimensional generalized Hu–Washizu variational formulation on the convex set of the displacements which satisfy unilateral contact restrictions. In this way Signorini's conditions are straightly recovered as a result of the variational approach and, by enforcing dual constraints on both strain and stress fields, the consistent reduction of the three-dimensional problem to two-dimensional laminate models is obtained, rationally reducing the contact problem at the same lower-dimensional manifold formulation. Clearly, contact constraints could be introduced after the reduction to the laminated plate model, but in this case Signorini's conditions should be a-priori postulated consistently with the problem dimension and not directly recovered as a consequence of the variational framework.

Generalizations to more complex contact and material contexts are possible (e.g., considering piezoelectric layers, and/or layer-wise approaches). Moreover, owing to its variational character, this formulation opens to the possibility to build up new consistent and refined laminate finite elements.

Acknowledgements The authors would like to thank Professor Paolo Bisegna for valuable suggestions and fruitful discussions on this paper. This work was developed within the framework of Lagrange Laboratory, a European research group comprising CNRS, CNR, the Universities of Rome "Tor Vergata", Calabria, Cassino, Pavia, and Salerno, Ecole Polytechnique, University of Montpellier II, ENPC, LCPC, and ENTPE.

References

1. Whitney, J.M., Pagano, N.J.: Shear deformation in heterogeneous anisotropic plates. J. Appl. Mech. Trans. ASME 37, 1031–1036 (1970)
2. Lo, K.H., Christensen, R.M., Wu, E.M.: A high-order theory of plate deformation. Part II: Laminated plates. J. Appl. Mech. Trans. ASME 44, 669–676 (1978)
3. Idlbi, A., Karama, M., Touratier, M.: Comparison of various laminated plate theories. Compos. Struct. 37, 173–184 (1997)
4. Reddy, J.N.: Mechanics of Laminated Composite Plates, Theory and Analysis. CRC Press, Boca Raton (1997)
5. Wang, C.M., Reddy, J.N., Lee, K.H.: Shear Deformable Beams and Plates. Relationships with Classical Solutions. Elsevier Science, Oxford (2000)
6. Auricchio, F., Sacco, E., Vairo, G.: A mixed FSDT finite element for monoclinic laminated plates. Comput. Struct. 84, 624–639 (2006)
7. Goldenveizer, A.L.: Derivation of an approximate theory of bending of a plate by the method of asymptotic integration of the equations of the theory of elasticity. J. Appl. Math. Mech. 26, 1000–1025 (1962)
8. Ciarlet, P.G., Destuynder, P.: A justification of a nonlinear model in plate theory. Comput. Methods Appl. Mech. 18, 227–258 (1979)
9. Ciarlet, P.G., Lods, V.: Asymptotic analysis of linearly elastic shells. I. Justification of membrane shell equations. Arch. Ration Mech. An. 136, 119–161 (1996)
10. Rodriguez, J.M., Viaño, J.M.: Asymptotic derivation of a general linear model for thin-walled elastic rods. Comput. Methods Appl. Mech. 147, 287–321 (1997)
11. Bisegna, P., Sacco, E.: A rational deduction of plate theories from the three-dimensional linear elasticity. Zeitschr. Angew. Math. Mech. 77, 349–366 (1997)
12. Podio-Guidugli, P.: An exact derivation of thin plates equations. J. Elasticity 22, 121–133 (1989)
13. Lembo, M., Podio-Guidugli, P.: Plate theory as an exact consequence of three-dimensional linear elasticity. Eur. J. Mech. A Solid 10, 485–516 (1991)
14. Bisegna, P., Sacco, E.: A layer-wise laminate theory rationally deduced from the three-dimensional elasticity. J. Appl. Mech. Trans. ASME 64, 538–545 (1997)
15. Lembo, M., Podio-Guidugli, P.: Internal constraints, reactive stresses, and the Timoshenko beam theory. J. Elasticity 65, 131–148 (2001)
16. Maceri, F., Bisegna, P.: Modellazione strutturale. In: Elio Giangreco – Ingegneria delle Strutture, Utet, Torino, vol. II, pp. 1–90 (2002) (in Italian)
17. Maceri, F., Vairo, G.: Anisotropic thin-walled beam models: A rational deduction from three-dimensional elasticity. J. Mech. Mater. Struct. 4(2), 371–394 (2009)
18. Vlasov, V.Z., Leont'ev, N.N.: Beams, Plates and Shells on Elastic Foundations. Israel Program for Scientific Translations, Jerusalem, Israel (1966)
19. Keer, L.M., Dundurs, J., Tsai, K.C.: Problems involving receding contact between a layer and a half-space. J. Appl. Mech. Trans. ASME 39, 1115–1120 (1972)
20. Popov, G.Y.: Plates on a linearly elastic foundation (a survey). Int. Appl. Mech. 8(3), 3–17 (1972)
21. Gladwell, G.M.L.: Contact Problems in the Classical Theory of Elasticity. Sijthoff and Noordhoff, The Netherlands (1980)
22. Westbrook, D.R.: Contact problems for the elastic beam. Comput. Struct. 15(4), 473–479 (1982)

23. Tishchenko, V.N.: Contact problems for thin-walled structures. Int. Appl. Mech. 21(10), 75–79 (1985)
24. Khludnev, A.M., Hoffmann, K.H.: A variational inequality in a contact elastoplastic problem for a bar. Adv. Math. Sci. Appl. 1(1), 127–136 (1992)
25. Han, W., Kuttler, K.L., Shillor, M., Sofonea, M.: Elastic beam in adhesive contact. Int. J. Solids Struct. 39(5), 1145–1164 (2002)
26. Galin, L.A.: Contact problems. In: Gladwell, G.M.L. (ed.) Contact Problems. The Legacy of L.A. Galin. Solid Mechanics and Its Applications, vol. 155, Springer, Dordrecht (2008)
27. Maceri, F., Vairo, G.: Beams comprising unilateral material in frictionless contact: A variational approach with constraints in dual spaces. LNACM. Springer, Berlin (to appear)
28. Bert, C.W.: Models for fibrous composites with different properties in tension and in compression. J. Eng. Mater. Tech. ASME 99, 344–349 (1977)
29. Bisegna, P., Maceri, F., Sacco, E.: On the fiber-governed bimodular constitutive models. In: Sih, G.C., et al. (eds.) Advanced Technology for Design and Fabrication of Composite Materials and Structures: Applications to the Automotive, Marine, Aerospace, and Construction Industry, pp. 113–128. Kluwer, Dordrecht (1995)
30. Maceri, F., Sacco, E.: A contribution to the mechanics of bimodular materials. Mater. Eng. 1, 189–199 (1990)
31. Reddy, J.N., Bert, C.W.: On the behaviour of plates laminated of bimodulus composite materials. Zeitschr. Angew. Math. Mech. 62, 213–219 (1982)
32. Soltz, M.A., Ateshian, G.A.: A conewise linear elasticity mixture model for the analysis of tension-compression nonlinearity in articular cartilage. J. Biomech. Eng. 122, 576–586 (2000)
33. Patel, B.P., Gupta, S.S., Sarda, R.: Free flexural vibration behavior of bimodular material angle-ply laminated composite plates. J. Sound Vibr. 286, 167–186 (2005)
34. Klisch, S.M.: A bimodular theory for finite deformations: Comparison of orthotropic second-order and exponential stress constitutive equations for articular cartilage. Biomech. Model Mechanobiol. 5, 90–101 (2006)
35. Curnier, A., He, Q.C., Zysset, P.: Conewise linear elastic materials. J. Elasticity 37(1), 1–38 (1995)
36. Antman, S.S., Marlow, R.S.: Material constraints, Lagrange multipliers, and compatibility. Applications to rod and shell theories. Arch. Ration. Mech. Anal. 116, 257–299 (1991)
37. Kikuchi, N., Oden, J.T.: Contact Problems in Elasticity: A Study of Variational Inequalities and Finite Element Methods. Studies in Applied Mathematics 8 (1988)
38. Signorini, A.: Sopra alcune questioni di elastostatica. Atti della Società Italiana per il Progresso delle Scienze (1933) (in Italian)
39. Laitinen, M., Lahtinen, H., Sjölind, S.G.: Transverse shear correction factors for laminates in cylindrical bending. Commun. Numer. Meth. Eng. 11, 41–47 (1995)
40. Pai, P.F.: A new look at the shear correction factors and warping functions of anisotropic laminates. Int. J. Solids Struct. 32, 2295–2313 (1995)
41. Auricchio, F., Sacco, E.: Partial-mixed formulation and refined models for the analysis of composite laminates within an FSDT. Compos. Struct. 46, 103–113 (1999)
42. Auricchio, F., Sacco, E., Vairo, G.: A mixed FSDT finite-element formulation for the analysis of composite laminates without shear correction factors. In: Maceri, F., Fremond, M. (eds.) Mechanical Modelling and Computational Issues in Civil Engineering. LNACM, vol. 23, pp. 345–358. Springer, Berlin (2005)

Contact Modelling in Structural Simulation – Approaches, Problems and Chances

Rolf Steinbuch

Abstract Nearly all structural problems deal to some extend with interacting bodies [1]. In consequence, contact between these bodies is one of the most often occurring problems in modelling. On the other hand, many contacts are either neglected in modelling or assumed to be represented by appropriate boundary conditions. Today most of the simulation codes used in structural mechanics allow to model contact. In consequence there is a need to qualify more users to apply contact models in everyday simulation practice. Modelling contact implies to consider the appropriate setting of the available contact parameters. Using the standard values proposed by the codes or unqualified selection of parameter combinations may produce simulation results which are either poor or even misleading. To stimulate the discussion about contact modelling some of the aspects to be considered and their non-unique consequences are discussed.

Choosing qualified element types surely will influence significantly the simulation velocity and quality. Adapting meshes to the current state of the problem may improve speed and quality essentially, but mapping the analysis results from old meshes to the new ones may cause loss of local quality. Reduced integration helps to speed up the computing time and avoid numerical stiffness, but the stress results are less qualified. Artificial stiffness introduced to prevent misleading effects like hourglasing produces potential sources of numerical errors.

The ways to detect and release contact in time and space influence significantly the course of the analysis. Simple approaches yield fast responses at the price of incorrect penetrations and incomplete definitions. Qualified algorithms may converge only after large numbers of recycles and time step reductions, so acceptable response times may not occur. Friction models suffer from the uncertainty of data and the large scatter of their input values, especially in transitions of stick to slip, high speed sliding and lubrication. The decision whether to use explicit or implicit time integration is not easy do be made in many cases.

Rolf Steinbuch
Department of Engineering, Reutlingen University, Alteburgstraße 150, D-72762 Reutlingen, Germany; e-mail: rolf.steinbuch@reutlingen-university.de

G. Zavarise & P. Wriggers (Eds.): Trends in Computational Contact Mechanics, LNACM 58, pp. 339–354.
springerlink.com

A qualified discussion of the interaction of the parameters in contact analysis will certainly not yield a simple scheme applicable in all cases required to be considered in structural simulation. Users will be faced with problems where the ideas they have been applying successfully in many studies fail, do not converge in appropriate time or even come up with misleading results without indicating the potential danger. So the most important parameter in engineering contact analysis is the experience of the people running the jobs. Some application examples underline the problems mentioned and sketch solution proposals for some parameter configurations. Critical regions are defined, workarounds proposed and experiences presented. Small parameter changes may cause large changes in system response. In consequence for contact problems robustness studies, indicating the sensitivity of the physical and numerical parameters should be performed.

In any case the qualification and care of the analyst applying contact methods in engineering remain the most important tools to solve contact problems successfully and to learn about the interaction of structures.

1 Contact in Structural Problems

Nearly all coherent structures like metal parts, stones, plants or furniture, to mention some, interact with their surrounding via surface contact. A table is standing on the floor, the ends of its legs contacting the surface of the floor of the room. The only exceptions of bodies not in structural contact with some neighbours are structural units flying through the air or the empty space like meteorites. So contact and the necessity to handle contact is not an exotic task, but one of the basic questions in structural mechanics. But as contact problems are often related to nonlinear phenomena many engineering approaches try to avoid contact modelling by the use of linear surrogates.

1.1 Boundary Conditions or Contact?

In mechanics the idea of free body models, replacing all interaction with the contacting surrounding by application of corresponding forces or moments is crucial to start the studies of static problems. Introducing contact models at that early state of education would increase the confusion in the novices' heads, who are struggling hard enough to enter the world of mechanics. So models like the one shown in Figure 1a are supposed to represent real support loading situations indicated in Figure 1b. Effects like local notches or sharp corners between the interacting bodies are ignored and believed to be of none or little influence. But when dealing with real components and their failure modes, we realise, that a good part of the problems are caused by these local disturbances we neglected before.

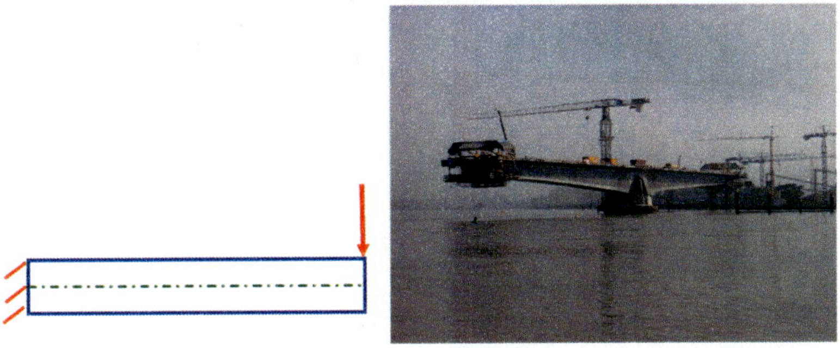

(a) Beam with boundary conditions (b) Build-in beam under real loading

Fig. 1 Idealised and real loading situation of beams.

1.2 Contact as Part of Manufacturing Processes

Contact, being present as a static boundary condition in most structural systems, is additionally defining some of the processes in mechanics. All the shaping of structural but deformable material including unwanted forming like crash is specified by some contact history. Segments of parts surfaces come into contact, the acting forces cause deformation of the surrounding material [2, 3].

New contacts are possible due to the fact that material preventing the contact has been removed by the contact process. Fig. 2.a demonstrates some steps in an axisymmetric deep drawing process, using different tools and finally covering the newly manufactured can by a seal [4].

During two steps of forming, the shape of the top of the can is manufactured (subfig. 1 and 2 of Figure 2a). Then the tampering head is introduced (subfig. 3 of Figure 2a). Finally some thermal load steps caused by temperature changes of the whole system are applied (subfig. 4 of Figure 2a). The analysis aimed to check if the sealing by the tampering head would be sufficient during the lifetime of the assembly. The contact study performed checks whether the manufacturing is possible, if problems may occur, e.g. large strains may indicate the failure of the sheet metal and other related quantities like the spring-back after the removal of the die. In addition including the deep drawing into the system's analysis helps to come up with a better evaluation of the structure. Figure 2b plots the sealing forces between the can and the tamping due to temperature changes. Obviously the study not including the deep drawing steps (w/o manufacturing) predicts a failure of the seal during the thermal loading (radial pressure $= 0$) while the more elaborate simulation (with manufact) predicts a sufficient tightness (radial pressure > 0).

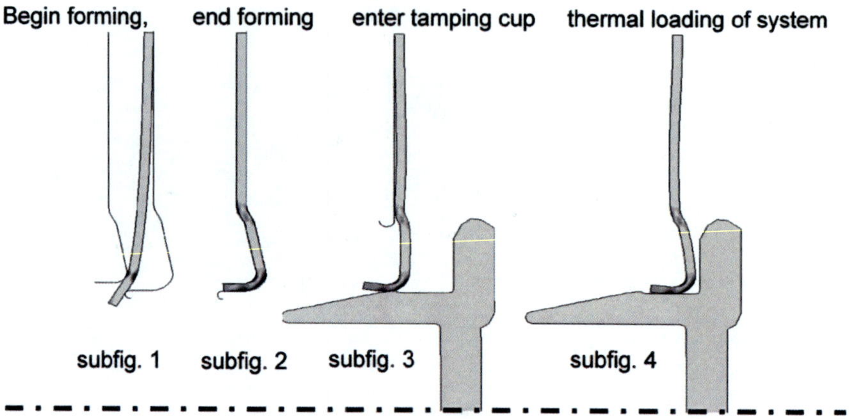

(a) Manufacturing, grey tones indicate plastic strain

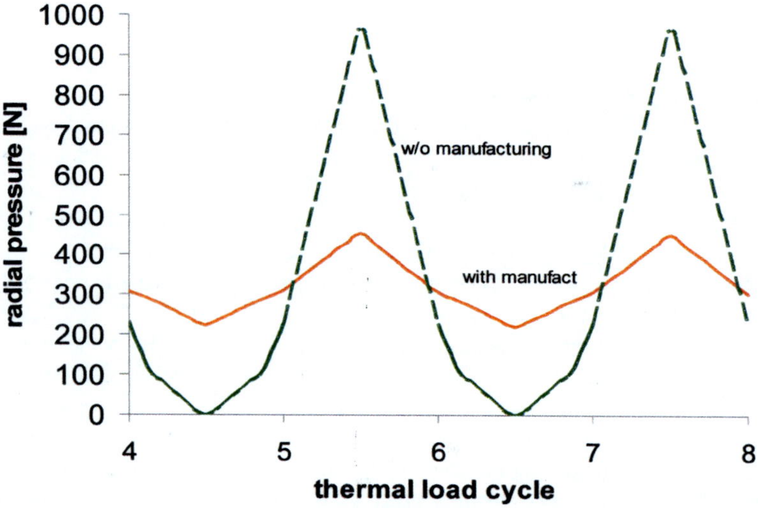

(b) Sealing pressure

Fig. 2 Some steps of a deep drawing of the top of a can (axisymmetric) and sealing.

1.3 Numerical Contact Studies

If two bodies come into contact there is a small region of first encounter, which will be increasing, if the bodies are continuing to move in the closing direction. Perhaps some of the early contact regions are released from contacting, while others support the load to be transmitted via the contact areas. A continuous process of establishing and releasing contacts will evolve. When studying more complex problems of continuum mechanics it is well known that there are no analytical or classical solu-

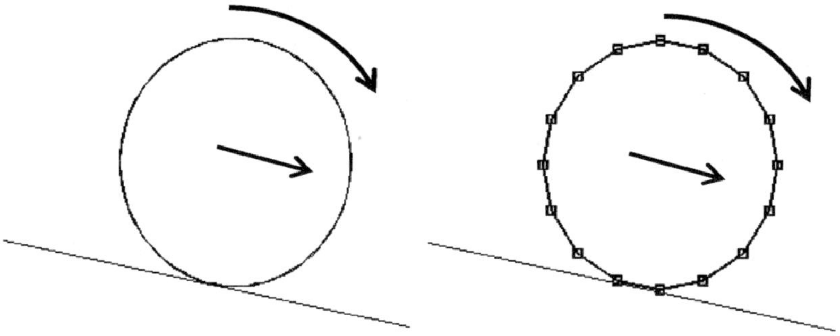

(a) Perfect real circular cylinder (b) FE discretisation of the cylinder

Fig. 3 Rolling of a cylinder and a FE-model of the cylinder on an inclined plane.

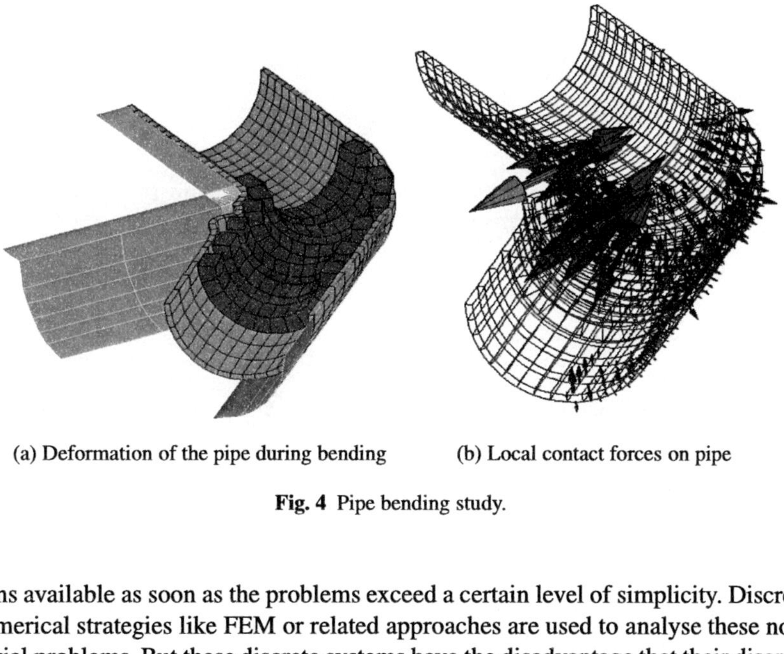

(a) Deformation of the pipe during bending (b) Local contact forces on pipe

Fig. 4 Pipe bending study.

tions available as soon as the problems exceed a certain level of simplicity. Discrete numerical strategies like FEM or related approaches are used to analyse these non-trivial problems. But these discrete systems have the disadvantage that their discrete schemes including discrete discovery and release of contact will fail to reproduce the soft initiation of continuous contact. Figure 3 compares the smooth rolling of a circular wheel with the inevitable rumbling of a discrete representation of this wheel. While Figure 4a depicts a pipe bending study, Figure 4b presents the local contact forces during this pipe bending study at a given time step. There are sharp changes in the contact state and forces at the Finite Element nodes, which is quite the opposite of the real smooth process in time.

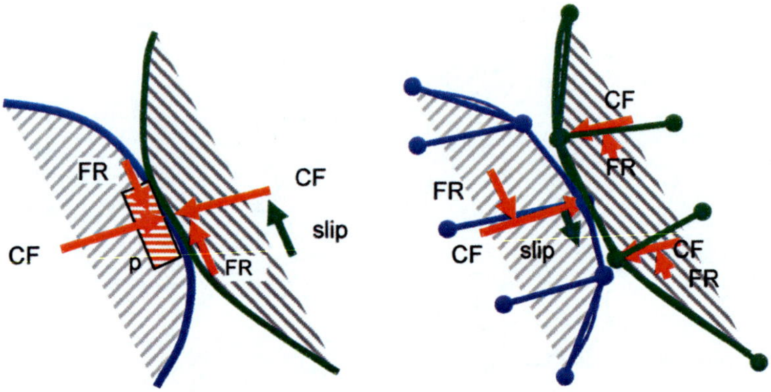

(a) Real contact (b) Discrete contact modelling

Fig. 5 Main contact parameters.

2 Terms and Definitions

Having mentioned some of the basic terms of contact modelling the need to clarify terms used in contact studies arises.

2.1 Real Contact

The following short definitions for contact between real parts hold and are generally accepted (see Figure 5a):

Contact:	Two bodies meet, are in structural interaction, the contact normal force is positive.
Release:	The contact ends, no more structural interaction, the contact normal force is zero.
Contact force:	Force normal to the contact area acting between the contacting bodies (CF).
Contact pressure:	Contact force/contact area (p).
Friction:	Force transverse to contact closing direction, opposing the sliding tendency (FR).
Stick:	No relative transverse movement, often due to friction.
Slip:	Sliding along contact surface overpowering the sticking friction forces.

Table 1 Some of the major contact modelling parameters.

Parameter	Values		
elements	linear	quadratic	higher order
adaptivity	refine	coarsen	remesh
element integration	reduced	full	–
deformation	elasto-plastic	rigid-plastic	–
contact type	master-slave	both partners	contact range
contact partners	flexible vs. stiff	both flexible	–
contact modelling	penalty	Lagrange multipliers	merging
shell contact	include thickness	neglect thickness	1 or 2 sided
corner handling	one side only	slip around	–
catching	range	after penetration	–
release	immediately	slowly	–
friction	Coulomb	shear	glue
friction definition	velocity	displacement	–
friction onset	stick-slip	continuous	–
time integration	explicit	implicit	–
time modelling	static	dynamic	large masses

2.2 Numerical Contact

The same terms used for real contact may be used for the discrete numerical modelling of contact. We are taking into account that no longer areas are interacting, but discrete nodes or points are checked, if they are penetrating opposite surfaces. All contact properties, especially normal and friction forces, stick or slip responses have to be transferred to the discrete nodes like indicated in Figure 5b.

2.3 Parameters Used in Contact Studies

Going deeper into the problem of modelling, the process so simple to describe experimentally exhibits a large number of parameters to be covered. Table 1 tries to identify some of them, including the most commonly ways used to handle them. Neither the completeness of the parameters list nor the proposed handling must be accepted by other contact modelling engineers or systems. Nevertheless Table 1 gives some idea that contact modelling is not a simple task. Furthermore it tries to motivate potential users not to trust their codes without counterchecking the default values proposed by the different simulation systems. The users should analyse their problems, compare the specific requirements with the possibilities the codes offer and search for reasonable solutions.

There is no commonly accepted agreement about the preferences to be made, the parameter combinations to be used or unwanted interactions of parameter choices to be avoided. In many cases extensive studies have to be performed before a reliable contact analysis may converge to a satisfactory result. Two examples may enlighten this warning.

Linear elements are simple to be used, but require small contact edges, while quadratic elements may yield faster results using essentially smaller node and element numbers. On the other hand, the local forces of quadratic elements are different at the corner and mid-side nodes. Larger local forces occur at the mid-side nodes than at the edge nodes. As a consequence the forces transmitted locally between contacting structures may not be correctly placed.

Explicit time integration is known to produce fast results at relatively small time and storage requirement, while implicit time integration needs large storage and lots of time for the matrix inversion.

3 Some Problems and Some Proposals

Some of the contact parameters and related problems should be looked at, to come up with an understanding of the necessity of sensible handling contact simulation.

3.1 Element Types and Contact Definition

Contact is (see Figure 5) the interaction between two bodies. The result of the modelled contact should not depend too much on the modelling approach. Figure 6 however indicates that local effects may vary significantly due to the element size and element type used. Only for very fine meshes, the desired symmetric and element type independent result occurs. This leads to the proposal to use very fine meshes or sensitive adaptive algorithms.

3.2 Contact State

The definition of the contact closing and the removing of penetrations may be done by many strategies, Lagrange Multipliers and Penalty Methods being the most popular ones. Some codes propose locally conforming meshes, merging the respective nodes or degrees of freedom. Others apply forces instead of displacement constraints to remove overlapping of the contacting parts. Even the definition when contact happens is not unique, master-slave and symmetric checking strategies show acceptable results, but sometimes yield different contact response.

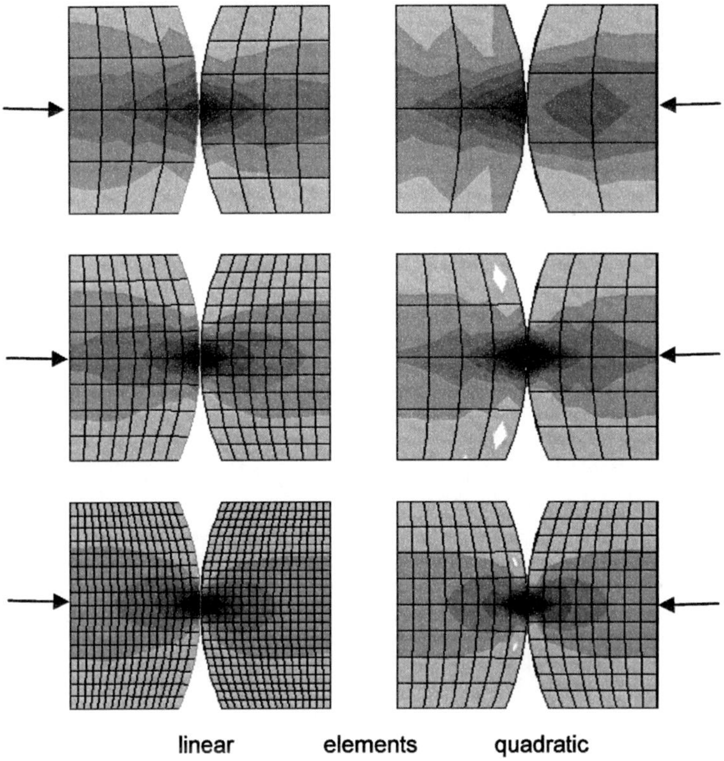

Fig. 6 Contact convergence, grey tones indicate contact normal stresses.

3.3 Friction

Friction, the force acting normal to the contact closure direction and parallel to the contacting surfaces is one of the most difficult and little understood phenomena in contact studies. In the microscopic scale local structures interact, build bridges, hooks, joints, are removed by sliding of the surfaces and rebuild at a new position. Molinari [5] presents impressive studies of local interaction yielding local forces and the resulting stresses. Simulation codes generally use coarse simplification of the friction effects by introducing friction coefficients, sometimes dependent on the local pressure and sliding velocity. Nevertheless even more qualified ideas are hardly able to represent the friction of real surfaces. Figure 7 demonstrates for two simple 2D-surfaces, how the local friction depends on the meshing parameters, even the refinement of the meshes produces no tendency of convergence.

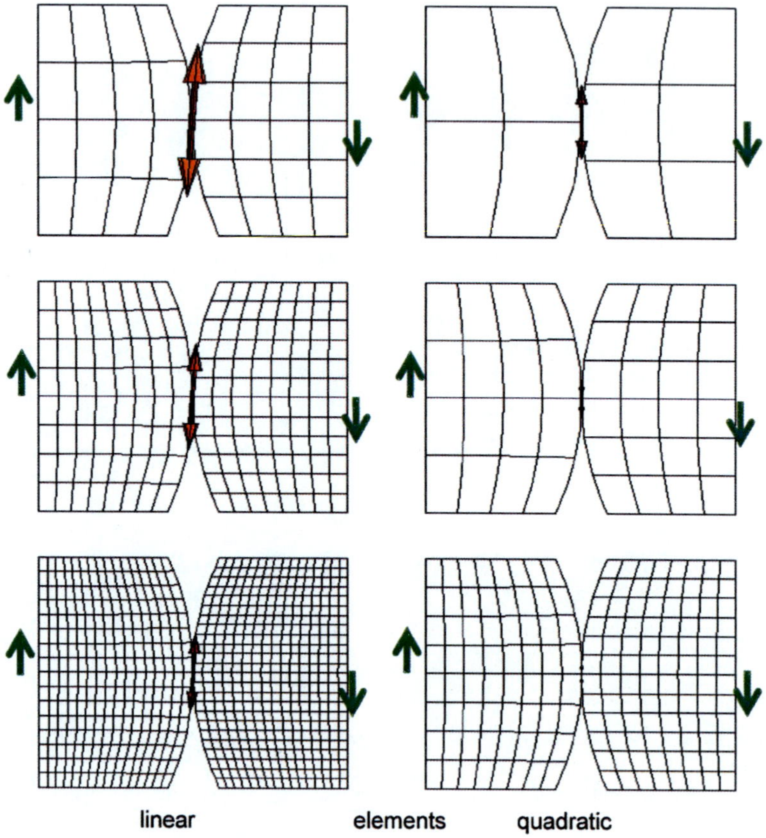

Fig. 7 Friction as function of element type and size, outside arrows indicate transverse displacement, inner arrows indicate local friction force.

3.4 Time History Integration

Contact is the result of the motion of different bodies, which initially had some distance or little common surface. During the contact process the area of interaction, the contact surfaces vary as well as the local and global forces. Regions being initially free are detected by free surfaces of the other contact partners, contacting regions are released or at least carry a reduced load. There are some promising approaches to use the final relative position of the bodies for the analysis [6, 7]. In most cases however it is a good idea to start with free bodies. Then one defines a motion, bringing into contact the bodies like indicated in Figure 8. From the initial independent state contact surfaces are evolving, showing a time dependent local response, again different for linear and quadratic elements.

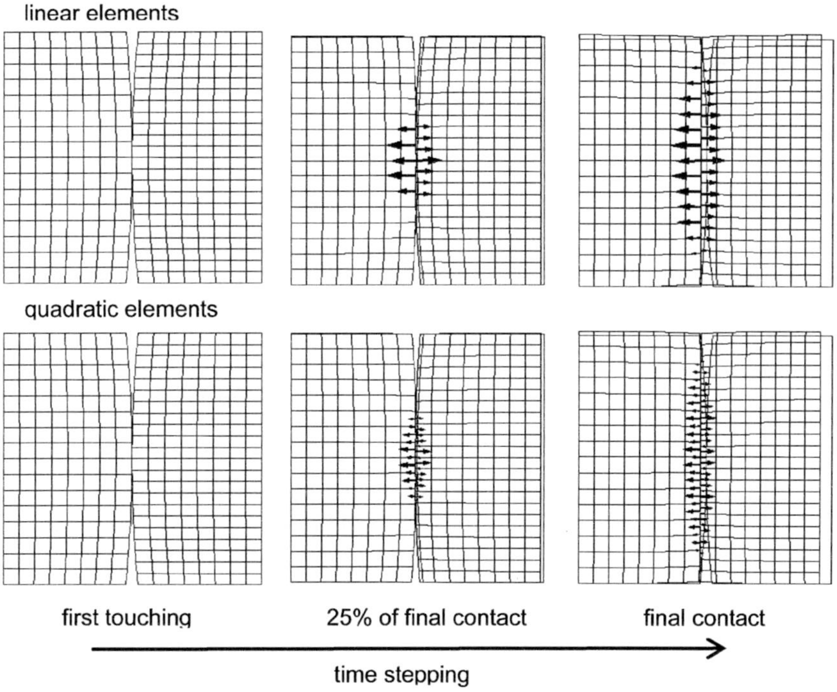

Fig. 8 Time dependent contact surfaces and local forces.

3.5 Time Integration Schemes

There are many discussions about the procedures of time integration. Since the first upcoming of LS-Dyna, [8] scientists discuss whether to use explicit or implicit time integration schemes. From the classical point of view there is no doubt, that the implicit solution

$$\mathbf{K}_n\mathbf{u}_n = \mathbf{F}_n \tag{1}$$

or its dynamic version

$$\mathbf{M}_n\ddot{\mathbf{u}}_n + \mathbf{C}_n\mathbf{u}_n + \mathbf{K}_n\mathbf{u}_n = \mathbf{F}_n \tag{2}$$

where the index n indicates the new time step, should be preferred, as it implies the equilibrium of forces at the new time step. In consequence relatively large time steps may still yield sufficient quality of the numerical prediction of the time history. Unfortunately for both static and dynamic problems the need to solve large linear equation systems requires large computation time and storage requirements. Especially in the case of metal forming or crash one needs to recycle the analysis at every time step until satisfactory convergence in terms of equilibrium of internal and external forces in equations (1) or (2) is achieved. This makes implicit time integration an expensive, in many cases an inacceptable tool. So the upcoming of explicit integ-

Table 2 Comparison of implicit and explicit resources used for pipe bending.

Integration	Processors	RAM [GByte]	CPU-Time [h]	Rel. velocity
implicit	16	24	8	1
explicit [9]	2	2	2	32

ration schemes, often related to the first versions of LS-Dyna since 1976, marked an essential step towards the feasibility of large contact studies. Assuming that in small time increments, things will not change essentially, the equation to be solved is

$$\mathbf{K}_o \Delta \mathbf{u} = \Delta \mathbf{F} \tag{3}$$

respectively,

$$\mathbf{M}_o + \ddot{\mathbf{u}}_o + \mathbf{C}_o \dot{\mathbf{u}}_o + \mathbf{K}_o \mathbf{u}_o = \mathbf{F}_o \tag{4}$$

the index o now indicating the old stiffness, mass and damping matrix, displacement and force. Substituting the finite difference approximations for the time derivates and using diagonal mass and damping matrices, equation (4) yields

$$\mathbf{M}_o \frac{\mathbf{u}_n - 2\mathbf{u}_o - \mathbf{u}_{o-1}}{\Delta t^2} + \mathbf{C}_o \frac{\mathbf{u}_n - \mathbf{u}_{o-1}}{2\Delta t} + \mathbf{K}_o \mathbf{u}_o = \mathbf{F}_o \tag{5}$$

$\mathbf{u}_n$ denoting the new displacement again, $\mathbf{u}_{o-1}$ standing for the past time step's displacement. From equation (5) we find the new displacement $\mathbf{u}_n$. Using equations (4) and (5) has the big advantage that the stiffness matrix has never to be assembled or inverted. So the CPU-time and storage requirements to perform one time step are reduced by large factors. On the other hand the time steps have to be reduced essentially to avoid to large an accumulation of errors due to the fact that the equilibrium at the old time step is used to predict the new displacements. The numerical limit to the time step enforces even smaller time increments.

To give some ideas about the impact of the time integration scheme we remember the pipe bending example demonstrated in Figure 4. The CPU-time and the storage requirements are compared in Table 2. Obviously the explicit code has large advantages. But on the other hand, the fact that explicit codes always yield results is not without drawbacks. When implicit codes show little stability or fail to converge, this may be indicate that physical problems arise. So the deficit in numerical performance may be an advantage, motivating the qualified analyst to improve his ideas, to check assumptions, to go deeper into the details of the problem posed.

3.6 Reliability of Codes

One of the most dangerous obstacles in simulation of contact is the reliability of the commercial or private codes. Nonlinearity and contact are included in most today

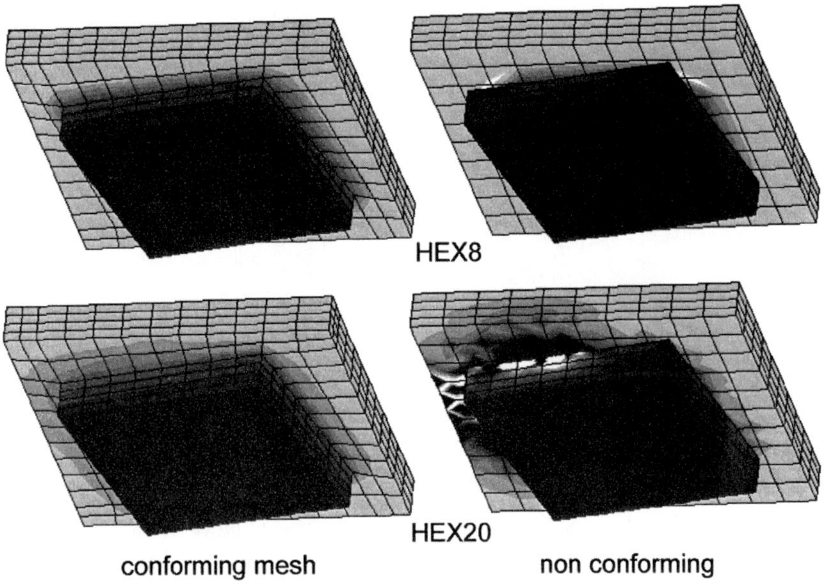

Fig. 9 Failing contact quality using HEX20 [11].

modelling systems, even if sometimes doubtful contact options are offered to the user. Unfortunately many of the options and parameters are neither clearly understood nor defined. In addition, the difficulty of contact modelling makes the software provider proposing many different parameter schemes. It is not uncommon, that the definition of contact input requires about 100 pages in the simulation system manuals (see e.g. [10]). Even high level quality assurance does not inhibit erroneous codes as indicated in Figure 9 [11]. The thermal contact between two bodies is modelled with conforming and nonconforming meshes using linear (HEX8) and quadratic (HEX20) elements. Obviously the contact definition of the higher order elements is wrong, at least unsatisfactory. This is even more surprising, as the HEX20-element of this code shows very good performance when used in structural contact problems as shown in Figure 4. So any contact used in commercial codes needs to be checked at some benchmark problems before using it in engineering or scientific application.

4 Scatter and Robustness

The increasing computing power improved the simulation of parts from single jobs of few variants to the study of the response of large sets of models. This may be necessary when we need to vary some or many of the parts parameters. A typical application is the modelling of the scatter of manufactured components in large series production. The term of *robustness* entered the world of virtual development,

Table 3 Most sensitive and influencing parameters in contact modelling.

Physical data	Numerical data
– friction	– gap, catch range
– surface quality	– mesh quality
– surface definition	– element definition
– loads	– initial contact
– supports	– penetration
– flexibility	– local stiffness
	– friction model

opening new sights, new possibilities and new problems to the simulation teams. At the onset of robustness studies due to the large number of parameter combinations it was considered to be available only for few elaborated problems. The introduction of statistically based approaches like the Latin Hypercube and the Response Surface allowed accelerations of the studies. So today we may handle many problems with scattering parameters in acceptable and affordable time. Nevertheless robustness studies, especially in the nonlinear regime need essentially more time and computing power. This is due to the fact that many variants have to be analysed over a certain time of the contact process.

4.1 Scatter in Contact Studies

Table 1 offered a selection of parameters and solution strategies used in contact modelling. The selected values of the parameters and their values influence the result of the contact analysis. In consequence robustness or sensitivity studies seem inevitable. Table 3 lists some of the most dangerous and scattering components of the physical contact and its numerical modelling. Once more neither completeness nor uniqueness is assumed. Other analysts with other experiences may add or remove entries to or from Table 3. In any case care should be taken that the range of possible states is covered by the range of numerical parameters.

4.2 Parameter Sensitivity

Figures 6 to 9 indicated some of the possible uncertainties encountered using specific contact models. The range of input definitions covering the aforementioned 100 pages is hardly understood by the majority of the engineers working with the different numerical codes. Figure 10 gives an idea how small input changes result in large differences in the results predicted by the simulation. As a consequence automated or semi-automated systems should provide the possibility to do a sensitivity

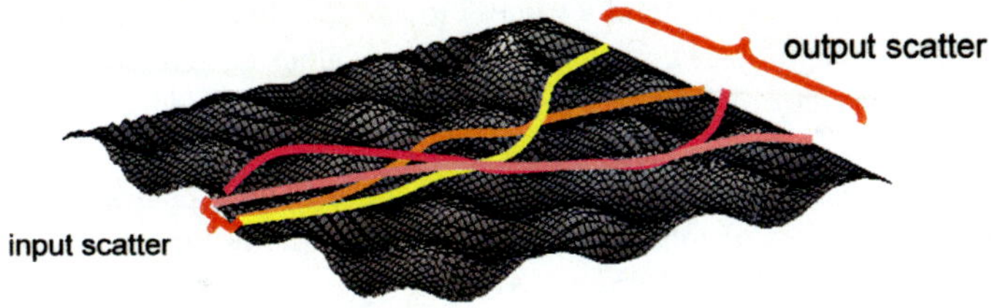

Fig. 10 Scatter and robustness.

study to understand whether physical or numerical effects influence the results of the nonlinear analysis.

Some typical steps may include:

- Check which input parameter changes cause significant changes in results.
- Cover possible parameter range.
- Find critical parameter combinations.
- Check reliability of parameters used.
- Compare results with experimental data, if available and reliable.

The outcome of these studies could result in an increased critical interpretation of the numerical results. This is always a good idea in nonlinear studies, especially in large and complex contact problems like crash and metal forming.

5 Conclusions

As a résumé, some statements may help to improve the awareness of problems, obstacles and traps in contact simulation.

- Contact studies are possible with commercial and private simulation codes and without deep understanding of the underlying principles. Regarding the questions mentioned in this article, we doubt that this is always positive. Sometimes there should be some fences preventing inexperienced users to enter fields where their simulation is in great danger of being at least not very meaningful. The definition of these fences requires large experience. They should be provided by the software suppliers.
- As many parameters for contact modelling are available in commercial simulation systems, many aspects have to be taken into account, many of the sometimes contradicting possibilities checked, interactions looked at, inconsistencies excluded.
- Often the meaning of parameters is not properly defined so that even the software supplier's people do not know what they are speaking about. Therefore a reproducible strategy of benchmarking and quality assurance needs to be established, improved and updated.

- The tendency of using the default values provided by the code includes the danger to run totally inappropriate studies. All default values have to be checked, questionable entries removed, large warning signs posted to avoid following nonsense roads.
- The large number of physical and numerical parameters in contact studies makes a consequent system of robustness and sensitivity studies inevitable. It should be included in the problem solving rules and never be neglected due to lack of time or money.
- The interpretation of nonlinear simulations in most cases is a difficult task. So experience, comparison with experiments, preceding studies, comparable results have to be included in the evaluation of simulation results. This holds not only but even more thoroughly when dealing with contact problems.
- Qualified users should not and cannot avoid doing contact studies. They should always be aware of the many possibilities of failing by ignoring or misinterpreting physical or numerical aspects. So the old rule holds, that

  "FEM makes a good engineer great, but a bad engineer dangerous."

References

1. Steinbuch, R.: Remarks on the Engineering Treatment of Nonlinear Simulation Problems. St. Petersburg, Russia (2003)
2. Kammler, G., Mauch, G., Steinbuch, R.: Automatische Neuvernetzung stark deformierter Strukturen. Berichtsband Des MARC-Benutzertreffens, München (1992) (in German)
3. Bogenschütz, H., Mohrhardt, B., König, H.G., Steinbuch, R.: Numerical analysis of the pedestrian-car accident. In: Proceedings of the 1st European MARC- Users Conference, Düsseldorf (1995)
4. Haak, T., Mück, M., Schultz, O., Steinbuch, R.: On the influence of manufacturing on the load carrying capacity of metal structures. In: Proceedings of the CADFEM Users' Meeting 2008, Darmstadt (October 23, 2008)
5. Molinari, J.F.: Multiscale modelling of nanotribology: Challenges and opportunities. Presented at ICCCM 2009, Lecce (2009)
6. Zavarise, G., De Lorenzis, L.: A strategy for contact problems with large initial penetrations. Presented at ICCCM 2009, Lecce (2009)
7. `http://www.altairhyperworks.com/pdfs/product_brochuresHW_HyperForm_Web.pdf`
8. Benson, D.J.: The history of LS-DYNA. University of California, San Diego, `http://blog.d3view.com/wp-content/uploads/2007/06/benson.pdf` (retrieved March 25, 2009)
9. Private communication with P. Vogel, Dynamore, Stuttgart, Germany
10. LS Dyna Keyword User Manual, Livermore Software Technolog Coperation, `http://www.lstc.com`
11. Private communication with Michael Lautsch, Lautsch Finite Elemente, Esslingen, Germany